计算理论

THEORY OF COMPUTATION

（双语版）

成科扬　周从华　李茂贞　编著

镇　江

图书在版编目(CIP)数据

计算理论：汉英对照 / 成科扬，周从华，李茂贞编著. — 镇江：江苏大学出版社，2024.1
ISBN 978-7-5684-1955-0

Ⅰ. ①计⋯ Ⅱ. ①成⋯ ②周⋯ ③李⋯ Ⅲ. ①计算机算法－理论－教材－汉、英 Ⅳ. ①TP301.6

中国国家版本馆 CIP 数据核字(2023)第 072518 号

计算理论
Jisuan Lilun

编　著 /	成科扬　周从华　李茂贞
责任编辑 /	杨海濒
出版发行 /	江苏大学出版社
地　　址 /	江苏省镇江市京口区学府路 301 号(邮编：212013)
电　　话 /	0511-84446464(传真)
网　　址 /	http://press.ujs.edu.cn
排　　版 /	镇江市江东印刷有限责任公司
印　　刷 /	苏州市古得堡数码印刷有限公司
开　　本 /	787 mm×1 092 mm　1/16
印　　张 /	22.5
字　　数 /	683 千字
版　　次 /	2024 年 1 月第 1 版
印　　次 /	2024 年 1 月第 1 次印刷
书　　号 /	ISBN 978-7-5684-1955-0
定　　价 /	69.00 元

如有印装质量问题请与本社营销部联系(电话：0511-84440882)

Preface

Throughout the thousands of years of human civilization, people have encountered many computational problems, such as calculating prices at the checkout, choosing routes when traveling, and even planning at the national level. As a result, various kinds of computations were studied and many algorithms were created, but the mathematical theory, which is based on the nature of the computation or algorithm itself, was not developed until the 1930s. The theory of computation is the mathematical theory used to study the process and efficacy of computation, and in 1936, experts in mathematical logic posed the problem of a model of computation to address the question of whether every problem has a solution. Turing, a logician, was the first to propose an abstract model of computation, the "Turing Machine", which laid the theoretical foundation for computers, and on this foundation came computer science. The theory of computation, which is the theoretical foundation of computer science, has been widely used in various fields of computer science. Therefore, learning the theory of computation is of great importance to our subsequent research in computer science and technology.

The theory of computation includes algorithmics, computational complexity theory, computability theory, automata theory, and formal language theory. The book divides them into six main parts, and combines a large number of diagrams, formulas, and codes to lead the reader to understand the theory of computation from the easy to the difficult and complicated, and at the end of each chapter there are a considerable number of exercises to help readers consolidate what they have learned. In addition, the book also includes two sections on algorithm analysis and system modeling and reasoning, which are applications of computational theory and can help readers better understand and apply the basic principles of computational theory. At the same time, the book is in bilingual version, with a bilingual two-column layout, which makes it easy for readers to read in both languages and develop their reading and writing skills in English.

With the development of science and technology, the theory of computation will be applied to other fields more often. We hope that this book will help readers master the knowledge and methods related to the theory of computation and lay a solid foundation for participating in innovative research and experimental work in the future.

During the writing process, we consulted various versions of discrete mathematics textbooks and related literature both domestically and internationally. We learned many good ideas and methods from these scholars and experts, and we would like to express our gratitude to them. Additionally, we would like to thank Yan Liuyang, Wan Hao, Shen Weijie, Zhou Hao, Chen Nan, Yu Yue and others for their contributions to the writing and proofreading of this book, as well as the financial support provided by the Graduate Textbook Development Project of Jiangsu University and the All-English Tanght Courses Textbook Development Project for International Students.

Due to our limited expertise, there may still be some shortcomings in this book. We welcome and appreciate feedback and criticism from our readers.

Authors
June 2023

前　言

在几千年的人类文明发展史中，人们遇到了许多计算类的问题，例如结账时的价格计算、出行时的路线选择，甚至国家层面的统筹规划，等等．于是，人们研究了各种各样的计算，创立了许许多多的算法，但以计算或算法本身的性质为研究对象的数学理论却是到20世纪30年代才发展起来的．计算理论正是用来研究计算的过程与功效的数学理论．1936年，数理逻辑专家们便提出了计算模型的问题，借以回答每个问题是否都有解．逻辑学家图灵最先提出计算的抽象模型"图灵机"，为计算机科学的发展奠定了理论方面的基础，在这个基础上计算机科学产生了．目前，作为计算机科学理论基础的计算理论已经广泛应用于计算机科学的各个领域．因此，学好计算理论对于我们后续研究计算机科学技术有着十分重大的意义．

计算理论主要包括算法学、计算复杂性理论、可计算性理论、自动机理论和形式语言理论等．本书将它们分为六大部分，并结合大量的图例、公式以及代码，带领读者由浅入深地了解计算理论，并在每一章节的最后设置相当数量的习题帮助读者巩固所学知识．除此之外，本书还包含算法分析和系统建模与推理两部分内容，将其作为计算理论知识的应用，帮助读者更好地理解和运用所学计算理论基本原理．同时，本书为双语版本，采用双语双栏的形式排版，方便读者对照阅读，从而培养读者的专业外语阅读能力．

随着科技的发展，计算理论会更多地应用于其他领域．希望本书能够帮助读者掌握计算理论的相关知识和方法，为将来参与创新性的研究和实验工作打下坚实的基础．

最后，在编写过程中，我们参考了国内外多种版本的计算理论教材和相关的文献资料，从中吸取了许多适用的思想与方法，在此一并向相关学者和专家表示感谢．此外，还要感谢严浏阳、万浩、沈维杰、周昊、陈楠、余悦等人在本书的编写、校对等方面做出的贡献，以及江苏大学研究生教材建设项目、来华留学全英文授课课程教材建设项目的资助．

由于编著者水平有限，书中不足之处在所难免，敬请读者批评、指正．

编著者
2023年6月

Contents

Part I Preliminary Knowledge

Chapter 1 Sets, Relations, and Languages
1.1 Sets ·············· 001
1.2 Relations and functions ·············· 005
1.3 Special types of binary relations ·············· 010
1.4 Finite and infinite sets ·············· 016
1.5 Three fundamental principle ·············· 019
1.6 Finite representations of languages ·············· 025
Exercise 1 ·············· 032

Part II Automata and Languages

Chapter 2 Finite Automata
2.1 Deterministic finite automata ·············· 037
2.2 Non-deterministic finite automata ·············· 042
2.3 Finite automata and regular expressions ·············· 044
2.4 Regular language and non-regular language ·············· 048
2.5 State minimization ·············· 052
2.6 Algorithmic aspects of finite automata ·············· 054
Exercise 2 ·············· 058

Chapter 3 Context-Free Languages
3.1 Context-free grammar (CFG) ·············· 061
3.2 Parse trees ·············· 062
3.3 Pushdown automata (PDA) ·············· 067
3.4 Context-free grammar and pushdown automata ·············· 068
3.5 Context-free and non-context-free languages ·············· 069
Exercise 3 ·············· 072

Part III Theory of Computability

Chapter 4 Turing Machine
- 4.1 Definition of Turing machine (TM) ... 073
- 4.2 Church-Turing thesis ... 073
- 4.3 Variants of Turing machine ... 084
- 4.4 Computing with Turing machine ... 092
- 4.5 Definition of algorithm ... 093
- Exercise 4 ... 095

Chapter 5 Decidability
- 5.1 Decidable languages ... 097
- 5.2 Halting problem ... 105
- Exercise 5 ... 115

Chapter 6 Undecidability
- 6.1 Undecidable problems from language theory ... 118
- 6.2 Undecidable problems about grammars ... 124
- 6.3 An undecidable tiling problem ... 125
- 6.4 Formal definition of mapping reducibility ... 127
- Exercise 6 ... 131

Part IV Theory of Computational Complexity

Chapter 7 Time Complexity
- 7.1 Measuring complexity ... 132
- 7.2 Class P ... 144
- 7.3 Class NP ... 154
- 7.4 NP-completeness ... 163
- 7.5 Typical NP-complete problems ... 178
- Exercise 7 ... 193

Chapter 8 Space Complexity
- 8.1 Savitch's theorem ... 196
- 8.2 Class $PSPACE$ ... 197
- 8.3 $PSPACE$-completeness ... 199
- 8.4 Class L and NL ... 200

第Ⅲ部分 可计算性理论

第4章 图灵机
- 4.1 图灵机(TM)定义 ... 073
- 4.2 丘奇-图灵论题 ... 073
- 4.3 图灵机的变形 ... 084
- 4.4 用图灵机进行计算 ... 092
- 4.5 算法的定义 ... 093
- 习题4 ... 095

第5章 可判定性
- 5.1 可判定性语言 ... 097
- 5.2 停机问题 ... 105
- 习题5 ... 115

第6章 不可判定性
- 6.1 语言理论中的不可判定问题 ... 118
- 6.2 与文法有关的不可判定问题 ... 124
- 6.3 不可判定的铺砖问题 ... 125
- 6.4 映射可归约性的形式定义 ... 127
- 习题6 ... 131

第Ⅳ部分 计算复杂性理论

第7章 时间复杂性
- 7.1 度量复杂性 ... 132
- 7.2 P 类 ... 144
- 7.3 NP 类 ... 154
- 7.4 NP 完全性 ... 163
- 7.5 典型的 NP 完全问题 ... 178
- 习题7 ... 193

第8章 空间复杂度
- 8.1 萨维奇定理 ... 196
- 8.2 $PSPACE$ 类 ... 197
- 8.3 $PSPACE$ 完全性 ... 199
- 8.4 L 类和 NL 类 ... 200

8.5 *NL*-completeness ················ 202
8.6 *NL* equals *coNL* ················ 204
Exercise 8 ······························ 207

Part V Algorithm Design and Analysis

Chapter 9 Divide and Conquer Strategy
9.1 Basic idea of divide and conquer
　　································· 209
9.2 Binary search ················ 212
9.3 Matrix multiplication ······ 215
9.4 Merge sort ···················· 218
9.5 Quick sort ···················· 222
Exercise 9 ······························ 229

Chapter 10 Dynamic Programming
10.1 Basic idea of dynamic programming
　　································· 231
10.2 The shortest path problem of multi-segment graphs ··············· 233
10.3 Multiplication problem of matrix chain
　　································· 238
10.4 The longest common subsequence problem ···················· 243
10.5 0/1 knapsack problem ······ 247
Exercise 10 ···························· 252

Chapter 11 Greedy Algorithm
11.1 Basic idea of greedy algorithm
　　································· 254
11.2 The single-source shortest path problem ···················· 258
11.3 The minimum spanning tree problem
　　································· 265
Exercise 11 ···························· 281

Chapter 12 Lower Bounds
12.1 Trivial lower bounds ······ 283
12.2 Decision tree model ······ 284
12.3 Algebraic decision tree model
　　································· 289
12.4 Linear time reductions ···· 291

8.5 *NL* 完全性 ··················· 202
8.6 *NL* 等于 *coNL* ··············· 204
习题 8 ·································· 207

第 V 部分 算法设计与分析

第 9 章 分治策略
9.1 分治法的基本思想 ········ 209

9.2 二分搜索 ···················· 212
9.3 矩阵乘法 ···················· 215
9.4 合并排序 ···················· 218
9.5 快速排序 ···················· 222
习题 9 ·································· 229

第 10 章 动态规划
10.1 动态规划的基本思想 ······ 231

10.2 多段图最短路径问题 ······ 233

10.3 矩阵连乘问题 ··············· 238

10.4 最长公共子序列问题 ······ 243

10.5 0/1 背包问题 ················ 247
习题 10 ································ 252

第 11 章 贪心算法
11.1 贪心算法的基本思想 ······ 254

11.2 单源最短路径问题 ········ 258

11.3 最小生成树问题 ··········· 266

习题 11 ································ 281

第 12 章 下界
12.1 平凡下界 ···················· 283
12.2 判定树模型 ················· 284
12.3 代数判定树模型 ··········· 289

12.4 线性时间归约 ··············· 291

Exercise 12 ·················· 296

Chapter 13　Backtracking Method

13.1　Basic idea of backtracking method
　·················· 297

13.2　n-queens problem ·········· 300

13.3　m-coloring problem of graphs
　·················· 302

13.4　0/1 knapsack problem ······ 304

Exercise 13 ·················· 309

Part Ⅵ　Modelling and Reasoning about Systems

Chapter 14　Model-based Verification

14.1　Formal verification and model checking
　·················· 310

14.2　Syntax and semantics of linear temporal logic ·········· 316

14.3　Syntax and semantics of computation tree logic ·········· 320

14.4　System verification technologies based on FSM ·········· 327

Exercise 14 ·················· 345

References ·················· 348

习题 12 ·················· 296

第 13 章　回溯法

13.1　回溯法的基本思想 ·········· 297

13.2　n 个皇后问题 ············ 300

13.3　图的 m 着色问题 ·········· 302

13.4　0/1 背包问题 ············ 304

习题 13 ·················· 309

第Ⅵ部分　系统建模与推理

第 14 章　基于模型的验证

14.1　形式化验证和模型检测 ······ 310

14.2　线性时态逻辑的语法与语义 ·· 316

14.3　分支时态逻辑的语法与语义 ·· 320

14.4　基于有穷状态机的系统验证技术
　·················· 327

习题 14 ·················· 345

参考文献 ·················· 348

Part I Preliminary Knowledge

Chapter 1 Sets, Relations, and Languages

1.1 Sets

It is said that mathematics is the language of science, it is certainly the language of the theory of computation, the scientific discipline we shall study in this book. And the language of mathematics deals with sets, and the complex ways in which they overlap, intersect, and in fact take part themselves in forming new sets.

A **set** is a collection of objects. For example, the collection of the four letters a, b, c, and d is a set, which may be named as L and denoted as $L = \{a, b, c, d\}$. The objects comprising a set are called its **elements** or **members**. For example, b is an element of the set L denoted as $b \in L$. Sometimes we simply say that b is in L, or that L contains b. On the other hand, z is not an element of L and denoted as $z \notin L$.

In a set we do not distinguish repetitions of the elements. Thus the set {red, blue, red} is a same set with {red, blue}. Similarly, the order of the elements is immaterial, for example, $\{3,1,9\}$, $\{9,3,1\}$, and $\{1,3,9\}$ are the same sets. In summary, two sets are equal (that is, the same) if and only if they have the same elements.

The elements of a set need not be related in any way (other than happening to be all members of the same set); for example, {3, red, {d, blue}} is a set with three elements, one of which is itself a set. A set may have only one element, it is then called a **singleton**. For example, {1} is the set with 1 as its only element, thus {1} and 1 are quite different. There is also a set with no element at all. Naturally, there can be only one such set, it is called a **empty set**, and is

第 I 部分 预备知识

第 1 章 集合、关系、语言

1.1 集合

数学是科学的语言,它当然也是我们在这本书中学习的科学学科——计算理论的语言. 数学语言包括集合和集合的重叠、相交,以及用集合构成新的集合的各种复杂方式.

集合是对象的汇集. 例如,4 个字母 a, b, c, d 的汇集是一个集合,叫作 L,记作 $L = \{a, b, c, d\}$. 构成集合的所有对象叫作它的**元素**或**成员**. 例如,b 是集合 L 的一个元素,表示为 $b \in L$,有时直接说 b 在 L 中,或 L 包含 b. 又如,z 不是 L 的元素,记作 $z \notin L$.

在集合中,不计重复的元素. 于是,集合{red, blue, red}与{red, blue}是相同的集合. 类似地,所有元素的顺序是无关紧要的. 例如,$\{3,1,9\}$,$\{9,3,1\}$和$\{1,3,9\}$是相同的集合. 总之,当且仅当它们有相同的元素时,两个集合相等(或相同).

一个集合的元素之间不需要有什么关系(除非它们都是同一个集合的元素). 例如,{3, red, {d, blue}}是有 3 个元素的集合,其中一个元素本身就是一个集合. 集合可能只有一个元素,这样的集合叫作**单元集**. 例如,{1}是以 1 作为它的唯一一个元素的集合,{1}与 1 是完全不同的. 还有根本没有元素的集合. 当然,只可能有一个这样

denoted by $\varnothing$. Any set other than the empty set is called nonempty.

So far we have defined sets by simply listing all their elements, separated by commas and included in braces. Some sets cannot be written in this way, because they are infinite. For example, the set **N** of natural numbers is infinite, it can be denoted by $\mathbf{N} = \{0,1,2,\ldots\}$, using the three dots to represent the elements of an infinite list intuitively. A set that is not infinite is finite.

Another way to specify a set is by referring to properties that other sets and elements may or may not have. if $I = \{1,3,9\}$ and $G = \{3,9\}$, G can be described as the set of elements of I that are greater than 2. This fact is written as follows.

$$G = \{x : x \in I \text{ and } x \text{ is greater than } 2\}.$$

In general, if a set A has been defined and P is a property that elements of A may or may not have, then we can define a new set

$$B = \{x : x \in A \text{ and } x \text{ has property } P\}.$$

As another example, the set of odd natural numbers is

$$O = \{x : x \in \mathbf{N} \text{ and } x \text{ is not divisible by } 2\}.$$

If every element of set A is an element of set B, then set A is a **subset** of set B, denoted as $A \subseteq B$. Thus $O \subseteq \mathbf{N}$, since each odd natural number is a natural number. Note that any set is a subset of itself. If A is a subset of B but A is not the same as B, we say that A is a **proper subset** of B and denoted as $A \subset B$. Also note that the empty set is a subset of any set. For if B is any set, then $\varnothing \subseteq B$, since each element of $\varnothing$ (of which there are no such elements) is also an element of B.

To prove that two sets A and B are equal, we can prove that $A \subseteq B$ and $B \subseteq A$. Therefore, every element of A must be an element of B and vice versa, so that A and B have the same elements and $A = B$.

Two sets can be combined to form a third set by various set operations, just as numbers are combined by arithmetic operations such as addition. **Union** is one set operation: the union of two sets is a set whose elements are in at least one of these two given sets or may be in both of the two given sets. We use the symbol $\cup$ to denote union, so that

$$A \cup B = \{x : x \in A \text{ or } x \in B\}.$$

For example,

$$\{1,3,9\} \cup \{3,5,7\} = \{1,3,5,7,9\}.$$

The **intersection** of two sets is the collection of all elements shared by the two sets, that is

$$A \cap B = \{x : x \in A \text{ and } x \in B\}.$$

For example,

$$\{1,3,9\} \cap \{3,5,7\} = \{3\},$$

and

$$\{1,3,9\} \cap \{a,b,c,d\} = \varnothing.$$

Finally, the **difference** of two sets A and B, denoted by $A-B$, is the set of all elements of A that are not elements of B, that is

$$A - B = \{x : x \in A \text{ and } x \notin B\}.$$

For example,

$$\{1,3,9\} - \{3,5,7\} = \{1,9\}.$$

Certain properties of the set operations follow easily from their definitions. For example, if A, B, and C are sets, the following laws hold.

Idempotency $\qquad A \cup A = A,$
$\qquad\qquad\qquad\quad A \cap A = A;$
Commutativity $\quad A \cup B = B \cup A,$
$\qquad\qquad\qquad\quad A \cap B = B \cap A;$
Associativity
$\qquad (A \cup B) \cup C = A \cup (B \cup C),$
$\qquad (A \cap B) \cap C = A \cap (B \cap C);$
Distributivity
$\qquad (A \cup B) \cap C = (A \cap C) \cup (B \cap C),$
$\qquad (A \cap B) \cup C = (A \cup C) \cap (B \cup C);$
Absorption $\qquad (A \cup B) \cap A = A,$

通过加法算术运算可以把数结合在一起. 同样, 可以用各种集合运算把两个集合结合成第三个集合. 这种集合运算就是**并**. 两个集合的并是一个集合, 它的元素至少在这两个给定集合的一个中, 也可能同时在这两个集合中. 并用符号 $\cup$ 表示, 记作

$$A \cup B = \{x : x \in A \text{ 或 } x \in B\}.$$

例如:

$$\{1,3,9\} \cup \{3,5,7\} = \{1,3,5,7,9\}.$$

两个集合的**交**是这两个集合共有的全部元素组成的集合, 用符号 $\cap$ 表示, 即

$$A \cap B = \{x : x \in A \text{ 且 } x \in B\}.$$

例如:

$$\{1,3,9\} \cap \{3,5,7\} = \{3\},$$

和

$$\{1,3,9\} \cap \{a,b,c,d\} = \varnothing.$$

最后, 两个集合 A 与 B 的**差**, 记作 $A-B$. 它是不在 B 中的 A 的所有元素组成的集合, 即

$$A - B = \{x : x \in A \text{ 且 } x \notin B\}.$$

例如:

$$\{1,3,9\} - \{3,5,7\} = \{1,9\}.$$

这几个集合运算的某些性质容易从它们的定义中得到. 例如, 设 A, B, C 是集合, 则下述算律成立:

幂等律 $\qquad A \cup A = A,$
$\qquad\qquad\quad A \cap A = A;$
交换律 $\qquad A \cup B = B \cup A,$
$\qquad\qquad\quad A \cap B = B \cap A;$
结合律
$\qquad (A \cup B) \cup C = A \cup (B \cup C),$
$\qquad (A \cap B) \cap C = A \cap (B \cap C);$
分配律
$\qquad (A \cup B) \cap C = (A \cap C) \cup (B \cap C),$
$\qquad (A \cap B) \cup C = (A \cup C) \cap (B \cup C);$
吸收律 $\qquad (A \cup B) \cap A = A,$

$(A\cap B)\cup A=A$;

De-Morgan's laws

$$A-(B\cup C)=(A-B)\cap(A-C),$$
$$A-(B\cap C)=(A-B)\cup(A-C).$$

Example 1.1.1 Let us prove the first of De-Morgan's laws. Let

$$L=A-(B\cup C),$$
$$R=(A-B)\cap(A-C);$$

we are to prove that $L=R$. We do this by proving (a) $L\subseteq R$ and (b) $R\subseteq L$.

(a) Let x be any element of L, then $x\in A$, but $x\notin B$ and $x\notin C$. Hence x is an element of both $A-B$ and $A-C$, and is thus an element of R. Therefore $L\subseteq R$.

(b) Let $x\in R$, then x is an element of both $A-B$ and $A-C$, and is therefore in A but in neither B nor C. Hence $x\in A$ but $x\notin B\cup C$, that is $x\in L$. So $R\subseteq L$.

It is proved that $L=R$.

Two sets are **disjoint** if they have no element in common, that is, their intersection is an empty set.

It is possible to form intersections and unions of two or more sets. If S is any collection of sets, we write $\cup S$ for the set whose elements are the elements of all the sets in S. For example, if $S=\{\{a,b\},\{b,c\},\{c,d\}\}$, then $\cup S=\{a,b,c,d\}$; if $S=\{\{n\}:n\in\mathbf{N}\}$, that is, the collection of all the singleton sets with natural numbers as elements, then $\cup S=\mathbf{N}$.

In general,

$$\cup S=\{x:x\in P \text{ for some set } P\subseteq S\}.$$

Similarly,

$$\cap S=\{x:x\in P \text{ for each set } P\subseteq S\}.$$

The collection of all subsets of a set A is itself a set, called the **power set** of A and denoted 2^A. For example, the subsets of $\{c,d\}$ are $\{c,d\}$ itself, the singletons $\{c\}$ and $\{d\}$ and the empty set $\varnothing$, so

$$2^{\{c,d\}}=\{\{c,d\},\{c\},\{d\},\varnothing\}.$$

$(A\cap B)\cup A=A$;

德摩根律

$$A-(B\cup C)=(A-B)\cap(A-C),$$
$$A-(B\cap C)=(A-B)\cup(A-C).$$

例 1.1.1 证明第一条德摩根律，令

$$L=A-(B\cup C),$$
$$R=(A-B)\cap(A-C);$$

要证 $L=R$，先要证（a）$L\subseteq R$ 和（b）$R\subseteq L$.

（a）设 x 是 L 的任一元素，则 $x\in A$ 但 $x\notin B$ 且 $x\notin C$. 因此 x 既是 $A-B$ 的元素，也是 $A-C$ 的元素，从而是 R 的元素. 所以 $L\subseteq R$.

（b）设 $x\in R$，则 x 是 $A-B$ 和 $A-C$ 的元素，即 x 在 A 中但不在 B 和 C 中. 因此 $x\in A$ 但 $x\notin B\cup C$，即 $x\in L$. 所以 $R\subseteq L$.

这就证明了 $L=R$.

如果两个集合没有共同的元素，即它们的交是空集，则称它们**不相交**.

可以定义两个及以上集合的并和交. 设 S 是任一集合的汇集，$\cup S$ 表示 S 中所有集合的全部元素组成的集合. 例如，如果 $S=\{\{a,b\},\{b,c\},\{c,d\}\}$，则 $\cup S=\{a,b,c,d\}$；如果 S 是所有以自然数作元素的单元集 $\{\{n\}:n\in\mathbf{N}\}$，则 $\cup S=\mathbf{N}$.

一般地，

$$\cup S=\{x:\text{对于某个集合 } P\subseteq S, x\in P\}.$$

类似地，

$$\cap S=\{x:\text{对于每一个集合 } P\subseteq S, x\in P\}.$$

集合 A 的所有子集的汇集也是一个集合，叫作 A 的**幂集**，记作 2^A. 例如，$\{c,d\}$ 的全部子集是 $\{c,d\}$ 本身、单元集 $\{c\}$ 和 $\{d\}$，以及空集 $\varnothing$. 因此

$$2^{\{c,d\}}=\{\{c,d\},\{c\},\{d\},\varnothing\}.$$

A **partition** of a nonempty set A is a subset Π of 2^A such that $\varnothing$ is not an element of Π and such that each element of A is in one and only one set in Π. That is, Π is a partition of A if Π is a set of subsets of A such that

(1) each element of Π is nonempty;

(2) distinct members of Π are disjoint;

(3) $\cup \Pi = A$, and Π is partition of A.

For example, $\{\{a,b\},\{c\},\{d\}\}$ is a partition of $\{a,b,c,d\}$, but $\{\{b,c\},\{c,d\}\}$ is not. The sets of even and odd natural numbers form a partition of $\mathbf{N}$, respectively.

1.2 Relations and functions

Mathematics deals with statements about objects and the relations between them. It is natural to say, for example, that "less than" is a relation between objects of a certain kind namely, numbers which holds between 4 and 7 but does not hold between 4 and 2, or between 4 and itself. But how can we express relations between objects in the only mathematical language we have available at this point —that is to say, the language of sets? We simply think of a relation as being itself a set. The objects that belong to the relation are, in essence, the combinations of individuals for which that relation holds in the intuitive sense. So the less-than relation is the set of all pairs of numbers such that the first number is less than the second.

But we have moved a bit quickly. In a pair that belongs to a relation, we need to be able to distinguish the two parts of the pair, and we have not explained how to do so. We cannot write these pairs as sets, since $\{4,7\}$ is the same thing as $\{7,4\}$. It is easiest to introduce a new device for grouping objects called an **ordered pair**.

The ordered pair of a and b is denoted as (a,b), a and b are called the **components** of the ordered pair (a,b). The ordered pair (a,b) is not the same as the set $\{a,b\}$. First, the order matters: (a,b) is different from (b,a), whereas $\{a,b\} = \{b,a\}$. Second, the two components of an ordered pair need not be distinct, $(7,7)$ is a valid ordered pair. Note that two ordered pairs (a,b) and (c,d) are equal only when $a=c$ and $b=d$.

The **Cartesian product** of two sets A and B, denoted by $A \times B$, is the set of all ordered pairs (a,b) with $a \in A$ and $b \in B$. For example,

$$\{1,3,9\} \times \{b,c,d\}$$
$$= \{(1,b),(1,c),(1,d),$$
$$(3,b),(3,c),(3,d),$$
$$(9,b),(9,c),(9,d)\}.$$

A binary relation on two sets A and B is a subset of $A \times B$. For example, $\{(1,b),(1,c),(3,d),(9,d)\}$ is a binary relation on $\{1,3,9\}$ and $\{b,c,d\}$. And $\{(i,j): i,j \in \mathbf{N}$ and $i<j\}$ is the less-than relation, it is a subset of $\mathbf{N} \times \mathbf{N}$ (two sets related by a binary relation are often identical).

More generally, let n be any natural number. Then if $a_1, \ldots, a_n$ are any n objects, not necessarily distinct, $(a_1, \ldots, a_n)$ is an **ordered tuple**; for each $i=1, \ldots, n$, a_i is the ith component of $(a_1, \ldots, a_n)$. An ordered m-tuple $(b_1, \ldots, b_m)$, where m is a natural number, is the same as $(a_1, \ldots, a_n)$ if and only if $m=n$ and $a_i = b_i$, for $i=1, \ldots, n$. Thus $(4,4)$, $(4,4,4)$, $((4,4),4)$, and $(4,(4,4))$ are all distinct. Ordered 2-tuples are the same as the ordered pairs discussed above, and ordered 3-, 4-, 5-, and 6-tuples are called **ordered triples**, **quadruples**, **quintuples**, and **sixtuples**, respectively. On the other hand, a **sequence** is an ordered n-tuple for some unspecified n (the **length** of the sequence). If $A_1, \ldots, A_n$ are any sets, then the n-**fold Cartesian product** $A_1 \times \ldots \times A_n$ is the set of all ordered n-tuples $(a_1, \ldots, a_n)$, with $a_i \in A_i$, for each $i=1, \ldots, n$. In case all the A_i are the same set A, the n-fold Cartesian product $A_1 \times \ldots \times A_n$ of A with itself is also written

(b,a)不一样,而$\{a,b\}=\{b,a\}$. 其次,有序对的两个分量不必不同,即$(7,7)$是有效的有序对. 注意: 当且仅当$a=c$且$b=d$时,两个有序对(a,b)和(c,d)相等.

两个集合A和B的**笛卡儿积**是所有有序对(a,b)构成的集合,$a \in A$且$b \in B$,记作$A \times B$. 例如,

$$\{1,3,9\} \times \{b,c,d\}$$
$$= \{(1,b),(1,c),(1,d),$$
$$(3,b),(3,c),(3,d),$$
$$(9,b),(9,c),(9,d)\}.$$

集合A和B上的二元关系是$A \times B$的子集. 例如,$\{(1,b),(1,c),(3,d),(9,d)\}$是$\{1,3,9\}$和$\{b,c,d\}$上的二元关系. $\{(i,j): i,j \in \mathbf{N}$且$i<j\}$是小于关系,它是$\mathbf{N} \times \mathbf{N}$的子集(用二元关系关联的两个集合常常是相同的).

更一般地,设n是任一自然数. 如果$a_1, \cdots, a_n$是任意n个对象,且它们不必是不相同的,那么$(a_1, \cdots, a_n)$是一个**有序组**. 对每一个$i=1, \cdots, n$, a_i是$(a_1, \cdots, a_n)$的第i个分量. 当且仅当$m=n$且$a_i=b_i, i=1, \cdots, n$时,有序m元组$(b_1, \cdots, b_m)$与$(a_1, \cdots, a_n)$相同. 因此,$(4,4)$, $(4,4,4)$, $((4,4),4)$和$(4,(4,4))$都是不相同的. 有序二元组就是前面讨论的有序对. 有序3、4、5、6元组分别叫作**有序三元组**、**四元组**、**五元组**、**六元组**. 此外,对于某个未指定的数n(该序列的长度),**序列**是一个有序n元组. 设$A_1, \cdots, A_n$是任意n个集合,n重笛卡儿积$A_1 \times \cdots \times A_n$是所有有序$n$元组$(a_1, \cdots, a_n)$的集合,其中对于每一个$i=1, \cdots, n$,都有$a_i \in A_i$. 当所有的$A_i$都是同一集合$A$时,$A$自身的$n$重笛卡儿积$A_1 \times \cdots \times A_n$也写成$A^n$.

as A^n itself is also written as A^n. For example, $\mathbf{N}^2$ is the set of ordered pairs of natural numbers. An n-**ary relation** on sets $A_1, \ldots, A_n$ is a subset of $A_1 \times \cdots \times A_n$; 1-, 2-, and 3-ary relations are called **unary**, **binary**, and **ternary relations**, respectively.

Another fundamental mathematical concept is the function. On the intuitive level, a function is an association of each object of one kind with a unique object of another kind. For example, people with their ages, dogs with their owners, numbers with their successors, and so on. Using the idea of regarding a binary relation as a set of ordered pairs, we can replace this intuitive idea by a concrete definition. A function from a set A to a set B is a binary relation R on A and B with the following special property: for each element $a \in A$, there is exactly one ordered pair in R with first component a. Use the following example to illustrate the definition, let C be the set of prefecture-level cities in China, and S be the set of provinces; and let

$R_1 = \{(x,y): x \in C, y \in S, \text{ and } x \text{ is a prefecture-level city in } y \text{ province}\}$,

$R_2 = \{(x,y): x \in S, y \in C, \text{ and } y \text{ is a prefecture-level city in } x \text{ province}\}$.

Then R_1 is a function, since each prefecture-level city is in one and only one province, but R_2 is not a function, since some provinces have more than one prefecture-level city.

In general, we use letters such as f, g, and h for functions and we use $f: A \mapsto B$ to denote that f is a function from A to B. We call A the **domain** of f. If a is any element of A, denote the element b of B by $f(a)$ such that $(a,b) \in f$; since f is a function, there is exactly one $b \in B$ with this property, so $f(a)$ denotes a unique object. The object $f(a)$ is called the **image** of a under f. To specify a function $f: A \mapsto B$, it suffices to specify $f(a)$ for each $a \in A$. For example, to specify the function R_1 above, it suffices to specify, for each prefecture-level city, the provinces in which

例如，$\mathbf{N}^2$ 是所有自然数有序对的集合. 集合 $A_1, \cdots, A_n$ 上的 n **元关系**是 $A_1 \times \cdots \times A_n$ 的一个子集. 1、2、3 元关系分别叫作**一元**、**二元**、**三元关系**.

另一个基本的数学概念是函数. 直观上，函数是把一类中的每一个对象关联到另一类中的一个唯一的对象. 例如，人与他们的年龄，狗与它们的主人，数与它们的后继，等等. 利用把二元关系看作有序对的集合的思想，可以给出这个直观概念的具体定义. 从集合 A 到集合 B 的函数是 A 和 B 上的具有下述特殊性质的二元关系 R: 对于元素 $a \in A$, 在 R 中恰好有一个有序对以 a 为第一个分量. 用下面的例子说明这个定义. 令 C 是中国的地级市集合，S 是省份集合，即设

$R_1 = \{(x,y): x \in C, y \in S, 且 x 是 y 省的一个地级市\}$,

$R_2 = \{(x,y): x \in S, y \in C, 且 y 是 x 省的一个地级市\}$.

那么，R_1 是一个函数，因为每一个地级市在且只在一个省内；R_2 不是函数，因为一个省有一个以上地级市.

一般地，用字母 f, g, h 等表示函数，用 $f: A \mapsto B$ 表示 f 是从 A 到 B 的函数. 把 A 叫作 f 的**定义域**. 设 a 是 A 的任一元素，用 $f(a)$ 表示 B 中使得 $(a,b) \in f$ 的元素 b. 因为 f 是一个函数，恰好存在一个 $b \in B$ 具有这个性质，所以 $f(a)$ 表示一个唯一的对象. 这个 $f(a)$ 叫作 a 在 f 下的**象**. 为了详细说明函数 $f: A \mapsto B$, 只要对每一个 $a \in A$ 指定 $f(a)$. 例如，要详细说明上述函数 R_1, 只要指出每一个地级市所在的省. 设 $f: A \mapsto B$, A' 是 A 的子集，定义 $f[A'] = $

it is located. If $f: A \mapsto B$ and A' is a subset of A, then we define $f[A'] = \{f(a) : a \in A'\}$ (that is, $\{b : b = f(a)$ for a certain $a \in A'\}$). We call $f[A']$ the **image** of A' under f. The **range** of f is the image of its domain.

Ordinarily, if the domain of a function is a Cartesian product, one set of parentheses is dropped. For example, if $f: \mathbf{N} \times \mathbf{N} \mapsto \mathbf{N}$ is defined so that the image under f of an ordered pair (m,n) is the sum of m and n, we would write $f(m,n) = m+n$ rather than $f((m,n)) = m+n$, simply as a matter of notational convenience.

If $f: A_1 \times \ldots \times A_n \mapsto B$ is a function, and $f(a_1, \ldots, a_n) = b$, where $a_i \in A_i$ for $i = 1, \ldots, n$ and $b \in B$, we sometimes call $a_1, \ldots, a_n$ the **arguments** of f and b the corresponding **value** of f. Thus the function f can be specified by giving its value for each n-tuple of arguments.

Certain kinds of functions are of special interest. A function $f: A \mapsto B$ is **one-to-one** if for any two distinct elements $a, a' \in A$, $f(a) \neq f(a')$. For example, Let C be the set of prefecture-level cities in China and S be the set of provinces, and $g: S \mapsto C$ is specified by

$g(s) = $ the capital of province s, for each $s \in S$

then g is one-to-one since no two provinces have the same capital. A function $f: A \mapsto B$ is **onto** B if each element of B is the image under f of some element of A. The function g just specified is not onto C, but the function R_1 defined above is onto S since each province contains at least one prefecture-level city. Finally a function $f: A \mapsto B$ is a **bijection** between A and B if it is both one-to-one and onto B. For example, if C_0 is a set of capital cities, then the function $g: S \mapsto C_0$ specified, as before, by

$g(s) = $ the capital of province s

g is a bijection between S and C_0.

The inverse of a binary relation R is the relation $\{(b,a) : (a,b) \in R\}$, denoted R^{-1}. Clearly, if $R \subseteq$

$\{f(a) : a \in A'\}$（即 $\{b$：对某个 $a \in A'$，$b = f(a)\}$），称 $f[A']$ 是 A' 在 f 下的**象**. f 的**值域**是它的定义域的象.

当函数的定义域是笛卡儿积时，通常省略一对圆括号. 例如，若定义 $f: \mathbf{N} \times \mathbf{N} \mapsto \mathbf{N}$ 使得有序对 (m,n) 在 f 下的象是 m 与 n 的和，则按照习惯记法写成 $f(m, n) = m+n$，而不写成 $f((m, n)) = m+n$.

设 $f: A_1 \times \cdots \times A_n \mapsto B$ 是一个函数，$f(a_1, \cdots, a_n) = b$，其中 $a_i \in A_i, i = 1, \cdots, n$ 且 $b \in B$，有时称 $a_1, \cdots, a_n$ 是 f 的**自变量**，而 b 是 f 对应的**值**. 于是，对自变量的每一个 n 元组给出 f 的值可以指定函数 f.

某些类型的函数具有特别的意义. 如果对任意两个不同的值 $a, a' \in A, f(a) \neq f(a')$，则称函数 $f: A \mapsto B$ 是**一对一**的. 例如，设 C 是中国的地级市集合，S 是省份集合，规定 $g: S \mapsto C$ 为

$g(s) = s$ 省的省会，每一个 $s \in S$

那么 g 是一对一的，因为没有两个省有相同的省会. 如果 B 的每一个元素是 A 的某个元素在 f 下的象，则称函数 $f: A \mapsto B$ **满射**到 B. 上面定义的函数 g 不是满射到 C 的，但函数 R_1 满射到 S，因为每一个省至少有一个地级市. 最后，如果映射 $f: A \mapsto B$ 既是一对一的又是满射到 B 的，则称 f 是 A 与 B 之间的**双射**. 例如，如果 C_0 是省会集合，如前，规定函数 $g: S \mapsto C_0$ 为

$g(s) = s$ 省的省会

则 g 是 S 与 C_0 之间的双射.

二元关系 R 的逆是关系 $\{(b,a)$：$(a,b) \in R\}$，记作 R^{-1}. 显然，若 $R \subseteq A \times$

$A \times B$, then $R^{-1} \subseteq B \times A$. For example, the relation R_2 defined above is the inverse of R_1. Thus, the inverse of a function need not be a function. In the case of R_1, its inverse fails to be a function since some provinces have more than one prefecture-level city, that is, there are distinct prefecture-level cities c_1 and c_2 such that $R_1(c_1) = R_1(c_2)$. Set $f: A \mapsto B$ is a function and there exists $b \in B$ such that for all $a \in A$, $f(a) \neq b$, then the inverse of the function f is also not a function. If $f: A \mapsto B$ is a bijection, then, neither of these eventualities can occur, and f^{-1} is a function, indeed, a bijection between B and A. Moreover $f^{-1}(f(a)) = a$ for each $a \in A$, and $f^{-1}(f(b)) = b$ for each $b \in B$.

When a particularly simple bijection between two sets has been specified, it is sometimes possible to view an object in the domain and its image in the range as virtually indistinguishable: the one may be viewed as a renaming or a way of rewriting the other. For example, singletons and ordered 1-tuples are, strictly speaking, different, but not much harm is done if we occasionally blur the distinction, since for each, singleton $\{a\}$, $f(\{a\}) = a$ is an obvious bijection. Such a bijection is called a natural isomorphism. Of course this is not a formal definition since what is "natural" and what distinctions can be blurred depend on the context. Some slightly more complex examples should make the point more clearly.

Example 1.2.1 For any three sets A, B, and C, there is a natural isomorphism of $A \times B \times C$ to $(A \times B) \times C$, namely for any $a \in A$, $b \in B$, and $c \in C$.
$$f(a,b,c) = ((a,b),c)$$

Example 1.2.2 For any sets A and B, there is a natural isomorphism φ from $2^{A \times B}$, that is, the set of all binary relations on A and B, to the set
$$\{f: f \text{ is a function from } A \text{ to } 2^B\}.$$
Namely, for any binary relation $R \subseteq A \times B$, let $\varphi(R)$ be the function $f: A \mapsto 2^B$ such that
$$f(a) = \{b: b \in B \text{ and } (a,b) \in R\}.$$

For example, if S is the set of provinces and $R\subseteq S\times S$ contains any ordered pair of provinces with a common border, then the naturally associated function $f:S\mapsto 2^S$ is specified by

$f(s)=\{s':s'\in S$ and s' shares a border with $s\}$.

Example 1.2.3 Sometimes we regard the inverse of a function $f:A\mapsto B$ as a function even when f is not a bijection. The idea is to regard $f^{-1}\subseteq B\times A$ as a function from B to 2^A, using the natural isomorphism described under Example 1.2.2. Thus $f^{-1}(b)$ is, for any $b\in B$, the set of all $a\in A$ such that $f(a)=b$. For example, if R_1 is defined as described earlier (for each prefecture, the value of this function is equal to the province in which it is located), then for each province s, $R_1^{-1}(s)$ is equal to the set of all prefecture level cities in province s.

Example 1.2.4 If Q and R are binary relations, then their composition $Q\circ R$, or simply QR, is the relation

$\{(a,b):$ for some $c,(a,c)\in Q$ and $(c,b)\in R\}$.

Note that, the composition of two functions $f:A\mapsto B$ and $g:B\mapsto C$ is a function h from A to C such that $h(a)=g(f(a))$ for each $a\in A$. For example, if f is the function that maps each dog to its owner, and g is a function that maps each person to his or her age, then $f\circ g$ maps each dog to the age of its owner.

1.3 Special types of binary relations

Binary relations will be found over and over again in these pages, it will be helpful to have convenient ways of representing them and some terminologies for discussing their properties. A completely "random" binary relation has no significant internal structure, but many relations we shall encounter arise out of specific contexts and therefore have important regularities. For example, the relation that holds between two prefecture-level cities if they belong to the same provinces has certain "symmetries" and other properties that are worth noting, discussing, and exploiting.

In this section we discuss binary relations that exhibit these similar regularities. We shall deal only with binary relations on a set and itself. Thus, let A be a set, and $R \subseteq A \times A$ be a binary relation on A. The relation R can be represented by a **directed graph**. Each element of A is represented by a small circle—what we call a node of the directed graph—and an arrow is drawn from a to b if and only if $(a,b) \in R$. The arrows are the **edges** of the directed graph. For example, the relation $R = \{(a,b),(b,a),(a,d),(d,c),(c,c),(c,a)\}$ is represented by the graph in Fig. 1-1. Note in particular the loop from c to itself, corresponding to the pair $(c,c) \in R$. From a node of a graph to another, there is either no edge, or one edge, "parallel arrows" are prohibited.

意、讨论和运用的性质.

本节讨论表示这些关系及具有类似规律的二元关系. 下面只讨论集合与集合自身的二元关系. 设 A 是一个集合, $R \subseteq A \times A$ 是 A 上的一个二元关系. 可以用一个**有向图**表示关系 R. A 的每一个元素用一个小圆圈表示(叫作有向图的**顶点**), 从 a 到 b 画一个箭头当且仅当 $(a,b) \in R$. 这些箭头是该有向图的**边**. 例如, 用图 1-1 表示关系 $R = \{(a,b),(b,a),(a,d),(d,c),(c,c),(c,a)\}$. 特别要注意, c 到自身的环对应有序对 $(c,c) \in R$. 从图的一个顶点到另一个顶点, 要么没有边, 要么只有一条边, 这里不允许出现"平行的箭头".

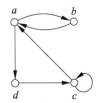

图 1-1　关系 $R = \{(a,b),(b,a),(a,d),(d,c),(c,c)(c,a)\}$
Fig. 1-1　Relation $R = \{(a,b),(b,a),(a,d),(d,c),(c,c)(c,a)\}$

There is no formal distinction between binary relations on a set A and directed graphs with nodes from A. We use the term directed graph when we want to emphasize that the set on which the relation is defined is of no independent interest to us, outside the context of this particular relation. Directed graphs, as well as the undirected graphs soon to be introduced, are useful as models and abstractions of complex systems (traffic and communication networks, computational structures and processes, etc.). We will discuss many interesting computational problems related to directed graphs in more detail in Section 1.6, especially in Chapter 7.

集合 A 上的二元关系与来自 A 的顶点有向图没有形式上的区别. 当我们想强调对上面定义的二元关系的集合没有兴趣时, 可使用有向图这个术语, 特定关系的情况除外. 有向图和将要介绍的无向图作为复杂系统(交通网与通信网、计算结构与过程等)的模型和抽象是有用的. 我们将在本章 1.6 节, 尤其是在第 7 章更加详细地讨论与有向图有关的许多有趣的计算问题.

For another example of a binary relation or directed graph, the less-than-or-equal-to relation ≤ defined on the natural numbers is illustrated in Fig. 1-2. Of course, the entire directed graph cannot be drawn, since it would be infinite.

二元关系或有向图的另一个例子是定义在自然数集合上的小于等于关系≤,如图 1-2 所示. 当然,不可能画出整个有向图,因为它是无穷的.

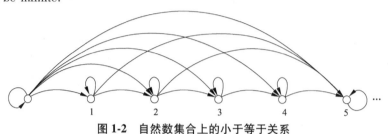

图 1-2　自然数集合上的小于等于关系

Fig. 1-2　The less than or equal to relation on the set of natural numbers

A relation $R \subseteq A \times A$ is **reflexive**, if $(a, a) \in R$ for each $a \in A$. The directed graph representing a reflexive relation has a loop from each node to itself. For example, the directed graph of Fig. 1-2 represents a reflexive relation, but that of Fig. 1-1 does not.

A relation $R \subseteq A \times A$ is **symmetric**, if $(b, a) \in R$ whenever $(a, b) \in R$. In the corresponding directed graph, whenever there is an arrow between two nodes, there are arrows between those nodes in both directions. For example, the directed graph of Fig. 1-3 represents a symmetric relation. This directed graph might depict the relation of "friendship" among six people, since whenever x is a friend of y, y is also a friend of x. The relation of friendship is not reflexive, since we do not regard a person as his or her own friend. Of course, a relation could be both symmetric and reflexive. For example, $\{(a, b) : a$ and b are persons with the same father$\}$ is such a relation.

若对于每一个 $a \in A, (a, a) \in R$,则称关系 $R \subseteq A \times A$ 是**自反**的. 表示一个自反关系的有向图从每一个顶点到它自身有一个环. 例如,图 1-2 中的有向图表示一个自反关系,而图 1-1 中的图所表示的关系不是自反的.

若只要$(a, b) \in R$ 就有$(b, a) \in R$,则称关系 $R \subseteq A \times A$ 是**对称**的. 在对应的有向图中,只要两个顶点之间有一个箭头,则在这两个顶点之间两个方向都有箭头. 例如,图 1-3 中的有向图表示一个对称关系. 这个有向图可以描述六个人中的"朋友"关系,因为只要 x 是 y 的朋友,y 也是 x 的朋友. 朋友关系不是自反的,因为我们不会把一个人看作他(她)自己的朋友. 当然,一个关系可以既是对称的又是自反的. 例如,$\{(a, b): a$ 和 b 的父亲是同一个人$\}$就是这样一个关系.

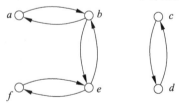

图 1-3　表示有向图的对称关系

Fig. 1-3　Symmetric relationship represented by a directed graph

A symmetric relation without pairs of the form (a,a) is represented as an **undirected graph**, or simply a **graph**. Graphs are drawn without arrowheads, combining pairs of opposite-direction arrows between two nodes into a single undirectional edge. For example, the relation shown in Fig.1-3 could also be represented by the graph in Fig.1-4.

把没有(a,a)形式的有序对的对称关系用**无向图**表示,无向图简称**图**. 画无向图不使用箭头,把两个顶点之间一对方向相反的箭头合并成一条不带箭头的边. 例如,图 1-3 给出的关系也可以用图 1-4 表示.

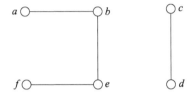

图 1-4　用无向图表示对称关系

Fig. 1-4　Symmetric relationship represented by an undirected graph

A relation R is **antisymmetric** if when $(a,b) \in R$ and a and b are distinct, then $(b,a) \notin R$. For example, let P be the set of all persons, then

$$\{(a,b):a,b \in P \text{ and } a \text{ is the father of } b\}$$

is antisymmetric. A relation may be neither symmetric nor antisymmetric. For example, the relation

$$\{(a,b):a,b \in P \text{ and } a \text{ is the brother of } b\}$$

and the relation represented in Fig. 1-1 are neither symmetric nor antisymmetric.

A binary relation R is **transitive** if whenever $(a,b) \in R$ and $(b,c) \in R$, then $(a,c) \in R$. For example, the relation

$$\{(a,b):a,b \in P \text{ and } a \text{ is an ancestor of } b\}$$

is transitive, since if a is an ancestor of b and b is an ancestor of c, then a is an ancestor of c. So is the less-than-or-equal relation. In terms of the directed graph representation, transitivity is equivalent to the requirement that whenever there is a sequence of arrows leading from an element a to an element z, there is an arrow directly from a to z. For example, the relation illustrated in Fig. 1-5 is transitive.

A relation that is reflexive, symmetric, and transitive is called an **equivalence relation**. The representation of an equivalence relation by an undirected

若当$(a,b) \in R$且a与b不同时,$(b,a) \notin R$,则称关系R是**反对称**的. 例如,设P是所有的人组成的集合,则

$$\{(a,b):a,b \in P \text{ 且 } a \text{ 是 } b \text{ 的父亲}\}$$

是反对称的. 一个关系可能既不是对称的,也不是反对称的. 例如,关系

$$\{(a,b):a,b \in P \text{ 且 } a \text{ 是 } b \text{ 的兄弟}\}$$

和图 1-1 表示的关系既不是对称的,也不是反对称的.

若只要$(a,b) \in R$且$(b,c) \in R$就有$(a,c) \in R$,则称二元关系R是**传递**的. 例如,关系

$$\{(a,b):a,b \in P \text{ 且 } a \text{ 是 } b \text{ 的前辈}\}$$

是传递的,因为若a是b的前辈且b是c的前辈,则a是c的前辈. 小于等于关系也是传递的. 用有向图表示传递的时候,要求只要有从元素a引导到元素z的箭头序列,就有从a直接到z的箭头. 例如,图 1-5 给出的关系是传递的.

把自反、对称和传递的关系叫作**等价关系**. 表示等价关系的无向图由若干集团组成,在每一个集团中的每一

graph consists of a number of clusters; within each cluster, each pair of nodes is connected by a line (Fig. 1-6). The "clusters" of an equivalence relation are called its **equivalence classes**. $[a]$ is usually used to denote the equivalence class containing an element a, when the equivalence relation R does not have to be specified according to the context. That is, $[a] = \{b : (a,b) \in R\}$, or, since R is symmetric, $[a] = \{b : (b,a) \in R\}$. For example, the equivalence relation in Fig. 1-6 has three equivalence classes, one with four elements, one with three elements, and one with one element.

对顶点由一条线连接(图1-6). 把等价关系的这种"集团"叫作**等价类**. 当根据上下文不必指明等价关系 R 时, 通常用 $[a]$ 表示包含元素 a 的等价类, 即 $[a] = \{b : (a,b) \in R\}$, 或者 $[a] = \{b : (b,a) \in R\}$ (因为 R 是对称的). 例如, 图1-6中的等价关系包含3个等价类. 一个等价类有4个元素, 一个等价类有3个元素, 还有一个等价类有1个元素.

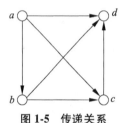

图 1-5 传递关系

Fig. 1-5 Transitive relationship

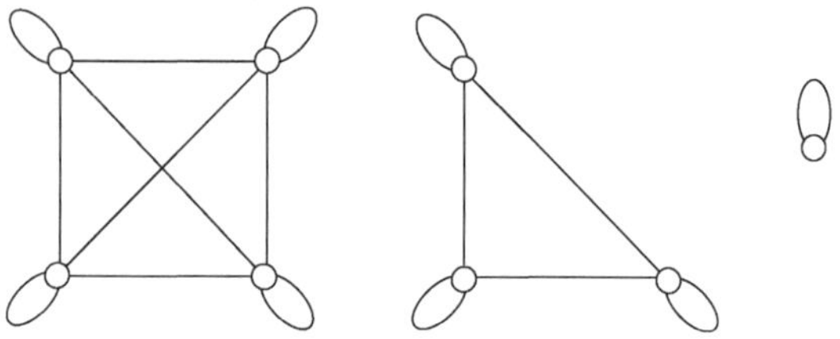

图 1-6 等价关系

Fig. 1-6 Equivalence relationship

Theorem 1.3.1 Let R be an equivalence relation on a nonempty set A. Then the equivalence classes of R constitute a partition of A.

Proof Let $\Pi = \{[a] : a \in A\}$. To prove that the sets in Π are nonempty, disjoint, and together equal A. All equivalence classes are nonempty, since $a \in [a]$ for all $a \in A$, by reflexivity. To prove that they are disjoint, consider any two distinct equivalence classes $[a]$ and $[b]$, and assume that $[a] \cap [b] \neq \emptyset$. Thus there is an element c such that $c \in [a]$ and $c \in [b]$. Hence $(a,c) \in R$ and $(c,b) \in R$; since R is transitive, $(a,b) \in R$; and since R is symmetric, $(b,a) \in R$. Now take any element $d \in [a]$, then $(d,a) \in R$ and, by transitivity, $(d,b) \in R$. Hence $d \in [b]$, so that $d \in [b]$. Likewise $[b] \subseteq [a]$, therefore $[a] = [b]$. But this contradicts the assumption that $[a]$ and $[b]$ are distinct.

定理 1.3.1 设 R 是非空集合 A 上的等价关系, 则 R 的等价类构成 A 的一个分割.

证 令 $\Pi = \{[a] : a \in A\}$. 要证明 Π 中的集合是非空不相交的, 并且合在一起等于 A. 根据自反性, 因为对于所有的 $a \in A, a \in [a]$, 所以所有的等价类非空. 欲证明它们是不相交的, 考虑任意两个不同的等价类 $[a]$ 和 $[b]$, 且假设 $[a] \cap [b] \neq \emptyset$. 于是, 存在元素 c 使得 $c \in [a]$ 和 $c \in [b]$. 从而 $(a,c) \in R$ 且 $(c,b) \in R$, 由 R 具有传递性得 $(a,b) \in R$. 又由 R 是对称的, 有 $(b,a) \in R$. 现在取任一元素 $d \in [a]$, 则 $(d,a) \in R$, 由传递性得 $(d,b) \in R$. 因此 $d \in [b]$, 从而 $[d] \in [b]$. 类似可证 $[b] \subseteq [a]$, 因此, $[a] = [b]$. 这与 $[a]$ 和 $[b]$ 是两个不相同的类矛盾.

As long as we notice the reflexivity, each element a of A is in a certain set in Π, that is, $a \in [a]$, it is easy to see that $\cup \Pi = A$.

Thus setting an equivalence relation R, we can always construct a corresponding partition Π. For example, if $R = \{(a,b) : a \text{ and } b \text{ are the people who have the same parents}\}$, then the equivalence classes of R are all groups of siblings. Note that the construction of Theorem 1.3.1 can be reversed: from any partition, we can construct a corresponding equivalence relation. Namely, if Π is a partition of A, then $R = \{(a,b) : a \text{ and } b \text{ belong to the same set of } \Pi\}$ is an equivalence relation. Thus there is a natural isomorphism between the set of equivalence relations on a set A and the set of partitions of A.

A relation that is reflexive, antisymmetric, and transitive is called a **partial order**. For example, $\{(a,b) : a,b \text{ are persons and } a \text{ is an ancestor of } b\}$ is a partial order (assuming that we consider each person to be an ancestor of himself or herself). Let $R \subseteq A \times A$ is a partial order, if for all of $b \in A, (b,a) \in R$ only when $b = a$, then $a \in A$ is said to be a **minimal element**. A finite partial order must have at least one minimal element, but an infinite partial order does not necessarily have a minimal element.

If $R \subseteq A \times A$ is a partial order, and for all $a, b \in A$, either $(a,b) \in R$ or $(b,a) \in R$, then R is said to be a **total order**. Thus the ancestor relation is not a total order since not any two people are ancestrally related (for example, siblings are not), but the less-than-or-equal-to relation on numbers is a total order. A total order cannot have two or more minimal elements.

A **path** in a binary relation R is a sequence $(a_1, \ldots, a_n)$ satisfying the following conditions, for $i = 1, \ldots, n-1$, $(a_i, a_{i+1}) \in R$, where $n \geq 1$, it is called a path from a_1 to a_n. The **length** of a path $(a_1, \ldots, a_n)$ is n. The path $(a_1, \ldots, a_n)$ is a **cycle**, if the a_i are all distinct and $(a_n, a_1) \in R$.

只要注意到由自反性,A 的每一个元素 a 在 Π 的某个集合中,即 $a \in [a]$,便容易看出 $\cup \Pi = A$.

于是,给定一个等价关系 R,总能够构造出对应的划分 Π. 例如,设 $R = \{(a,b) : a \text{ 与 } b \text{ 是两个人且有相同的双亲}\}$,则 R 的等价类是同胞兄弟姐妹. 注意:定理 1.3.1 的构造能够反过来说,即任给一个划分,可以构造一个对应的等价关系. 也就是说,设 Π 是 A 的一个划分,则 $R = \{(a,b) : a \text{ 和 } b \text{ 在 } \Pi \text{ 的同一个集合中}\}$ 是一个等价关系. 于是,在集合 A 上的等价关系与集合 A 的划分之间存在一个自然同构.

自反、反对称和传递等关系叫作**偏序**. 例如,$\{(a,b) : a \text{ 和 } b \text{ 是两个人且 } a \text{ 是 } b \text{ 的前辈}\}$就是一个偏序(假定把每一个人看作自己的前辈). 设 $R \subseteq A \times A$ 是一个偏序,若对所有的 $b \in A$,当且仅当 $b = a$ 时,$(b,a) \in R$,则称 $a \in A$ 是**极小元**. 有穷的偏序一定至少有一个极小元,无穷的偏序不一定有极小元.

若 $R \subseteq A \times A$ 是一个偏序,且对所有的 $a,b \in A$,或者 $(a,b) \in R$ 或者 $(b,a) \in R$,则称 R 是一个**全序**. 前辈关系不是全序,因为不是任意两人都有前辈关系(例如,也可能是兄弟姐妹),而自然数上的小于等于关系是一个全序. 全序不可能有两个或两个以上极小元.

二元关系 R 中的一条**通路**是一个满足下述条件的序列 $(a_1, \cdots, a_n)$,对于 $i = 1, \cdots, n-1$,$(a_i, a_{i+1}) \in R$,其中 $n \geq 1$,称这是一条从 a_1 到 a_n 的通路. 通路 $(a_1, \cdots, a_n)$ 的**长度**为 n. 如果所有的 a_i 都不相同且 $(a_n, a_1) \in R$,则通路 $(a_1, \cdots, a_n)$ 是一个**圈**.

1.4 Finite and infinite sets

A basic property of a finite set is its size, that is, the number of elements it contains. Some facts about the sizes of finite sets are so obvious they hardly need proof. For example, if $A \subseteq B$, then the size of A is less than or equal to that of B; the size of A is zero if and only if A is the empty set.

However, difficulties arise if we attempt to generalize the concept of "large" and "small" to infinite sets by our intuition. Is the set of multiples of 17 $\{0, 17, 34, 51, 68, \ldots\}$ more than the set of perfect squares $\{0, 1, 4, 9, 16, \ldots\}$? You can guess which one is more, but experience has shown that the only satisfactory convention is to regard these sets as having the same size.

We call two sets A and B **equinumerous** if there is a bijection $f: A \mapsto B$. Recall that if there is a bijection $f: A \mapsto B$, then there is a bijection $f^{-1}: B \mapsto A$, hence equinumerosity is a symmetric relation. In fact, as is easily shown, it is an equivalence relation. For example, $\{8, \text{red}, \{\varnothing, b\}\}$ and $\{1, 2, 3\}$ are equinumerous; let $f(8) = 1$, $f(\text{red}) = 2$, $f(\{\varnothing, b\}) = 3$. So are the multiples of 17 and the perfect squares. A bijection is given by $f(17n) = n^2$ for each $n \in \mathbf{N}$.

In general, we call a set **finite**, if intuitively, it is equinumerous with $\{1, 2, \ldots, n\}$ for some natural number n (For $n = 0$, $\{1, \ldots, n\}$ is the empty set, the empty set is equinumerous with itself, so the empty set is finite). If A and $\{1, \ldots, n\}$ are equinumerous, then we say that the cardinality of A (denoted as $|A|$) is n. The cardinality of a finite set is thus the number of elements in it. A set is **infinite**, if it is not finite. For example, the set $\mathbf{N}$ of natural numbers is infinite, so are sets such as the set of integers, the set of reals, and the set of perfect squares. However, not all infinite sets are equinumerous.

1.4 有穷集合与无穷集合

有穷集合的一个基本性质是它的大小,即它包含的元素个数.关于有穷集合的大小,某些事实是非常明显的,几乎不需要证明.例如,若$A \subseteq B$,则A的大小小于等于B的大小;当且仅当A是空集时A的大小等于0.

然而,如果我们打算凭直觉把"大"和"小"的概念推广到无穷集合上就比较困难.17的倍数的集合$\{0, 17, 34, 51, 68, \cdots\}$比完全平方数的集合$\{0, 1, 4, 9, 16, \cdots\}$大吗?可以猜想哪一个大.经验表明,只有认为这两个集合大小相同才是唯一合理的规定.

若存在双射$f: A \mapsto B$,则称集合A与B**等势**.想一想,若存在双射$f: A \mapsto B$,则存在双射$f^{-1}: B \mapsto A$.因此等势性是一种对称关系.事实上,容易证明它是一个等价关系.例如,$\{8, \text{red}, \{\varnothing, b\}\}$与$\{1, 2, 3\}$等势,可以令$f(8) = 1$, $f(\text{red}) = 2$, $f(\{\varnothing, b\}) = 3$.所有17的倍数与完全平方数也是等势的.对于每一个$n \in \mathbf{N}$,令$f(17n) = n^2$,则f是双射.

通常,直观上如果对于某个自然数n,一个集合与$\{1, 2, \cdots, n\}$等势,则称这个集合是**有穷**的(当$n = 0$时,$\{1, \cdots, n\}$是空集,空集与它自身等势,所以空集是有穷的).如果A与$\{1, \cdots, n\}$等势,则称A的基数(记作$|A|$)为n.有穷集合的基数就是它的元素个数.如果集合不是有穷的,则称它是**无穷**的.例如,自然数集合$\mathbf{N}$是无穷的,整数集合、实数集合和完全平方数集合等都是无穷的.但是,不是所有的无穷集合都等势.

A set is said to be **countably infinite** if it is equinumerous with $\mathbf{N}$, and **countable** if it is finite or countably infinite. A set that is not countable is **uncountable**. To prove that a set A is countably infinite we must exhibit a bijection f between A and $\mathbf{N}$; equivalently, we need to suggest a way in which A can be enumerated as

$$A = \{a_0, a_1, a_2, \ldots\}$$

and so on, since such an enumeration immediately suggests a bijection: $f(0) = a_0, f(1) = a_1, \ldots$.

For example, we can prove that the union of any finite number of countably infinite sets is countably infinite. Let us only illustrate the proof for the case of three pairwise disjoint, countably infinite sets, the proof for the general case can be done similarly. Let these 3 sets be A, B, and C. The sets can be listed as above: $A = \{a_0, a_1, \ldots\}$, $B = \{b_0, b_1, \ldots\}$, $C = \{c_0, c_1, \ldots\}$. Then their union can be listed as $A \cup B \cup C = \{a_0, b_0, c_0, a_1, b_1, c_1, a_2, \ldots\}$. This listing method is a way of "visiting" all the elements in $A \cup B \cup C$ by alternating between different sets, as illustrated in Fig. 1-7. This method of enumerating several sets alternated is called "dovetailing" (any carpenter can tell the origin of this name after looking at Fig. 1-7).

与 $\mathbf{N}$ 等势的集合是**可数无穷**的,有穷的或可数无穷的集合是**可数**的. 不是可数的集合称作**不可数**集合. 要证明集合 A 是可数无穷的,必须给出 A 与 $\mathbf{N}$ 之间的一个双射 f;等价地,只需给出一种方式,能够用这种方式把 A 枚举成

$$A = \{a_0, a_1, a_2, \cdots\}$$

这样的枚举立即给出一个双射: $f(0) = a_0, f(1) = a_1, \cdots$.

例如,可以证明任意有穷个可数无穷集合的并是可数无穷的. 下面给出对于3个两两不相交的可数无穷集合的证明,一般情况的证明可类似进行. 设这3个集合是 A, B 和 C. 可以像上面那样把它们列成 $A = \{a_0, a_1, \cdots\}$, $B = \{b_0, b_1, \cdots\}$, $C = \{c_0, c_1, \cdots\}$. 于是,它们的并可以列成 $A \cup B \cup C = \{a_0, b_0, c_0, a_1, b_1, c_1, \cdots\}$. 这种列表方法是通过在不同集合之间交替"访问" $A \cup B \cup C$ 中所有元素的一种方法,如图 1-7 所示. 这种交替枚举几个集合的方法叫作"制作榫头"(任何木工在看了图 1-7 之后都能说出这个名字的由来).

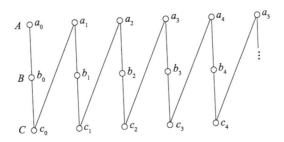

图 1-7 交替枚举法

Fig. 1-7 Alternating enumeration method

The same idea can be used to prove that the union of a countably infinite collection of countably infinite sets is countably infinite. For example, prove that **N**×**N** is countably infinite. Note that **N**×**N** is the union of {0}×**N**, {1}×**N**, {2}×**N**, and so on, that is, the union of a countably infinite collection of countably infinite sets. Dovetailing must here be more subtle than in the example above. It cannot be the same as in the previous example, where we access the second element of the first set before accessing the first element of each set, because with infinitely sets to visit, we could never even finish the first round! Instead we proceed as follows (Fig. 1-8).

可以用同样的思想来证明可数无穷个可数无穷集合的并是可数无穷的. 例如,可以证明 **N**×**N** 是可数无穷的. 注意 **N**×**N** 是 {0}×**N**,{1}×**N**,{2}×**N**,…的并,这是可数无穷个可数无穷集合的并. 在这里,制作榫头必须比上面的例子更巧妙一些. 不能与上例一样,在访问第一个集合的第二个元素之前访问每一个集合的第一个元素,因为有无穷多个集合要访问,根本不可能结束第一轮访问! 可换成如下方式进行访问(图 1-8).

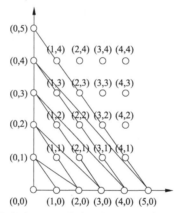

图 1-8 交替枚举法证明无穷个可数无穷集合的并是可数无穷的
Fig. 1-8 The proof of the countability of the union of an infinite collection of infinte sets using the method of alternating enumeration method

(1) In the first round, we visit the first element (0,0) from the first set.

(2) In the second round, we visit the next element (0,1) from the first set, and also the first element (1,0) from the second set.

(3) In the third round, we visit the next unvisited elements (0,2) and (1,1) of the first and second sets, and also the first element (2,0) of the third set.

(4) In general, in the nth round, we visit the nth element of the first set, the $(n-1)$th element of the second set, ... , and the first element of the nth set.

(1) 在第一轮,访问第一个集合的一个元素(0,0).

(2) 在第二轮,访问第一个集合的下一个元素(0,1)和第二个集合的第一个元素(1,0).

(3) 在第三轮,访问第一个和第二个集合的下一个未访问过的元素(0,2)和(1,1),还有第三个集合的第一个元素(2,0).

(4) 通常,在第 n 轮,访问第一个集合的第 n 个元素,第二个集合的第 $n-1$ 个元素,……,第 n 个集合的第一个元素.

Another way of viewing this use of dovetailing is to observe that the pair (i, j) is visited mth, where $m = \frac{1}{2}[(i+j)^2 + 3i + j]$; that is to say, the function $f(i, j) = \frac{1}{2}[(i+j)^2 + 3i + j]$ is a bijection from $\mathbf{N} \times \mathbf{N}$ to $\mathbf{N}$.

At the end of the next section, we present a technique for proving that two infinite sets are not equinumerous.

1.5　Three fundamental principle

Every proof is different, since every proof is designed to establish a different result. Like games of chess or baseball, patterns, experiences, skills can be exploited through a lot of observation. The main purpose of this section is to introduce three fundamental principles: **mathematical induction**, **pigeonhole principle**, and **diagonalization**. They appear in many proofs in various forms.

Principle of Mathematical Induction　Let A be a set of natural numbers such that

(1) $0 \in A$,

(2) for each natural number n, $if \{0, 1, \ldots, n\} \subseteq A$, then $n+1 \in A$, then $A = \mathbf{N}$.

In less formal terms, the principle of mathematical induction states that any set of natural numbers containing zero, if it has the following property: it contains $n+1$ as long as it contains all natural numbers less than or equal to n, then it is actually the set of all natural numbers. Intuitively the principle is obvious: every natural number must wind up in A since it can be "reached" from zero in a finite succession of steps by adding one each time. Another way to argue the same idea is by contradiction. Assume that (1) and (2) hold, but $A \neq \mathbf{N}$. Then there is a natural number not in A. In particular, let n be the first number among $0, 1, 2, \ldots$ that is not in A. Then n cannot be zero, since $0 \in A$ by (1); and since $\{0,1,\ldots,n-1\}$ $\subseteq A$ by the choice of n, then $n \in A$ by (2), which is a contradiction.

In practice, induction is used to prove assertions of the following form: "For all natural numbers n, property P is true." The principle of mathematical induction is applied to the set $A = \{n : P \text{ is true of } n\}$ in the following way.

(1) In the **basis step**, we prove that $0 \in A$, that is, for 0, P holds.

(2) The **induction hypothesis** is the assumption that for any fixed $n \geq 0$, P holds for each natural number $0, 1, \ldots, n$.

(3) In the **induction step**, using the induction hypothesis to show that P holds for $n+1$. By the induction principle, A is equal to $\mathbf{N}$, that is, P holds for every natural number.

Example 1.5.1 Proof: for any $n \geq 0$,
$$1+2+\ldots+n = \frac{1}{2}(n^2+n).$$

Basis Step Let $n = 0$, then the sum on the left is zero. Since there is nothing to add, the expression on the right is also zero.

Induction Hypothesis Suppose that, for some $n(n \geq 0)$, when $m \leq n$,
$$1+2+\ldots+m = \frac{1}{2}(m^2+m).$$

Induction Step
$$1+2+\ldots+n+(n+1)$$
$$= (1+2+\ldots+n)+(n+1)$$
$$= \frac{1}{2}(n^2+n)+(n+1) \text{ (by the induction hypothesis)}$$
$$= \frac{1}{2}(n^2+n+2n+2)$$
$$= \frac{1}{2}[(n+1)^2+(n+1)]$$

that's exactly what to prove.

n 的取法,$\{0,1,\cdots,n-1\} \subseteq A$,故由 (2),$n \in A$,矛盾.

在实践中,归纳法用来证明下述形式的论断:"对于所有的自然数 n,性质 P 成立."用下述方式把数学归纳法原理应用于集合 $A = \{n: 对于 n, P 为真\}$:

(1) 在**基础步骤**中证明 $0 \in A$,即对于 0,P 成立.

(2) **归纳假设**是假设对于任意固定的 $n \geq 0$,P 对于每一个自然数 0,$1,\cdots,n$ 成立.

(3) 在**归纳步骤**中,利用归纳假设证明 P 对于 $n+1$ 成立.于是,根据自然归纳法原理,A 等于 $\mathbf{N}$,即 P 对于每一个自然数成立.

例 1.5.1 证明:对于任意的 $n \geq 0$,
$$1+2+\cdots+n = \frac{1}{2}(n^2+n).$$

基础步骤 令 $n = 0$,左边的和等于零.因为没有要加的数,右边的表达式也等于零.

归纳假设 假设对于某个 $n(n \geq 0)$,当 $m \leq n$ 时,
$$1+2+\cdots+m = \frac{1}{2}(m^2+m).$$

归纳步骤
$$1+2+\cdots+n+(n+1)$$
$$= (1+2+\cdots+n)+(n+1)$$
$$= \frac{1}{2}(n^2+n)+(n+1)(由归纳假设)$$
$$= \frac{1}{2}(n^2+n+2n+2)$$
$$= \frac{1}{2}[(n+1)^2+(n+1)]$$

这正是要证明的结论.

Example 1.5.2 For any finite set A, $|2^A| = 2^{|A|}$, that is, the cardinality of the power set of A is equal to the the cardinality power of A of 2. We shall prove this statement by induction on the cardinality of A.

Basis Step Let A be a set of cardinality $n = 0$. Then $A = \varnothing$, and $2^{|A|} = 2^0 = 1$; on the other hand, $2^A = \{\varnothing\}$, and $|2^{|A|}| = |\{\varnothing\}| = 1$.

Induction Hypothesis Let $n \geq 0$, and assume that $|2^A| = 2^{|A|}$, when $|A| \leq n$.

Induction Step Let $|A| = n+1$. Since $n+1 > 0$, A contains at least one element a. Let $B = A - \{a\}$, then $|B| = n$. By the induction hypothesis, $|2^B| = 2^{|B|} = 2^n$. Now the power set of A can be divided into two parts, those sets containing the element a and those sets not containing a. The latter part is just 2^B, and the former part is obtained by introducing a into each member of 2^B. Thus
$$2^A = 2^B \cup \{C \cup \{a\} : C \in 2^B\}.$$
This division in fact partitions 2^A into two disjoint equinumerous parts, so the cardinality of the whole is twice $2^{|B|}$. By the induction hypothesis, we get $|2^A| = 2 \cdot 2^n = 2^{n+1}$, that's exactly what to prove.

We next use induction to prove our second fundamental principle, the pigeonhole principle.

Pigeonhole Principle If A and B are finite sets and $|A| > |B|$, then there is no one-to-one function from A to B. In other words, if we attempt to pair off the elements of A (the "pigeons") with elements of B (the "pigeonholes"), sooner or later, we will have to put more than one pigeon in a pigeonhole.

Basis Step Assume $|B| = 0$, that is, $B = \varnothing$. Then there is no function from A to B, let alone a one-to-one function.

Induction Hypothesis Assume that f is not one-to-one, if $f : A \mapsto B$, $|A| > |B|$, and $|B| \leq n$, where $n \geq 0$.

Induction Step Suppose that $f : A \mapsto B$ and $|A| >$

$|B|=n+1$, Choose a $a \in A$ (since $|A|>|B|=n+1 \geq 1$, A is nonempty, and therefore such a choice is possible). If there is another element of A, a', such that $f(a)=f(a')$, then obviously f is not a one-to-one function, and the proof is over. So, assume that a is the only element mapped by f to $f(a)$. Consider then the sets $A-\{a\}$, $B-\{f(a)\}$, and the function g from $A-\{a\}$ to $B-\{f(a)\}$. g is the same as f in $A-\{a\}$. Because $B-\{f(a)\}$ has n elements, and $|A-\{a\}|=|A|-1>|B|-1=|B-\{f(a)\}|$. According to the induction hypothesis, there are two distinct elements of $A-\{a\}$ that are mapped by g (thus also by f) to the same element of $B-\{f(a)\}$, and hence f is not one-to-one.

This simple fact is of use in a surprisingly large variety of proofs. Only a simple application is given here, and it will be pointed out in later chapters when it is used.

Theorem 1.5.1 Let R be a binary relation on a finite set A, and $a, b \in A$. If there is a path from a to b in R, then there is a path of length at most $|A|$.

Proof Assume that $(a_1, a_2, \ldots, a_n)$ is the shortest path from $a_1 = a$ to $a_n = b$, that is, the path with the smallest length, and assume that $n > |A|$. By the pigeonhole principle, there is an element of A that repeats on the path, for example, $a_i = a_j$, $1 \leq i < j \leq n$. Thus, $(a_1, a_2, \ldots, a_i, a_{j+1}, \ldots, a_n)$ is a shorter path from a to b, contradicting our assumption that $(a_1, a_2, \ldots, a_n)$ is the shortest path from a to b.

Finally, we come to our third basic proof technique, the diagonalization principle. Although it is not as widely used in mathematics as the other two principles we have discussed, it seems particularly well-suited for proving certain important results in the theory of computation.

Diagonalization Principle Let R be a binary relation on a set A, $D=\{a: a \in A \text{ and } (a,a) \notin R\}$ is

$n+1$，取一个 $a \in A$（因为 $|A|>|B|=n+1 \geq 1$，A 非空，故此做法是可行的）. 如果有 A 中的另一个元素 a' 使得 $f(a)=f(a')$，则显然 f 不是一对一的函数，证明结束. 因此，假设 a 是被 f 映射到 $f(a)$ 的唯一元素. 现在考虑集合 $A-\{a\}$ 和 $B-\{f(a)\}$，以及从 $A-\{a\}$ 到 $B-\{f(a)\}$ 的函数 g. 在 $A-\{a\}$ 上 g 与 f 相同. 因为 $B-\{f(a)\}$ 有 n 个元素，$|A-\{a\}|=|A|-1>|B|-1=|B-\{f(a)\}|$. 根据归纳假设，$A-\{a\}$ 有两个不同的元素被 g（因而也被 f）映射到 $B-\{f(a)\}$ 中的同一个元素，从而 f 不是一对一的.

在很多各种各样的证明中要用到这个简单的事实. 这里只给出一个简单的应用，在后面的各章当用到它时会指出来.

定理 1.5.1 设 R 是有穷集合 A 上的二元关系，$a,b \in A$. 如果在 R 中有一条从 a 到 b 的通路，则有一条长度不超过 $|A|$ 的通路.

证 假设 $(a_1,a_2,\cdots,a_n)$ 是从 $a_1=a$ 到 $a_n=b$ 的最短的通路，即长度最短的通路，又假设 $n>|A|$. 由鸽巢原理，A 有一个元素重复出现在这条通路上，比如说 $a_i=a_j$，$1 \leq i<j \leq n$. 于是，$(a_1,a_2,\cdots,a_i,a_{j+1},\cdots,a_n)$ 是一条更短的从 a 到 b 的通路，与假设 $(a_1,a_2,\cdots,a_n)$ 是从 a 到 b 的最短通路矛盾.

最后，介绍第三个基本原理——对角化原理. 虽然这个原理在数学中不像前面讨论的两个原理使用得那样广泛，但它似乎特别适合证明计算理论中的某些重要结果.

对角化原理 设 R 是集合 A 上的二元关系，则 $D=\{a: a \in A \text{ 且 } (a,a) \notin R\}$

called, the diagonal set of R. For each $a \in A$, let $R_a = \{b : b \in A \text{ and } (a,b) \in R\}$. Then D is distinct from each R_a.

If A is a finite set, then R can be drawn as a square array. The rows and columns are labeled with the elements of A, and there is a cross in the box with row labeled a and column labeled b just in case $(a,b) \in R$. The diagonal set D corresponds to the complement of the sequence of boxes along the main diagonal, boxes with crosses replaced by boxes without crosses, and boxes without crosses replaced by boxes with crosses. The sets R_a corresponds to the rows of the array. The diagonalization principle can then be stated as: the complement of the diagonal is different from each row.

Example 1.5.3 Let us consider the relation $R = \{(a,b), (a,d), (b,b), (b,c), (c,c), (d,b), (d,c), (d,e), (d,f), (e,e), (e,f), (f,a), (f,c), (f,d), (f,e)\}$; notice that $R_a = \{b,d\}$, $R_b = \{b,c\}$, $R_c = \{c\}$, $R_d = \{b,c,e,f\}$, $R_e = \{e,f\}$ and $R_f = \{a,c,d,e\}$. All in all, R may be drawn like Fig. 1-9.

	a	b	c	d	e	f
a		×		×		
b		×	×			
c			×			
d		×	×		×	×
e					×	×
f	×		×	×	×	

Fig. 1-9 Relation R

The sequence of boxes along the diagonal is

The complement of the given boxes sequence is

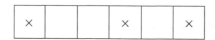

which corresponds to the diagonal set $D = \{a, d, f\}$. Indeed, D is different from each row of the array. For D, because of the way it is constructed, it differs from the first row in the first position, and differs from the second row in the second position, and so on.

The diagonalization principle holds for infinite sets as well, for the same reason: on the question of whether α is an element, the diagonal set is always different from the set R_a. So for any a, D cannot be the same as R_a.

Next, we will illustrate the use of diagonalization by a classic theorem of Georg Cantor (1845—1918).

Theorem 1.5.2 The set 2^N is uncountable.

Proof: Assume that 2^N is countably infinite. That is, we assume there is a way of enumerating all members of 2^N as

$$2^N = \{R_0, R_1, R_2, \ldots\}$$

(notice that R are the sets R_a in the statement of the diagonalization principle, when consider the relation $R = \{(i, j) : j \in R_i\}$.) Now consider the set

$$D = \{n \in \mathbf{N} : n \notin R_n\} \quad (D \text{ is the diagonal set})$$

D is a set of natural numbers, and therefore it should appear somewhere in the enumeration $\{R_0, R_1, R_2, \ldots\}$. But D cannot be R_0, because it differs from it with respect to containing 0 (it does if and only if R_0 does not); and it cannot be R_1 because D and R_1 differ in whether they contain 1; and so on. We can conclude that D does not appear on the enumeration at all, and this is a contradiction.

To restate the argument a little more formally, assume that $D = R_k$ for some $k \geq 0$ (since D is a set of natural numbers, and suppose $\{R_0, R_1, R_2, \ldots\}$ enumerates all subsets of $\mathbf{N}$, such a k must exist). We obtain a contradiction by asking whether $k \in R_k$:

(a) Assume the answer is yes, $k \in R_k$. Since $D = \{n \in \mathbf{N} : n \notin R_n\}$, it follows that $k \notin D$, but $D = R_k$, a contradiction.

(b) Assume the answer is no, $k \notin R_k$, then $k \in D$. But D is R_k, so $k \in R_k$, another contradiction.

We arrived at this contradiction starting from the assumption that $2^{\mathbf{N}}$ is countably infinite, and continuing by otherwise impeccably rigorous mathematical reasoning, we must therefore conclude that this assumption was in error. Hence $2^{\mathbf{N}}$ is uncountable.

In a similar way, we can prove that the set of all real numbers on the interval $[0,1]$ is uncountable. See Exercise 1, question 21.

1.6 Finite representations of languages

A central issue in the theory of computation is the representation of languages by finite specifications. Naturally, any finite language is amenable to finite representation by exhaustive enumeration of all the strings in the language. The issue becomes challenging only when infinite languages are considered.

Let us discuss about the notion of "finite representation of a language" more precisely. The first point to be made is that any such representation itself must be a string, a finite sequence of symbols over some alphabet Σ. Second, we certainly want different languages to have different representations, otherwise the term representation could hardly be considered appropriately. But these two requirements already imply that the possibilities for finite representation are severely limited. For the set Σ^* of strings over an alphabet Σ is countably infinite, so the number of possible representations of languages is countably infinite (This would remain true even if we were not bound to use a particular alphabet Σ, so long as the total number of available symbols was countably infinite). On the other hand, since $2^{\mathbf{N}}$ is not countably infinite, and thus the power set of any countably infinite set is not

(a) 假设 $k \in R_k$. 因为 $D = \{n \in \mathbf{N} : n \notin R_n\}$, 所以 $k \notin D$, 而 $D = R_k$, 矛盾.

(b) 假设 $k \notin R_k$, 则 $k \in D$. 而 $D = R_k$, 故 $k \in R_k$, 也矛盾.

我们从假设 $2^{\mathbf{N}}$ 是可数无穷的开始, 经过无懈可击的严格数学推理得出这个矛盾, 从而得出结论: 这个假设不成立. 因此, $2^{\mathbf{N}}$ 是不可数的.

用类似的方法可以证明区间 $[0,1]$ 上全体实数的集合是不可数的(见习题21).

1.6 语言的有穷表示

计算理论中的一个核心问题是用有穷的规定表示语言. 自然地, 任何有穷语言可以通过穷举该语言中的所有字符串给出它的有穷表示. 只有当考虑无穷语言时这个问题才是有争议的.

让我们更精确地讨论一下"语言的有穷表示"这个概念. 第一, 任何这样的表示本身必须是一个字符串, 即某个字母表 Σ 上的一个有穷的符号序列. 第二, 要求不同的语言有不同的表示, 否则很难认为所表示的这个词是合适的. 但是, 这两点要求已经意味着有穷表示的可能性是极其有限的. 一方面, 由于字母表 Σ 上的字符串集 Σ^* 是可数无穷的, 因此可能的语言表示也是可数无穷的(甚至不限制使用一个具体的字母表 Σ, 这仍然是对的, 只要所有可供使用的符号是可数无穷的). 另一方面, 由于 $2^{\mathbf{N}}$ 不是可数无穷的, 从而任何可数无穷集合的幂集不是可数无穷的, 因此在一个给定的字母表 Σ 上的所有可能的语言的集合 2^{Σ^*}

countably infinite, the set 2^{Σ^*} of all possible languages on a given alphabet Σ is uncountably infinite. There are only countable representations, but there are uncountable things to represent, we are unable to represent all languages finitely. Thus, the most we can hope is to find some kind of finite representation for at least some of the languages we are more interested in.

This is our first result in the theory of computation: no matter how powerful are the methods we use for representing languages, only a countable number of languages can be represented, so long as the representations themselves are finite. There are an uncountable number of languages in all, the vast majority of them will inevitably be missed under any finite representational scheme.

Of course, this is not the last thing we shall have to say along these lines. We shall describe several ways of describing and representing languages, each is more powerful than the previous in the following senses: each is capable of describing languages the previous one cannot. This hierarchy does not contradict the fact that all these finite representational methods are inevitably limited in scope for the reasons just explained.

We shall also want to derive ways of exhibiting particular languages that cannot be represented by the various representational methods we study. We know that the world of languages is inhabited by vast numbers of such unrepresentable specimens. But it's exceedingly difficult to catch one, put it on display, provide proof for it. This is probably a strange thing. Diagonalization arguments will eventually assist us here.

To begin our study of finite representations, we consider expressions, a string of symbols, that describe how languages can be built up by using the operations described in the previous section.

是不可数无穷的. 只有可数个表示, 但是有不可数个要表示的东西, 我们不可能有穷地表示所有的语言. 因此, 我们顶多可以希望至少为某些我们更感兴趣的语言找到某种有穷表示.

这是我们在计算理论中的第一个结果: 不论用来表示语言的方法怎样有力, 只要表示的本身是有穷的, 就只有可数个语言能够被表示. 总共有不可数个语言, 在任何有穷的表示方案下它们中的绝大多数将不可避免地被遗漏.

当然, 这不是我们根据这些线索要说的最后一件事情. 我们将要叙述几种描述和表示语言的方式, 在下述意义下每一个比前一个更有力: 每一个能够描述前一个不能描述的语言. 这个层次与所有这些有穷的表示方法由于刚才的解释而不可避免地被限制在某个范围内是不矛盾的.

我们还打算导出一些方法用来展示不能用我们所研究的各种表示方法表示的具体语言. 我们知道, 在语言世界中有大量的这种不可表示的奇怪的事物. 但是, 要抓住一个, 把它拿出来展览并且为它提供证明是非常困难的. 这大概是一件奇怪的事情. 对角化论证最终将帮助我们.

下面开始研究有穷表示. 首先考虑一种表达式, 即一串符号, 描述怎样用上一节的运算构造出语言.

Example 1.6.1 Let $L=\{w\in\{0,1\}^*: w$ has two or three occurrences of 1, the first and second of which are not consecutive$\}$. This language can be described using only singletons and the symbols $\cup$, $\circ$, and * as

$$\{0\}^*\circ\{1\}\circ\{0\}^*\circ\{0\}\circ\{1\}\circ\{0\}^*\circ((\{1\}\circ\{0\}^*)\cup\varnothing^*).$$

It is not hard to see that the language represented by the above expression is precisely the language L defined above. The important thing to notice is that the only symbols used in this representation are the braces $\{$ and $\}$, the parentheses(and), $\varnothing$, 0, 1, *, $\circ$, and $\cup$. In fact, we may dispense with the braces and $\circ$ and write simply

$$L=0^*10^*010^*(10^*\cup\varnothing^*).$$

Roughly speaking, an expression such as the one for L in Example 1.6.1 is called a regular expression. That is, a regular expression describes a language exclusively by means of single symbols and $\varnothing$, combined perhaps with the symbols $\cup$ and *, possibly with the aid of parentheses.

But in order to keep correct the expressions about which we are talking and the "mathematical vernacular" we are using for discussing them, we must tread rather carefully. Instead of using $\cup$, *, and $\varnothing$, which are the names in this book for certain operations and sets, we introduce special symbols $\cup$, *, and $\odot$ which should be regarded for the moment as completely free of meaning, just like the symbols a, b, and 0 used in earlier examples. The regular expressions over an alphabet Σ^* are all strings over the alphabet $\Sigma\cup\{(,),\odot,\cup,^\star\}$ that can be obtained as follows:

(1) Each member of $\odot$ and Σ is a regular expression.

(2) If α and β are regular expressions, then so is $(\alpha\beta)$.

(3) If α and β are regular expressions, then so is $(\alpha\cup\beta)$.

(4) If α is a regular expression, then so is $\alpha^\star$.

(5) Nothing is a regular expression unless it follows from (1) through (4).

Every regular expression represents a language, according to the symbols $\cup$ and * are interpreted as unions of sets and Kleene star respectively, and concatenations of expressions are interpreted as connections. Formally, the relation between regular expressions and the languages they represent is established by a function L, such that if α is any regular expression, then $L(\alpha)$ is the language represented by α. That is, L is a function from strings to languages. The function L is defined as follows:

(1) $L(\odot) = \varnothing$, and $L(\alpha) = \{\alpha\}$ for each $\alpha \in \Sigma$.

(2) If α and β are regular expressions, then $L((\alpha\beta)) = L(\alpha)L(\beta)$.

(3) If α and β are regular expressions, then $L((\alpha \cup \beta)) = L(\alpha) \cup L(\beta)$.

(4) If α is a regular expression, then $L(\alpha^\star) = L(\alpha)^*$.

Statement (1) defines $L(\alpha)$ for each regular expression α that consists of a single symbol. When the length of a regular expression α is greater than 1, (2) to (4) define $L(\alpha)$ in terms of $L(\alpha')$ for one or two regular expressions α' of shorter length. Thus, every regular expression is associated in this way with some language.

Example 1.6.2 What is $L(((a \cup b)^\star a))$? We have the following:

$L(((a \cup b)^\star a)) = L((a \cup b)^\star)L(a)$ by (2)
$= L((a \cup b)^\star)\{a\}$ by (1)
$= L((a \cup b))^*\{a\}$ by (4)
$= (L(a) \cup L(b))^*\{a\}$ by (3)
$= (\{a\} \cup \{b\})^*\{a\}$ by (1) twice
$= \{a, b\}^*\{a\}$
$= \{w \in \{a, b\}^* : w \text{ ends with an } a\}$

(4) 若 α 是正则表达式,则 $\alpha^\star$ 也是正则表达式.

(5) 除由(1)~(4)得到的正则表达式之外,没有任何别的正则表达式.

把符号 $\cup$ 和 * 分别解释成集合的并和 Kleene 星号,把表达式的并列解释成连接,每一个正则表达式表示一种语言. 形式上,用函数 L 建立起正则表达式与它们表示的语言之间的关系,如果 α 是任意一个正则表达式,则 $L(\alpha)$ 是 α 表示的语言. 也就是说,L 是从字符串到语言的函数. 函数 L 的定义如下:

(1) $L(\odot) = \varnothing$,且对每一个 $\alpha \in \Sigma, L(\alpha) = \{\alpha\}$.

(2) 若 α 和 β 是正则表达式,则 $L((\alpha\beta)) = L(\alpha)L(\beta)$.

(3) 若 α 和 β 是正则表达式,则 $L((\alpha \cup \beta)) = L(\alpha) \cup L(\beta)$.

(4) 若 α 是正则表达式,则 $L(\alpha^\star) = L(\alpha)^*$.

对于每一个由单个符号组成的正则表达式 α,(1)中定义了 $L(\alpha)$. 当正则表达式 α 的长度大于 1 时,(2)~(4)用一个或两个长度短一些的正则表达式 α' 的函数 $L(\alpha')$ 定义 $L(\alpha)$. 于是,每一个正则表达式以这种方式与某个语言相关联.

例 1.6.2 $L(((a \cup b)^\star a))$ 是什么?计算如下:

$L(((a \cup b)^\star a)) \xrightarrow{(2)} L((a \cup b)^\star)L(a)$
$\xrightarrow{(1)} L((a \cup b)^\star)\{a\}$
$\xrightarrow{(4)} L((a \cup b))^*\{a\}$
$\xrightarrow{(3)} (L(a) \cup L(b))^*\{a\}$
$\xrightarrow{2\text{次}(1)} (\{a\} \cup \{b\})^*\{a\}$
$= \{a, b\}^*\{a\}$
$= \{w \in \{a, b\}^* : w \text{ 以 } a \text{ 结束}\}$

Example 1.6.3 What language is represented by $(c^\star(a \cup (bc^\star))^\star)$? This regular expression represents the set of all strings over $\{a, b, c\}$ that do not have the substring ac. Clearly, no string in $L((c^\star(a \cup (bc^\star))^\star))$ can contain the substring ac, since each occurrence of a in such a string is either at the end of the string, or is followed by another occurrence of a, or is followed by an occurrence of b. On the other hand, let w be a string with no substring ac. Then w begins with zero or more c. If they are removed, the result is a string with no substring ac and not beginning with c. Any such string is in $L((a \cup (bc^\star))^\star)$; for it can be read, left to right, as a sequence of a, b, and c, with any blocks of c immediately following b (not following a, and not at the beginning of the string). Therefore $w \in L((c^\star(a \cup (bc^\star))^\star))$.

Example 1.6.4 $(0^\star \cup (((0^\star(1 \cup (11)))((00^\star)(1 \cup (11)))^\star)0^\star))$ represents the set of all strings over $\{0,1\}$ that do not have the substring 111.

Every language that can be represented by a regular expression can be represented by an infinite number of regular expressions. For example, α and $(\alpha \cup \odot)$ always represent the same language, so do $((\alpha \cup \beta) \cup \gamma)$ and $(\alpha \cup (\beta \cup \gamma))$. Since set union and concatenation are associative operations, that is, for all L_1, L_2 and L_3, there are $(L_1 \cup L_2) \cup L_3 = L_1 \cup (L_2 \cup L_3)$ and $(L_1 \circ L_2) \circ L_3 = L_1 \circ (L_2 \circ L_3)$, we normally omit the extra symbols (and) in regular expressions. For example, we treat $a \cup b \cup c$ as a regular expression even though "officially" it is not. For another example, the regular expression of Example 1.6.4 might be rewritten as

$$0^\star \cup 0^\star(1 \cup 11)(00^\star(1 \cup 11))^\star 0^\star.$$

Moreover, now that we have shown that regular expressions and the languages they represent can be defined formally and unambiguously. We feel free, when no confusion can result, to blur the distinction between the regular expressions and the "mathematical vernacular" we are using for talking about languages. Thus we may say at one point that a^*b^* is the set of all strings consisting of some number of as followed by some number of bs. To be precise, it should be written as $\{a\}^* \circ \{b\}^*$. At another point, we might say that a^*b^* is a regular expression representing that set. In this case, to be precise, it should be written as $(a^\star b^\star)$.

The class of **regular languages** over an alphabet Σ is defined to consist of all languages L such that $L = L(\alpha)$, where α is any regular expression over Σ. That is, regular languages are all languages that can be described by regular expressions. Alternatively, regular languages can be thought of in terms of closures. The class of regular languages over Σ is precisely the closure of the set of languages

$$\{\{\sigma\} : \sigma \in \Sigma\} \cup \{\varnothing\}$$

with respect to the functions of union, concatenation, and Kleene star.

We have already seen that regular expressions describe some nontrivial and interesting languages. Unfortunately, we can not describe by regular expressions some languages that have very simple descriptions by other methods. For example, $\{0^n1^n : n \geq 0\}$ will be proved in Chapter 2 not to be regular. Surely any theory of the finite representation of languages will have to accommodate at least such simple languages as this. Thus regular expressions are an inadequate specification method in general.

In search of a general method for finitely specifying languages, we might return to our general scheme

$$L = \{w \in \Sigma^* : w \text{ has property } P\}.$$

But which properties P should be needed? For example, what makes the preceding properties "w consists of a number of 0's followed by an equal number of 1's" and "w has no occurrence of 111" clearly acceptable? The reader may ponder about the right answer,

白话". 于是, 我们可以在一个地方说 a^*b^* 是由若干个 a 后面跟着若干个 b 构成的所有字符串的集合. 确切地讲, 应该写成 $\{a\}^* \circ \{b\}^*$. 在另一个地方我们还可以说 a^*b^* 是表示这个集合的正则表达式. 确切地讲, 这时应该写成 $(a^\star b^\star)$.

定义字母表 Σ 上的**正则语言**类由所有可写成 $L = L(\alpha)$ 的语言 L 组成, 其中 α 是 Σ 上的任一正则表达式. 即正则语言是所有能够用正则表达式描述的语言. 换一种说法, 可以认为正则语言是闭包. Σ 上的正则语言类恰好是语言集合

$$\{\{\sigma\} : \sigma \in \Sigma\} \cup \{\varnothing\}$$

关于并、连接和 Kleene 星号函数的闭包.

我们已经看到正则表达式描述了某些不平凡的、有趣的语言. 不幸的是, 不能用正则表达式描述某些可以用别的方法简单地描述的语言. 例如, 在第 2 章将证明 $\{0^n1^n : n \geq 0\}$ 不是正则的. 任何语言的有穷表示理论确实应该至少能接纳这么简单的语言. 因而, 正则表达式大体上是一种不充分规范的说明方法.

在寻找有关有穷地说明语言的一般方法时, 我们可以回到一般的方式

$$L = \{w \in \Sigma^* : w \text{ 具有 } P \text{ 性质}\}.$$

但是, P 应该具备什么样的性质? 例如, 是什么使得前面的性质 "w 由若干个 0 后面跟着个数相同的 1" 和 "w 中不出现 111" 明显是可以接受的? 读者可以去考虑正确的答案, 但是我们现

but we now accept and only accept the algorithmic properties. That is, for a property P of strings to be admissible as a specification of a language, there must be an algorithm for deciding whether a given string belongs to the language. An algorithm that is specifically designed, for some language L, to answer questions of the form "Is string w a member of L?" will be called a language recognition device. For example, a device for recognizing the language

$$L = \{w \in \{0,1\}^* : w \text{ does not have 111 as a substring}\}$$

may operate as follows: read the string from left to right, a symbol at a time. There is a counter, which starts at zero and is set back to zero every time a 0 is encountered in the input; add one every time a 1 is encountered in the input; stop with a "No" answer if the count ever reaches three, and stop with a "Yes" answer if the whole string is read without the count reaching three.

A completely different method for specifying a language is to describe how a generic specimen in the language is generated. For example, a regular expression such as $(e \cup b \cup bb)(a \cup ab \cup abb)^*$ my be viewed as a way of generating members of a language:

To generate a member of L, first write down either nothing, or b, or bb; then write down a or ab, or abb, and do this any number of times, including zero. All and only members of L can be generated in this way.

Such **language generators** are not algorithms, since they are not designed to answer questions and are not completely explicit about what to do (how to choose which of a, ab, or abb is to be written down?). But it is also an important and useful method of representing languages.

在接受并且只接受算法性质. 也就是说,为了使字符串 P 的性质可以接受为语言的规范,必须有算法能够确定给予的字符串是否属于该语言. 专门为某个语言 L 设计的,用来回答"字符串 w 是 L 的成员吗?"这种形式的问题的算法称作语言识别装置. 例如,识别语言

$$L = \{w \in \{0,1\}^* : w \text{ 不含子串 } 111\}$$

的装置可以进行如下运算:从左到右读字符串,每次读一个. 有一个计数器,开始时为 0,每次在输入中遇到 0 时将它置回到 0;每次在输入中遇到 1 时加 1;若计数已经达到 3,则停止计算并且回答"No";若读完整个字符串后计数没有达到 3,则停止计算并且回答"Yes".

说明语言的另一种完全不同的方法是描述如何生成语言中的一般样品. 例如,可以把正则表达式 $(e \cup b \cup bb)(a \cup ab \cup abb)^*$ 看作生成语言的成员的一种方法:

为了生成 L 的一个成员,第一步什么都不写,或者写 b,或者写 bb;然后写 a,或者 ab,或者 abb,包括 0 次在内,可以这样做任意次. L 中的所有和唯一的成员可以这样生成.

这种**语言生成器**不是算法,因为它们不是用来回答问题的,也不完全清楚做什么(怎样从 a,ab 和 abb 中选择一个要写的?)但它同样是一种重要和有用的表示语言的方法.

Exercise 1

1. Determine whether each of the following is true or false:
   (a) $\varnothing \in \varnothing$;
   (b) $\varnothing \subseteq \varnothing$;
   (c) $\varnothing \subseteq \{\varnothing\}$;
   (d) $\varnothing \in \{\varnothing\}$;
   (e) $\{a,b\} \in \{a,b,c,\{a,b\}\}$;
   (f) $\{a,b\} \subseteq \{a,b,\{a,b\}\}$;
   (g) $\{a,b\} \subseteq 2^{\{a,b,\{a,b\}\}}$;
   (h) $\{a,b\} \in 2^{\{a,b,\{a,b\}\}}$;
   (i) $\{a,b,\{a,b\}\} - \{a,b\} = \{a,b\}$.

2. Write the following sets using braces, commas, and numbers:
   (a) $(\{1,3,5\} \cup \{3,1\}) \cap \{3,5,7\}$;
   (b) $\cup \{\{3\},\{3,5\},\cap\{\{5,7\},\{7,9\}\}\}$;
   (c) $(\{1,2,5\} - \{5,7,9\}) \cup (\{5,7,9\} - \{1,2,5\})$;
   (d) $2^{\{7,8,9\}} - 2^{\{7,9\}}$;
   (e) $2^{\varnothing}$.

3. Prove each of the following equation:
   (a) $A \cup (B \cap C) = (A \cup B) \cap (A \cup C)$;
   (b) $A \cap (B \cup C) = (A \cap B) \cup (A \cap C)$;
   (c) $A \cap (A \cup B) = A$;
   (d) $A \cup (A \cap B) = A$;
   (e) $A - (B \cap C) = (A - B) \cup (A - C)$.

4. Write each of the following explicitly.
   (a) $\{1\} \times \{1,2\} \times \{1,2,3\}$;
   (b) $\varnothing \times \{1,2\}$;
   (c) $2^{\{1,2\}} \times \{1,2\}$.

5. Let $R = \{(a,b),(a,c),(c,d),(a,a),(b,a)\}$. What is $R \circ R$, the composition of R with itself? What is R^{-1}, the inverse of R? Is R, $R \circ R$, or R^{-1} a function?

6. Let $f: A \mapsto B$ and $g: B \mapsto C$. Let $h: A \mapsto C$ be their composition. In each of the following cases state h has the following properties if and only if f and g

习题 1

1. 判断下列各题是真是假:
   (a) $\varnothing \in \varnothing$;
   (b) $\varnothing \subseteq \varnothing$;
   (c) $\varnothing \subseteq \{\varnothing\}$;
   (d) $\varnothing \in \{\varnothing\}$;
   (e) $\{a,b\} \in \{a,b,c,\{a,b\}\}$;
   (f) $\{a,b\} \subseteq \{a,b,\{a,b\}\}$;
   (g) $\{a,b\} \subseteq 2^{\{a,b,\{a,b\}\}}$;
   (h) $\{a,b\} \in 2^{\{a,b,\{a,b\}\}}$;
   (i) $\{a,b,\{a,b\}\} - \{a,b\} = \{a,b\}$.

2. 用花括号、逗号和数字写出下列集合:
   (a) $(\{1,3,5\} \cup \{3,1\}) \cap \{3,5,7\}$;
   (b) $\cup \{\{3\},\{3,5\},\cap\{\{5,7\},\{7,9\}\}\}$;
   (c) $(\{1,2,5\} - \{5,7,9\}) \cup (\{5,7,9\} - \{1,2,5\})$;
   (d) $2^{\{7,8,9\}} - 2^{\{7,9\}}$;
   (e) $2^{\varnothing}$.

3. 证明下列等式:
   (a) $A \cup (B \cap C) = (A \cup B) \cap (A \cup C)$;
   (b) $A \cap (B \cup C) = (A \cap B) \cup (A \cap C)$;
   (c) $A \cap (A \cup B) = A$;
   (d) $A \cup (A \cap B) = A$;
   (e) $A - (B \cap C) = (A - B) \cup (A - C)$.

4. 写出下列各式的所有元素:
   (a) $\{1\} \times \{1,2\} \times \{1,2,3\}$;
   (b) $\varnothing \times \{1,2\}$;
   (c) $2^{\{1,2\}} \times \{1,2\}$.

5. 设 $R = \{(a,b),(a,c),(c,d),(a,a),(b,a)\}$. R 与自身的合成 $R \circ R$ 是什么? R 的逆 R^{-1} 是什么? R、$R \circ R$ 和 R^{-1} 是函数吗?

6. 设 $f: A \mapsto B, g: B \mapsto C$, 而 $h: A \mapsto C$ 是它们的合成. 叙述当且仅当 f 和 g 满足什么条件时 h 具有下述各条

satisfy what conditions:

(a) Onto;

(b) One-to-one;

(c) A bijection.

7. Let $R = \{(a,c),(c,e),(e,e),(e,b),(d,b),(d,d)\}$. Draw directed graphs representing each of the following types of relationships:

(a) R;

(b) R^{-1};

(c) $R \cup R^{-1}$;

(d) $R \cap R^{-1}$.

8. Draw directed graphs representing relations of the following types:

(a) Reflexive, transitive, and antisymmetric;

(b) Reflexive, transitive, and neither symmetric nor antisymmetric.

9. Let A be a nonempty set and let $R \subseteq A \times A$ be an empty set. Which properties does R have?

(a) Reflexivity;

(b) Symmetry;

(c) Antisymmetry;

(d) Transitivity.

10. Let $f: A \mapsto B$. Prove that the following relation R is an equivalence relation on A: $(a,b) \in R$ if and only if $f(a) = f(b)$.

11. Let $R \subseteq A \times A$ be a binary relation as defined below. In which cases is R a partial order? A total order?

(a) A is all positive integers; $(a,b) \in R$ if and only if b is divisible by a.

(b) $A = N \times N$; $((a,b),(c,d)) \in R$ if and only if $a \leq c$ or $b \leq d$.

(c) $A = N$; $(a,b) \in R$ if and only if $b = a$ or $b = a+1$.

(d) A is the set of all English words; $(a,b) \in R$ if and only if a is no longer than b.

(e) A is the set of all English words; $(a,b) \in R$ if and only if a is the same as b or occurs more frequently than b in the present book.

12. Let R_1 and R_2 be any two partial orders on the same set A. Prove that $R_1 \cap R_2$ is a partial order.

13. (a) Prove that if S is any collection of sets, then $R_s = \{(A, B) : A, B \in S \text{ and } A \subseteq B\}$ is a partial order.

(b) Let $S = 2^{\{1,2,3\}}$. Draw a directed graph representing the partial order R_S defined in (a). Which are the minimal elements of R_S?

14. Prove that any function from a finite set to itself contains a cycle.

15. Explicitly give bijections between each of the following pairs.

(a) $\mathbf{N}$ and the odd natural numbers.

(b) $\mathbf{N}$ and the set of all integers.

(c) $\mathbf{N}$ and $\mathbf{N} \times \mathbf{N} \times \mathbf{N}$.

(We are looking for formulas that are as simple as possible and involve only such operations as addition and multiplication.)

16. Let C be a set of sets defined as follows,

(1) $\varnothing \in C$.

(2) If $S_1 \in C$ and $S_2 \in C$, then $\{S_1, S_2\} \in C$.

(3) If $S_1 \in C$ and $S_2 \in C$, then $S_1 \times S_2 \in C$.

(4) The elements of C can be obtained from (1), (2) and (3).

(a) Explain carefully how to use (1) ~ (4) to get $\{\varnothing, \{\varnothing\}\} \in C$.

(b) Give a set S of ordered pairs such that $S \in C$, and $|S| > 1$.

(c) Does C contain any infinite sets? Explain.

(d) Is C countable or uncountable? Explain.

17. Prove that the dovetailing method of Fig. 1-8 visits the pair (i, j) at the m-th time, where:

$$m = \frac{1}{2}[(i+j)^2 + 3i + j]$$

书中出现的次数比 b 多.

12. 设 R_1 和 R_2 是同一个集合 A 上的任意两个偏序. 证明 $R_1 \cap R_2$ 是一个偏序.

13. (a) 证明:如果 S 是任意集合的汇集,则 $R_s = \{(A, B) : A, B \in S$ 且 $A \subseteq B\}$ 是一个偏序.

(b) 设 $S = 2^{\{1,2,3\}}$. 画出表示(a)中定义的偏序 R_S 的有向图. R_S 的极小元是什么?

14. 证明:有穷集合到它自身的任意函数包含一个圈.

15. 明确给出下列每一对集合之间的双射:

(a) $\mathbf{N}$ 与全体奇自然数.

(b) $\mathbf{N}$ 与所有整数的集合.

(c) $\mathbf{N}$ 与 $\mathbf{N} \times \mathbf{N} \times \mathbf{N}$.

(我们正在寻找尽可能简单的、只含有加法和乘法等运算的公式)

16. 设 C 是集合的集合,定义如下:

(1) $\varnothing \in C$.

(2) 若 $S_1 \in C$ 且 $S_2 \in C$,则 $\{S_1, S_2\} \in C$.

(3) 若 $S_1 \in C$ 且 $S_2 \in C$,则 $S_1 \times S_2 \in C$.

(4) C 的元素均可由(1),(2) 和(3)得到.

(a) 详细地说明怎么由(1) ~ (4) 得到 $\{\varnothing, \{\varnothing\}\} \in C$.

(b) 给出一个有序对的集合 S,使得 $S \in C$ 且 $|S| > 1$.

(c) C 包含无穷集合吗? 说明之.

(d) C 是可数的还是不可数的? 说明之.

17. 证明:图 1-8 所示的制作榫头的方法在第 m 次访问有序对 (i, j),其中,

$$m = \frac{1}{2}[(i+j)^2 + 3i + j]$$

18. Point out the errors in the "proof" below. The proof claims that all horses are the same color.

Proof by induction on the number of horses:

Basis Step: There is only one horse. Then clearly all horses have the same color.

Induction Hypothesis: In any group of up to n horses, all horses have the same color.

Induction Step: Consider a group of $n+1$ horses. Discard one horse, by the induction hypothesis, all the remaining horses have the same color. Now put that horse back and discard another. All the remaining horses have the same color. So all the horses have the same color as the ones that were not discarded either time, and so they all have the same color.

19. Prove that, if A and B are any finite sets, then there are $|B|^{|A|}$ functions from A to B.

20. Prove that in any group of at least two people, there are at least two persons that have the same number of acquaintances within the group. (Hint: Use the pigeonhole principle)

21. Prove that the set of all real numbers in the interval $[0, 1]$ is uncountable. (Hint: it is well known that each such number can be written in binary notation as an infinite sequence of 0s and 1s, such as 0.0110011100000…. Assume that an enumeration of these sequences exists, and create a "diagonal" sequence by "flipping" the ith bit of the ith sequence.)

22. What language is represented by the regular expression $(((a^\star a)b) \cup b)$?

23. Rewrite each of these regular expressions as a simpler regular expression representing the same set:

(a) $\odot^\star \cup a^\star \cup b^\star \cup (a \cup b)^\star$;

(b) $((a^\star b)^\star (b^\star a^\star)^\star)^\star$;

(c) $(a^\star b)^\star \cup (b^\star a)^\star$;

(d) $(a \cup b)^\star a (a \cup b)^\star$.

24. Which of the following are true? Explain.

(a) $baa \in a^\ast b^\ast a^\ast b^\ast$;

18. 指出下述"证明"中的错误. 该证明声称:所有马的颜色都相同.

对马匹数进行归纳证明.

基础步骤:只有一匹马. 于是显然所有的马是同一种颜色.

归纳假设:任何数量不超过 n 的一群马都是同一种颜色.

归纳步骤:考虑 $n+1$ 匹马. 牵走一匹马,根据归纳假设,所有留下的马是同一种颜色. 现在把牵走的马牵回来, 再牵走另一匹马. 留下的所有马也是同一种颜色. 于是,所有的马都与没有牵走过的马的颜色相同,从而所有的马是同一种颜色.

19. 证明:设 A 和 B 是任意两个有穷集合,则有 $|B|^{|A|}$ 个从 A 到 B 的函数.

20. 证明:在至少有两个人的任意一群人中,至少有两个相识的人数相同的人. (提示:利用鸽巢原理证明)

21. 证明:区间 $[0, 1]$ 上所有实数的集合是不可数的. (提示:众所周知, 每一个实数都可以用二进制表示写成 0 和 1 的无穷序列,如 0.0110011100000…. 假设能够枚举这些序列,用"翻转"第 i 个序列的第 i 位生成一个"对角线"序列.)

22. 正则表达式 $(((a^\star a)b) \cup b)$ 表示的是什么语言?

23. 把下列正则表达式重写成表示同样集合的更简单的正则表达式:

(a) $\odot^\star \cup a^\star \cup b^\star \cup (a \cup b)^\star$;

(b) $((a^\star b)^\star (b^\star a^\star)^\star)^\star$;

(c) $(a^\star b)^\star \cup (b^\star a)^\star$;

(d) $(a \cup b)^\star a (a \cup b)^\star$.

24. 下列各题中哪些是成立的? 并说明之.

(a) $baa \in a^\ast b^\ast a^\ast b^\ast$;

(b) $b^*a^* \cap a^*b^* = a^* \cup b^*$;
(c) $a^*b^* \cap b^*c^* = \varnothing$;
(d) $abcd \in (a(cd)^*b)^*$.

习题 1 答案
Key to Exercise 1

Part II Automata and Languages

Chapter 2 Finite Automata

2.1 Deterministic finite automata

A finite automata(**FA**) is a good model for computers with extremely limited storage. In the initial stage of describing the mathematical theory of finite automata, only abstract descriptions are made, and no specific applications are involved. Fig. 2-1 depicts a finite automaton, called M_1.

第 II 部分 自动机与语言

第 2 章 有穷自动机

2.1 确定型有穷自动机

有穷自动机(**FA**)是关于存储量极其有限的计算机的很好的模型. 在叙述有穷自动机的数学理论的初期, 只做抽象的描述, 不涉及具体的应用. 图 2-1 描述了一个有穷自动机, 称之为 M_1.

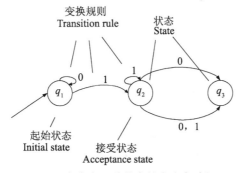

图 2-1 一台含有 3 种状态的有穷自动机 M_1
Fig. 2-1 A finite automaton M_1 with three states

Fig. 2-1 is called the **state diagram** of M_1. It has 3 states, denoted as q_1, q_2 and q_3. The initial state q_1 is marked with an arrow pointing to itself, with no starting point, and the acceptance state q_2 has a double circle. The arrow from one state to another is called a transition.

Example:
010: reject
11: accept
010100100100100: accept
010000010010: reject
ξ: reject

Use this automaton to test various output strings, and learn that it accepts strings 1, 01, 11 and 01010101. In fact, it accepts:

图 2-1 叫作 M_1 的**状态图**. 它有 3 个状态, 分别记作 q_1,q_2 和 q_3. 起始状态 q_1 用一个指向它的无出发点的箭头标示, 接受状态 q_2 带有双圈. 从一个状态到另外一个状态的箭头叫作转移.

例:
010: reject
11: accept
010100100100100: accept
010000010010: reject
ξ: reject

用这台自动机对各式各样的输出串进行试验, 得知它接受字符串 1, 01, 11 和 01010101. 事实上, 它接受:

(1) any string of 0 or 1 ending in 1;

(2) in any string of 0 or 1 with an even number of 0s after the last 1;

(3) reject other strings of 0 or 1.

The finite automaton first includes a set of finite states, and also includes the transition from one state to another. The finite automaton looks like a directed graph, where the states are the nodes of the graph, and the state transitions are the edges of the graph. In addition, there must be an initial state and at least one accepting state among these states. This is the **formal definition of finite automata**.

Example 2.1.1 Given the state diagram of the finite automaton M_2 (Fig. 2-2). Please give a formal description and determine the language it can recognize.

(1) 以 1 结尾的任何 0,1 串;

(2) 在最后一个 1 的后面有偶数个 0 的任何 0,1 串;

(3) 拒绝其他串的 0,1 串.

有穷自动机首先包含一个有限状态的集合,还包含从一个状态到另外一个状态的转移. 有穷自动机看上去就像是一个有向图,其中状态是图的结点,而状态转换则是图的边. 此外这些状态中还必须有一个初始状态和至少一个接受状态. 这就是**有穷自动机的形式定义**.

例 **2.1.1** 给定有穷自动机 M_2 的状态图(图 2-2). 请给出形式化的描述,并确定其能识别的语言.

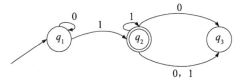

图 2-2 有穷自动机 M_2

Fig. 2-2 The finite automaton M_2

M_2 can be written formally as $M_2 = (Q, \Sigma, \delta, q_1, F)$, where

(1) Q represents the state set, $Q = \{q_1, q_2, q_3\}$.

(2) Σ stands for the alphabet, $\Sigma = \{0, 1\}$.

(3) δ is the transfer rule (transfer function), which is given in Tab. 2-1.

可以把 M_2 形式地写成 $M_2 = (Q, \Sigma, \delta, q_1, F)$,其中:

(1) Q 代表状态集合,$Q = \{q_1, q_2, q_3\}$.

(2) Σ 代表字母表,$\Sigma = \{0, 1\}$.

(3) δ 是转移规则(转移函数),由表 2-1 给出.

表 2-1 M_2 的 δ

Tab. 2-1 The δ of M_2

状态 State	输入 Input	
	0	1
q_1	q_1	q_2
q_2	q_3	q_2
q_3	q_2	q_2

(4) q_1 is the initial state.

(5) F is the acceptance state set, $F = \{q_2\}$.

(4) q_1 是起始状态.

(5) F 是接受状态集,$F = \{q_2\}$.

If A is the set of all character strings accepted by machine M, then A is the language of machine M, denoted as $L(M) = A$, also known as **M recognizing A or M accepting A**.

$L(M_2) = \{\omega | \omega$ has at least one 1 and an even number of 0s after the last 1$\}$

Each step of a finite automaton is determined, so it can be called a **deterministic finite automaton (DFA)**. Deterministic finite automaton means that when a state faces an input symbol, it transitions to a uniquely deterministic state. Its characteristic is that only one edge with a certain symbol can be sent from each state. That is to say, the same symbol cannot appear on the two edges of the same state.

Example 2.1.2 Given the state diagram of the finite automaton M_3 (Fig. 2-3). please give a formal description and determine the language it can recognize.

$M_3 = (\{q_1, q_2\}, \{0, 1\}, \delta, q_1, \{q_1\})$

The δ of M_3 is given in Table 2-2.

若 A 是机器 M 接受的全部字符串集,则称 A 是机器 M 的语言,记作 $L(M) = A$,又称 **M 识别 A 或 M 接受 A**.

$L(M_2) = \{\omega | \omega$ 至少有一个 1 并且在最后一个 1 后面有偶数个 0$\}$.

有穷自动机的每一步操作都是确定的,因此可称为**确定型有穷自动机(DFA)**. 确定型有穷自动机就是当一个状态面对一个输入符号时,它所转换到的是一个唯一确定的状态. 其特点是从每一个状态只能发出一条具有某个符号的边. 也就是说,同一个符号不能出现在同一状态发出的两条边上.

例 2.1.2 给定有穷自动机 M_3 的状态图(图 2-3). 请给出形式化的描述并确定其能识别的语言.

$M_3 = (\{q_1, q_2\}, \{0, 1\}, \delta, q_1, \{q_1\})$

M_3 的 δ 由表 2-2 给出.

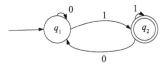

图 2-3 M_3 的状态图

Fig. 2-3 State diagram of M_3

表 2-2 M_3 的 δ

Tab. 2-2 The δ of M_3

状态 State	输入 Input	
	0	1
q_1	q_1	q_2
q_2	q_1	q_2

The tranfer function δ is

$L(M_3) = \{\omega | \omega$ ends with 1$\}$

Example 2.1.3 Given the state diagram of the finite automaton M_4 (Fig. 2-4). Please give a formal description and determine the language it can recognize.

$M_4 = (Q, \Sigma, \delta, q_0, F)$ where $Q = \{s, q_1, q_2, r_1, r_2\}, \Sigma = \{a, b\}$

转移函数 δ 为

$L(M_3) = \{\omega | \omega$ 以 1 结束$\}$

例 2.1.3 给定有穷自动机 M_4 的状态图(图 2-4). 请给出形式化的描述,并确定其能识别的语言.

$M_4 = (Q, \Sigma, \delta, q_0, F)$,其中 $Q = \{s, q_1, q_2, r_1, r_2\}, \Sigma = \{a, b\}$

The transfer function δ is given in Table 2-3.　　　转移函数 δ 由表 2-3 给出.

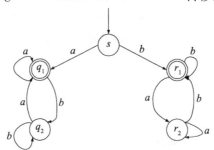

图 2-4　有穷自动机 M_4
Fig. 2-4　Finite automaton M_4

表 2-3　M_4 的 δ
Table 2-3　The δ of M_4

状态 State	输入 Input	
	a	b
s	q_1	r_1
q_1	q_1	q_2
q_2	q_1	q_2
r_1	r_2	r_1
r_2	r_2	r_1

$F = \{q_1, r_1\}$

M_4 accepts all strings with the same start and end symbols.

Example 2.1.4　Given the state diagram of the finite automaton M_5 (Fig. 2-5). Please give a formal description and determine the language it can recognize.

$F = \{q_1, r_1\}$

M_4 接受开始和结束符号相同的所有字符串.

例 2.1.4　给定有穷自动机 M_5 的状态图(图 2-5). 请给出形式化的描述, 并确定其能识别的语言.

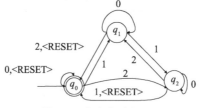

图 2-5　有穷自动机 M_5
Fig. 2-5　Finite automaton M_5

$\Sigma = \{0, 1, 2, <\text{RESET}>\}$

We use <RESET> as a symbol.

M_5 records the sum of the numbers it reads in the input string in modulo 3. Each time it reads the symbol <RESET>, it rests the count to 0. If the sum

$\Sigma = \{0, 1, 2, <\text{RESET}>\}$

我们把<RESET>作为一个符号.

M_5 以模 3 的方式记录它在输入串中读到的数字之和. 每次读到符号<RESET>, 它将计数重新置于 0. 如果

modulo 3 is equal to 0, in other words, the sum is a multiple of 3, then M_5 accepts.

The formal definition of finite automata is introduced above, and the design of finite automata is introduced below.

The most important factors in designing finite automata are two elements:

(1) **State**: Something that needs to be remembered.

(2) **Transfer**: According to the input symbol, transfer from one state to another state.

Example 2.1.5 $L(E_1) = \{\omega | \omega$ has an odd number of $1\}$, $\Sigma = \{0,1\}$.

First, we determine two states, odd q_{even} and even q_{odd}, as shown in Fig. 2-6.

Fig. 2-6 The two states q_{even} and q_{odd}

Then we join the transition, read 0 in the case of even number 1, keep even number unchanged; in the case of even number 1, read 1, become odd, enter the odd state; read 0 in the odd state of 1, the state does not change, read 1, becomes an even number (Fig. 2-7).

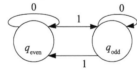

Fig. 2-7 q_{even} and q_{odd} after adding the transfer

Finally, the initial state and the acceptance state are added, as shown in Fig. 2-8.

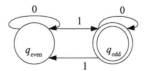

Fig. 2-8 Add initial state and acceptance state

2.2 Non-deterministic finite automata

Non-deterministic finite automata(NFA), which allows multiple edges with the same sign to be emitted from one state, and even edges marked with the ε (denoting empty) symbol, that is to say, NFA can automatically switch to the next state along the ε edge without entering any character.

The difference between deterministic finite automata (abbreviated as DFA) and non-deterministic finite automata (abbreviated as NFA) is obvious.

(1) Each state of DFA always has exactly one transition arrow shot out for each symbol in the alphabet. The non-deterministic finite automaton shown in Fig. 2-9 breaks this rule. State q_1 has one arrow shot for 0 and two arrows shot for 1. q_2 has an arrow for 0 and no arrow for 1. In NFA, a state may have 0, 1, or more arrows emitted for each symbol in the alphabet.

(2) In DFA, the label on the transfer arrow is a symbol taken from the alphabet. N_1 is an arrow with the label ε. Generally speaking, NFA arrows can mark symbols or ε in the alphabet. It is possible to shoot 0, 1, or more arrows with the label ε from a state.

Non-deterministic calculation: In the non-deterministic calculation, it is divided into ε movement and multiple options, and different backups are generated. When it cannot be moved, the backup disappears. When there is a backup accepted, the entire calculation is accepted.

2.2 非确定型有穷自动机

非确定型有穷自动机(NFA)允许从一个状态发出多条具有相同符号的边,甚至允许发出标有 ε(表示空)符号的边.也就是说,NFA 可以不输入任何字符就自动沿 ε 边转移到下一个状态.

确定型有穷自动机(DFA)和非确定型有穷自动机(NFA)之间的区别是显而易见的.

(1) DFA 的每一个状态对于字母表中的每一个符号总是恰好有一个转移箭头射出.图 2-9 中给出的非确定型有穷自动机打破了这个规则.状态 q_1 对于 0 有一个射出的箭头,而对于 1 有两个射出的箭头. q_2 对于 0 有一个箭头,而对于 1 没有箭头.在 NFA 中,一个状态对于字母表中的每一个符号可能有 0 个、1 个或者多个射出的箭头.

(2) 在 DFA 中,转移箭头上的标号是取自字母表的符号. N_1 是一个带有标号 ε 的箭头.一般来说,NFA 的箭头可以标记字母表中的符号或 ε. 从一个状态可能射出 0 个、1 个或多个带有标号 ε 的箭头.

非确定型计算:在非确定型计算中分为 ε 移动和多种选择,产生不同的备份.当无法移动时,该备份消失.当有一个备份接受时,整个计算就接受.

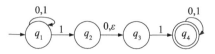

图 2-9 非确定型有穷自动机 N_1

Fig. 2-9 Non-deterministic finite automaton N_1

Example 2.2.1 $L(N_2) = \{\omega \mid$ the 3rd letter from the bottom of ω is $1\}$, $\Sigma = \{0,1\}$.

We do it in two ways: deterministic and non-deterministic:

(1) Non-deterministic: guess the 3rd letter from bottom (Fig. 2-10).

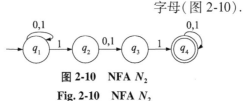

Fig. 2-10　NFA N_2

(2) Deterministic: memorize the last three letters.

Each NFA can be converted into an equivalent DFA, but sometimes that DFA may have many states. NFA can describe a language more simply than DFA, which is the advantage of NFA (Fig. 2-11).

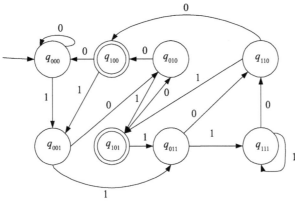

Fig. 2-11　DFA

The non-deterministic finite automaton is a five-tuple $(Q, \Sigma, \delta, q_0, F)$, where

Q: finite state set

Σ: input alphabet, ($\Sigma_\varepsilon = \Sigma \cup \{\varepsilon\}$)

δ: $Q \times \Sigma_\varepsilon \to P(Q)$, transfer function

$q_0 \in Q$: initial state

$F \subseteq Q$: receiving state (end state)

This is the **formal definition of non-deterministic finite automata.**

Example 2.2.2 $N_1 = (Q, \Sigma, \delta, q_0, F)$; $Q = \{q_1, q_2, q_3, q_4\}$; $\Sigma = \{0,1\}$; $F = \{q_4\}$; δ (Tab. 2-4):

Tab. 2-4 The δ of N_1

State	Input		
	0	1	ε
q_1	$\{q_4\}$	$\{q_1, q_2\}$	$\varnothing$
q_2	$\{q_3\}$	$\varnothing$	$\{q_3\}$
q_3	$\varnothing$	$\{q_4\}$	$\varnothing$
q_4	$\{q_4\}$	$\{q_4\}$	$\varnothing$

Fig. 2-12 NFA N_1

The formal definition of NFA calculation is also similar to that of DFA calculation. Let $N = (Q, \Sigma, \delta, q_0, F)$ be an NFA, and ω is a character string on the alphabet Σ. If ω can be written as $\omega = y_1 y_2 \cdots y_m$, where each y_i is a member of Σ_ε, and the state sequence $r_0, r_1, \ldots, r_m$ existing in Q satisfies the following three conditions:

(1) $r_0 = q_0$;
(2) $r_{i+1} \in \delta(r_i, y_{i+1})$, $i = 0, 1, \ldots, m-1$;
(3) $r_m \in F$,

It is said that N accepts ω.

2.3 Finite automata and regular expressions

In arithmetic, the basic object is the number, and the tool is the operation of the number, such as " + " and " × ". In computational theory, the object is language, and tools include operations specifically designed for processing languages. The three operations that define a language are called **regular operations**, and these operations are used to study the properties of regular languages.

Assume A and B are two languages, and define regular operations **union**, **connection**, and **asterisk** as follows:

union: $A \cup B = \{x \mid x \in A \text{ or } x \in B\}$
connection: $A \circ B = \{xy \mid x \in A \text{ and } y \in B\}$
asterisk: $A^* = \{x_1 x_2 \ldots x_k \mid k \geq 0, \text{ and each } x_i \in A\}$

In arithmetic, expressions can be constructed using the operators + and ×. Similarly, regular operators can be used to construct expressions describing languages, which are called **regular expressions**. For example,

$$(0 \cup 1) 0^*$$

The value of a regular expression is a language. The value of this regular expression is a language composed of all strings of a 0 or a 1 followed by any number of 0. Regular expressions play an important role in computer science applications. Regular expressions are often used to retrieve and replace text that conforms to a pattern (rule). Many programming languages support string operations using regular expressions. For example, a powerful regular expression engine is built in Perl. The concept of regular expression was first popularized by tool software in Unix (such as sed and grep). Regular expressions are usually abbreviated as "regex". The singular includes regexp and regex, and the plural includes regexps, regexes and regexen.

The characteristics of regular expressions are as follows:

(1) They are very flexible, logical and functional.

(2) They can quickly achieve the complex control of string in a very simple way.

(3) It is rather obscure and difficult for people who have just come into contact.

Because the main application objects of regular expression is text, it is used in various text editors, ranging from the famous editor EditPlus, to large edi-

设 A 和 B 是两个语言，定义正则运算**并**、**连接**和**星号**如下：

并：$A \cup B = \{x \mid x \in A \text{ 或 } x \in B\}$
连接：$A \circ B = \{xy \mid x \in A \text{ 且 } y \in B\}$
星号：$A^* = \{x_1 x_2 \cdots x_k \mid k \geq 0 \text{ 且每一个 } x_i \in A\}$

在算术中可以用运算符"+"和"×"构建表达式.类似地，可以用正则运算符构建描述语言的表达式，称作**正则表达式**.例如，

$$(0 \cup 1) 0^*$$

正则表达式的值是一种语言，这个正则表达式的值是由一个 0 或一个 1 后面跟着任意个 0 的所有字符串组成的语言.正则表达式在计算机科学应用中起着重要的作用.正则表达式通常被用来检索、替换那些符合某个模式(规则)的文本.许多程序设计语言都支持利用正则表达式的字符串操作.例如，在 Perl 中就内建了一个功能强大的正则表达式引擎.正则表达式这个概念最初是由 Unix 中的工具软件(如 sed 和 grep)普及开的.正则表达式通常缩写成"regex"，单数有 regexp、regex，复数有 regexps、regexes、regexen.

正则表达式的特点如下：

(1) 有很强的灵活性、逻辑性和功能性.

(2) 可以用极简单的方式迅速地达到对复杂的字符串的控制.

(3) 对于刚接触的人来说，比较晦涩难懂.

由于正则表达式主要应用对象是文本，因此它被应用于各种文本编辑器，小到著名的编辑器 EditPlus，大到 Microsoft word、Visual Studio 等大型编

tors such as Microsoft word and Visual Studio, all can use regular expressions to process text content.

Let R be a regular expression. If R is

(1) a, where a is an element of the alphabet Σ;

(2) ε;

(3) $\emptyset$;

(4) $(R_1 \cup R_2)$, where R_1 and R_2 are regular expressions;

(5) $(R_1 \circ R_2)$, where R_1 and R_2 are regular expressions;

(6) (R_1^*), where R_1 is a regular expression.

This is the **formal definition of a regular expression**, when you want to clearly distinguish the regular expression R and the language it represents, write the language represented by R as $L(R)$.

Regular expressions and finite automata are equivalent in terms of their descriptive capabilities. Although finite automata and regular expressions are very different on the surface, any regular expression can be converted into a finite automaton that recognizes the language it describes. The regular language is a language recognized by a finite automaton.

Theorem 2.3.1 A language is regular if and only if it can be described by regular expressions.

This theorem has two directions. Each direction is stated as a separate lemma.

Lemma 2.3.1 (1) If a language can be described by regular expressions, then it is regular.

(2) If a language is regular, it can be described by regular expressions.

Proof idea of lemma (1): Assume that the regular expression R describes the language A. It is necessary to explain how to convert R into an NFA that recognizes A. If an NFA recognizes A, then A is regular.

Proof idea of lemma (2): It must be proved that if language A is regular, there is a regular expression

辑器,都可以使用正则表达式来处理文本内容.

设 R 是一个正则表达式,如果 R 是

(1) a,这里 a 是字母表 Σ 中的一个元素;

(2) ε;

(3) $\emptyset$;

(4) $(R_1 \cup R_2)$,这里 R_1 和 R_2 是正则表达式;

(5) $(R_1 \circ R_2)$,这里 R_1 和 R_2 是正则表达式;

(6) (R_1^*),这里 R_1 是正则表达式.

这就是**正则表达式的形式定义**. 想要明确区分正则表达式 R 和它所表示的语言,可把 R 表示的语言写成 $L(R)$.

正则表达式和有穷自动机就其描述能力而言是等价的. 尽管有穷自动机和正则表达式在表面上看非常不同,但是任何正则表达式都能转换成识别它所描述语言的有穷自动机. 正则语言是被有穷自动机识别的语言.

定理2.3.1 一种语言是正则的,当且仅当可以用正则表达式描述它.

这个定理有两个方向. 每个方向都被叙述成一条单独的引理.

引理2.3.1 (1) 如果一种语言可以用正则表达式描述,那么它是正则的.

(2) 如果一种语言是正则的,那么它可以用正则表达式描述.

引理(1)的证明思路如下:假设正则表达式 R 描述语言 A,要说明怎么样把 R 转换成一台识别 A 的 NFA. 如果一台 NFA 识别 A,那么 A 是正则的.

引理(2)的证明思路如下:必须证明如果语言 A 是正则的,那么有一个

describing it. Since A is regular, it is accepted by a DFA.

Structure of NFA: The core of NFA is actually three states (Fig. 2-13—Fig. 2-15).

(1) $r = s|t$.

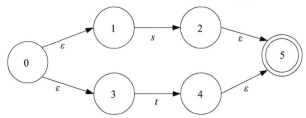

图 2-13　状态 1
Fig. 2-13　State one

(2) $r = st$.

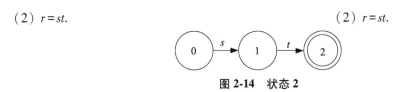

图 2-14　状态 2
Fig. 2-14　State two

(3) $r = s^*$.

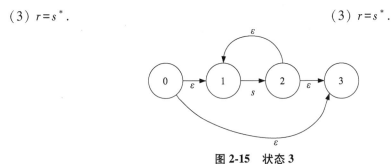

图 2-15　状态 3
Fig. 2-15　State three

Method: All the remaining expressions are a combination of the above three types, so what we have to do is two points.

(1) Remember the three expressions clearly.

(2) Make a clear distinction between which state and which state is consistent when collocation.

Example 2.3.1　$r = (s|t)^*$.

This is obviously a combination of two of them. The parentheses take precedence. Let's draw the state inside of the parentheses and analyze the state outside as well (Fig. 2-16).

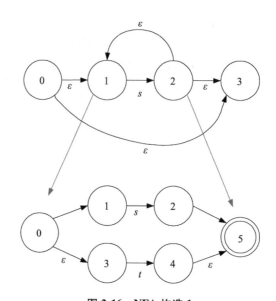

图 2-16 NFA 构造 1
Fig. 2-16 NFA structure one

It can be found that the following 0 and 5 are the beginning and ending states of the outermost layer of $s|t$, which are the beginning and ending states of the innermost layer of $(s|t)$, so the two are correspondingly combined into one (Fig. 2-17).

可以发现,下面的 0 和 5 作为 $s|t$ 最外层的始末状态,就是 $(s|t)$ 最内层的始末状态,所以两者合成一个状态(图 2-17).

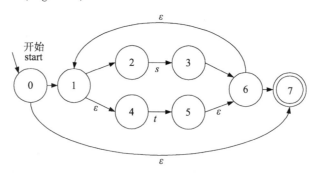

图 2-17 NFA 构造 2
Fig. 2-17 NFA structure one

2.4 Regular language and non-regular language

In Section 1, we have learned about the recognition of finite automata. If a language is recognized by a finite automaton (this automaton can be a deterministic finite automaton or a non-deterministic finite automaton), it is called a **regular language.**

2.4 正则语言与非正则语言

在 2.1 节中我们已经了解了有穷自动机的识别. 如果一个语言被一台有穷自动机识别(这个自动机可以是确定型有穷自动机也可是非确定型有穷自动机),那么称它是**正则语言**.

In Section 3, three operations of regular languages are introduced, which are union operation, connection operation and asterisk operation. The essence of calculation is cycle, and the essence of calculation in reality is also to keep repeating the cycles for calculation. Automata calculation is in the finite automata, you can do cyclic calculations, and use asterisk operations to achieve cyclic calculations.

Regular language is closed. The set of regular language is closed in union operation, connection operation and asterisk operation.

Closure description: A and B are both regular languages. A can find an automaton to recognize the language, and B can also find an automaton to recognize the language. Then an automaton must be found, which can respectively recognize: $A \cup B$, $A \circ B$, A^* language. (Three automata respectively recognize three languages.) If A and B are both regular languages, then $A \cup B$, $A \circ B$ and A^* are all regular languages.

The finite automata also have limitations, and some languages cannot be recognized by the finite automata. For example, let language $B = \{0^n 1^n \mid n \in \mathbf{N}\}$. If you want to find a DFA that recognizes B, you will find that this machine seems to need to remember how many 0s it reads in the input. Since there is no limit to the number of 0s, the machine will have to remember infinite possibilities. But this cannot be done with finite automata. However, such judgments are not rigorous, such as $C = \{\omega \mid \text{count}(0) = \text{count}(1)\}$ and $D = \{\omega \mid \text{count}(01) = \text{count}(10)\}$ (where $\text{count}(x)$ denotes counting the number of occurrences of the string x in ω), these two languages seem to be very similar, but C is not a regular language, and D is a regular language.

So we introduce a **pumping lemma** theorem. The theorem states that all regular languages have a specific value—**pumping length**. All strings in the language can be "pumped" as long as their length is not less than the pumping length. The following is a formal

在2.3节中介绍了正则语言的三种运算,分别是并运算、连接运算和星号运算.计算的本质就是循环,在现实中,其计算的本质也是不停地循环、计算.自动机计算就是在有穷自动机中,做循环计算,使用星号运算实现循环计算.

正则语言具有封闭性.正则语言组成的集合,在并运算、连接运算和星号运算中都是封闭的.

封闭性描述:A 和 B 都是正则语言,A 可以找到一个自动机识别该语言,B 也可以找到一个自动机识别该语言,那么一定可以找到一个自动机,分别可以识别 $A \cup B$,$A \circ B$,A^* 语言.(3个自动机分别识别3种语言.)若 A,B 都是正则语言,那么 $A \cup B$,$A \circ B$ 和 A^* 都是正则语言.

有穷自动机也是有其局限性的,有些语言不能被有穷自动机识别.比如,设语言 $B = \{0^n 1^n \mid n \in \mathbf{N}\}$.如果想找一台识别 B 的 DFA,那么这台机器需要记住它在输入中读到了多少个0,由于0的个数没有限制,因此机器将不得不记住无穷多种可能.但用有穷自动机无法做到这一点.不过这样的评判并不严谨,如 $C = \{\omega \mid \text{count}(0) = \text{count}(1)\}$ 和 $D = \{\omega \mid \text{count}(01) = \text{count}(10)\}$(其中 $\text{count}(x)$ 表示统计字符串 x 在 ω 中出现的次数),这两个语言看似很像,但是 C 不是正则语言,而 D 是正则语言.

因此我们引入了**泵引理**.该引理指出,所有的正则语言都有一个特定值——**泵长度**,语言中的所有字符串只要长度不小于泵长度,就可以被"抽取".下面是泵引理的形式化表述:

expression of the pumping lemma:

If A is a regular language, there is a number p (pumping length) such that if s is any string in A with a length not less than p, then s can be divided into three segments, $s=xyz$, satisfying the following conditions:

(1) $xy^iz \in A, \forall i \in \mathbf{N}$;

(2) $|y|>0$;

(3) $|xy| \leqslant p$,

where, $|\ |$ denotes the length of the string, and y^i denotes consecutive i ys.

Example 2.4.1 Let B be the language $\{0^n1^n \mid n \geqslant 0\}$. We use the pumping lemma to prove that B is not regular. The proof adopts the contradiction method. Assume the opposite, that is, B is regular. Let p be the pumping length given by the pumping lemma. Select s as the string 0^p1^p. Because s is a member of B and the length of s is greater than p, the pumping lemma guarantees that s can be divided into three segments $s=xyz$, so that for any $i \geqslant 0$, xy^iz is in B. Consider three situations below to show that this conclusion is impossible.

(1) The string y only contains 0. In this case, there are more 0s than 1s in the string $xyyz$, so it is not a member of B. Violation of pumping lemma condition 1, contradiction.

(2) The string y contains only 1, which also gives contradictions.

(3) The string y contains both 0 and 1. In this case, the number of 0 and 1 in the string $xyyz$ may be equal, but their order is messed up, and 1 appears before 0. Therefore, $xyyz$ is not a member of B, contradiction.

Therefore, if B is assumed to be regular, the contradiction is inevitable. Therefore, B is not regular.

In this example, it is easy to find the string s, because any string of length p or greater than p in B will work.

若 A 是一个正则语言,则存在一个数 p(泵长度),使得如果 s 是 A 中任一长度不小于 p 的字符串,那么 s 可以被分成三段,$s=xyz$,满足下述条件:

(1) $xy^iz \in A, \forall i \in \mathbf{N}$;

(2) $|y|>0$;

(3) $|xy| \leqslant p$,

其中,$|\ |$ 表示字符串的长度,y^i 表示连续的 i 个 y.

例 2.4.1 设 B 是语言 $\{0^n1^n \mid n \geqslant 0\}$. 用泵引理证明 B 不是正则的.

采用反证法. 假设相反, 即 B 是正则的. 令 p 是由泵引理给出的泵长度. 选择 s 为字符串 0^p1^p. 因为 s 是 B 的一个成员且 s 的长度大于 p,所以泵引理保证 s 可以分成 3 段 $s=xyz$,使得对于任意的 $i \geqslant 0, xy^iz$ 在 B 中. 下面考虑 3 种情况,说明这个结论是不可能的.

(1) 字符串 y 只包含 0. 在这种情况下,字符串 $xyyz$ 中的 0 比 1 多,从而不是 B 的成员,与泵引理的条件(1)矛盾.

(2) 字符串 y 只包含 1,这种情况同样存在矛盾.

(3) 字符串 y 既包含 0 也包含 1. 在这种情况下,字符串 $xyyz$ 中 0 和 1 的个数可能相等,但是它们的顺序乱了,在 0 的前面出现 1. 因此, $xyyz$ 不是 B 的成员,矛盾.

于是,如果假设 B 是正则的,那么矛盾就是不可避免的. 因此,B 不是正则的.

在这个例子中,很容易找到字符串 s,因为在 B 中任何长度为 p 或大于 p 的字符串都可作为 s.

Example 2.4.2 Let $F = \{\omega\omega | \omega \in \{0,1\}^*\}$. Use the pumping lemma to show that F is non-regular.

Assume the opposite, that is, F is regular. Let p be the sub-length given by the pumping lemma. Let s be the string $0^p 1 0^p 1$. Because s is a member of F, and the length of s is greater than p, the pumping lemma guarantees that s can be divided into three segments, $s = xyz$, which satisfies the three conditions in the lemma. It is impossible to prove this result.

Condition 3 is indispensable. Without this, let x and z be empty strings, we can pump s. When condition 3 is satisfied, the proof is as follows: because y must only consist of 0, $xyyz \notin F$.

Choose $s = 0^p 1 0^p 1$, and notice that this is a string that shows the "essence" of F's irregularity. Conversely, for example, the string $0^p 0^p$ cannot. Although $0^p 0^p$ is a member of F, it can be pumped, so no contradiction can be given.

Example 2.4.3 When using the pumping lemma, sometimes "pump" is used. Use the pumping lemma to prove that $E = \{0^i 1^j | i > j\}$ is not regular. The proof adopts the contradiction method.

Assume that E is regular. Let p be the pumping length with respect to E given by the pumping lemma. Let $s = 0^{p+1} 1^p$. Then s can be divided into xyz and satisfy the conditions of the pumping lemma. According to condition 3, y only contains 0. Let's check the string $xyyz$ to see if it can be in E. Adding a y increases the number of 0s. And E contains all the strings in $0^* 1^*$ that are more than 1, so increasing the number of 0 will still give the string in E, and there is no contradiction. We need to try other ways.

The pumping lemma states that when $i = 0$, there is also $xy^i z \in E$, so we consider the string $xy^0 z = xz$. Deleting y reduces the number of 0s, and there is only one more 0 than 1 in s. Therefore, there can be no more 0s in xz than 1, so it cannot be a member of E. So there is a contradiction.

例 2.4.2 令 $F = \{\omega\omega | \omega \in \{0,1\}^*\}$. 用泵引理说明 F 是非正则的.

假设相反,即 F 是正则的. 令 p 是泵引理给出的分长度. 设 s 是字符串 $0^p 1 0^p 1$. 因为 s 是 F 的一个成员,并且 s 的长度大于 p,所以泵引理保证 s 能够被划分成 3 段, $s = xyz$, 满足引理中的 3 个条件. 证明这个结果是不可能的.

条件(3)是不可缺少的. 如果没有这一条件,令 x 和 z 为空串,我们就能够抽取 s. 当满足条件(3)时,因为 y 一定仅由 0 组成,故 $xyyz \notin F$.

选择 $s = 0^p 1 0^p 1$,注意这是一个能显示 F 的非正则性的"本质"的字符串. 相反地,字符串 $0^p 0^p$ 就不能显示 F 的正则性. 虽然 $0^p 0^p$ 是 F 的一个成员,但是它能被抽取,所以不能给出矛盾.

例 2.4.3 在运用泵引理时,有时要使用"抽取". 用泵引理证明 $E = \{0^i 1^j | i > j\}$ 不是正则的.

采用反证法. 假设 E 是正则的. 设 p 是泵引理给出的关于 E 的泵长度. 令 $s = 0^{p+1} 1^p$. 于是 s 能够被划分成 xyz,且满足泵引理的条件. 根据条件(3), y 仅包含 0. 检查字符串 $xyyz$,看它是否在 E 中. 添加一个 y 使 0 的数目增加. 而 E 包含 $0^* 1^*$ 中所有 0 多于 1 的字符串,因而增加 0 的数目仍给出 E 中的字符串,没有矛盾. 我们需要试试别的办法.

泵引理指出,当 $i = 0$ 时也有 $xy^i z \in E$,因此考虑字符串 $xy^0 z = xz$. 删去 y 使 0 的数目减少,而 s 中 0 只比 1 多一个. 因此, xz 中的 0 不可能比 1 多,从而它不可能是 E 的一个成员. 于是得到矛盾.

2.5 State minimization

The most simplified DFA: This DFA has no redundant states, nor two equivalent states. A DFA can be transformed into a finite automaton with its equivalent minimum state by eliminating useless states and merging equivalent states.

Useless state: The state that is not reachable by any input string from the starting state of automation, or this state has no path to reach the final state.

Equivalent state: Two states recognize the same string, and the results are both correct or wrong. These two states are equivalent.

Difference state: Not equivalent state.

The simplification of DFA is the division of the state set according to the equivalence class:

(1) Make the states in any two different subsets distinguishable, and any state in the same state is equivalent.

(2) Any two subsets are disjoint.

(3) Finally, each subset retains a state.

The simplification process of DFA:

According to ε, the state set of DFA can be divided into two subsets of **final state** and **non-final state**, forming a basic division Π.

Assuming that at a certain time, Π contains m subsets, that is, $\Pi = \{I^{(1)}, I^{(2)}, \ldots, I^{(m)}\}$, check whether each subset in Π can be further divided by the following methods:

For a certain $I^{(i)}$, let $I^{(i)} = \{s_1, s_2, \ldots, s_n\}$, if there is an input character a such that $I_a^{(i)}$ is not included in a certain subset $I^{(j)}$ of any current Π, then $I^{(i)}$ should be at least divided into two parts (if s_1 and s_2 arrive at two different state sets after the character a is entered, and the existence of the character a can distinguish the two state sets, so the character a can distinguish the states s_1 and s_2, that is, s_1 and s_2 are not equivalent).

2.5 状态最小化

最简化的 DFA：这个 DFA 没有多余状态，也没有两个相互等价的状态. 一个 DFA 可以通过消除无用状态、合并等价状态而转换成一个与之等价的最小状态的有穷自动机.

无用状态：从自动机开始状态出发，任何输入串也无法到达的那个状态，或者这个状态没有可达最终状态的通路.

等价状态：两个状态识别相同的串，结果都同为正确或错误的，这两种状态就是等价的.

区别状态：不是等价状态.

DFA 的化简即是按等价类对状态集的划分：

（1）任何两个不同子集中的状态是可区分的，而同一状态中的任意状态是等价的.

（2）任何两个子集均不相交.

（3）每个子集保留一个状态.

DFA 的化简过程如下：

据 ε 可以将 DFA 的状态集分为**终态**和**非终态**两个子集，形成基本划分 Π.

假定某个时候，Π 包含 m 个子集，即为 $\Pi = \{I^{(1)}, I^{(2)}, \cdots, I^{(m)}\}$，检查 Π 中的每个子集是否可以进一步划分，检查方法如下：

对某个 $I^{(i)}$，设 $I^{(i)} = \{s_1, s_2, \cdots, s_n\}$，若存在一个输入字符 a 使得 $I_a^{(i)}$ 不包含在任何一个现行 Π 的某个子集 $I^{(j)}$ 中，则 $I^{(i)}$ 至少应该分为两部分（若 s_1 与 s_2 在字符 a 输入后，到达两个不同的状态集，而存在的字符 a 可以区分这两个状态集，则字符 a 是可以区分状态 s_1 与 s_2 的，即 s_1 与 s_2 不等价）.

Next, divide $I^{(i)}$ by the character a, component contains two parts $I^{(i1)}$ and $I^{(i2)}$ respectively containing s_1 and s_2:

(1) $I^{(i1)}$ contains s_1, $I^{(i1)} = \{s \mid s \in I^{(i)}$, and s and s_1 arrive subset of the same current Π via arc $a\}$;

(2) $I^{(i2)}$ contains s_2, $I^{(i2)} = I^{(i)} - I^{(i1)}$.

The new subset separated by $I^{(i1)}$ should be added to the inspection queue to determine whether it can proceed further division.

Generally speaking, for a certain character a and subset $I^{(i)}$, if $I_a^{(i)}$ falls into N different subsets of the current Π, then $I^{(i)}$ should be divided into N disjoint subsets according to whether the subsets into which the received character a falls are the same or not. After the division is completed, a state in each subset I is selected to represent the subset. The state selected from the subset containing the original initial state is the new initial state, and the state selected from the subset containing the original final state is the new final state.

Example 2.5.1 Simplify the following DFA (Fig. 2-18).

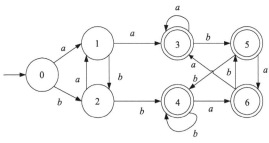

Fig. 2-18 DFA

$I = \{0,1,2,3,4,5,6\}$, where the final state is $\{3,4,5,6\}$. So the initial division is $I^{(1)} = \{3,4,5,6\}$, $I^{(2)} = \{0,1,2\}$, the subsets waiting to be divided are $I^{(1)}$, $I^{(2)}$, $\Pi = \{I^{(1)}, I^{(2)}\}$.

Check whether $I^{(1)}$ can be divided by character a or b, $I_a^{(1)}$ is included in $I^{(1)}$, and $I_b^{(1)}$ is also included in $I^{(1)}$, that is, the states in $I^{(1)}$ are equivalent, do not need to be divided. The subsets waiting to be divided are $I^{(2)}$, $\Pi = \{I^{(1)}, I^{(2)}\}$.

Check whether $I^{(2)}$ can be divided by the character a, $I_a^{(2)} = \{1, 3\}$, which fall into $I^{(1)}$ and $I^{(2)}$ respectively, so $I^{(2)}$ is divided into $I^{(21)} = \{0, 2\}$, $I^{(22)} = \{1\}$.

The subsets waiting to be divided are $I^{(21)}$, $I^{(22)}$, $\Pi = \{I^{(1)}, I^{(21)}, I^{(22)}\}$. Check whether $I^{(21)}$ can be divided by the character a, $I_a^{(21)}$ is included in $I^{(22)}$, check whether $I^{(21)}$ can be divided by the character b, $I_b^{(2)} = \{2, 4\}$, respectively fall into $I^{(21)}$ and $I^{(1)}$, so $I^{(21)}$ is divided into $I^{(211)} = \{0\}$, $I^{(212)} = \{2\}$.

The subsets waiting to be divided are $I^{(22)}$, $I^{(211)}$, $I^{(212)}$, $\Pi = \{I^{(1)}, I^{(22)}, I^{(211)}, I^{(212)}\}$. The size of $I^{(22)}$ is 1, and there is no need to continue to divide. The size of $I^{(211)}$ is 1, and there is no need to continue to divide. The size of $I^{(212)}$ is 1, and there is no need to continue to divide.

The division is complete, $\Pi = \{I^{(1)}, I^{(22)}, I^{(211)}, I^{(212)}\}$ $I^{(1)} = \{3, 4, 5, 6\}$, $I^{(22)} = \{1\}$, $I^{(211)} = \{0\}$, $I^{(212)} = \{2\}$, then the states 0, 1, 2, and 3 are reserved, where 0 is the initial state and 3 is the final state. According to the conversion relationship, the simplified result is as follows (Fig. 2-19):

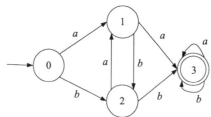

Fig. 2-19 Simplified NFA

2.6 Algorithmic aspects of finite automata

The process status and switching in the computer operating system can be used as an approximate understanding of the DFA algorithm. As shown in Fig. 2-20 below, the ellipse represents the state, the connection between the states represents the event, the state of the process and the event are all determinable, and can be exhaustively listed.

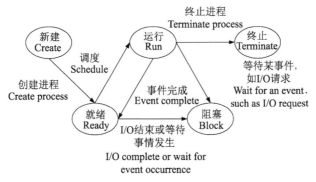

Fig. 2-20　Approximate understanding of DFA algorithm

The DFA algorithm has a variety of applications, here we first introduce the application in the field of matching keywords.

(1) **Matching keywords**: We can use each text segment as a state, for example, "matching keywords" can be divided into "match" "matching" "matching key" "matching keyword" and "matching keywords" five text fragments (Fig. 2-21).

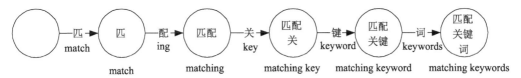

Fig. 2-21　Matching keywords

Procedure:

① The initial state is empty, and when the event "match" is triggered, it will switch to the state "match";

② When the event is triggered "ing", it will switch to the state "matching";

③ By analogy, until the switch "matching keywords" for the last state.

Let us consider the case of multiple keywords, such as "matching algorithm" "matching keywords" and "information extraction" (Fig. 2-22).

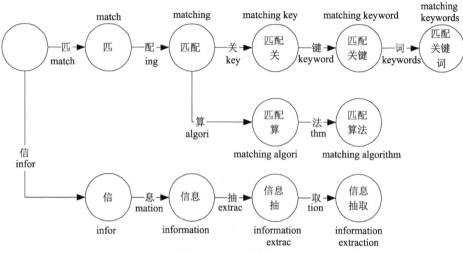

图 2-22 匹配关键词 2
Fig. 2-22 Matching keywords 2

You can see that the state diagram in the above figure is similar to a tree structure. It is precisely because of this structure that the DFA algorithm is faster than the keyword iteration method "for loop" in keyword matching. The time complexity of the tree structure is less than the time complexity of the for loop.

for loop:

keyword_list = [ ]

for keyword in ["Matching Algorithm", "Matching Keywords", "Information Extraction"]:

 if keyword in "DFA algorithm matching keywords":

  keyword_list.append(keyword)

The "for" loop needs to traverse the keyword table. With the expansion of the keyword table, the time will be longer and longer.

DFA algorithm: when a "match" is found, it will only go to a specific sequence according to the event, such as "matching keyword", not to the "matching algorithm", so the number of traversal is less than that of the for loop.

(2) **Sensitive word filtering**: among the algorithms for text filtering, DFA is the only better algorithm. DFA is Deterministic Finite Automaton. It ob-

可以看到上图的状态图结构类似树形结构,也正是因为这个结构,使得 DFA 算法在关键词匹配方面要快于关键词迭代方法(for 循环).树形结构的时间复杂度要小于 for 循环的时间复杂度.

for 循环:

keyword_list = [ ]

for keyword in ["匹配算法""匹配关键词""信息抽取"]:

 if keyword in "DFA 算法匹配关键词":

  keyword_list.append(keyword)

for 循环需要遍历关键词表,随着关键词表的扩充,所需的时间也会越来越长.

DFA 算法:找到"匹"时,只会按照事项走向特定的序列,例如"匹配关键词",而不会走向"匹配算法",因此遍历的次数要小于 for 循环.

(2) **敏感词过滤**:在实现文字过滤的算法中,DFA 是唯一比较好的实现算法. DFA 即 Deterministic Finite

tains the next state through event and current state, that is, event + state = nextstate. The following figure shows the transition of its state (Fig. 2-23).

Automaton, 也就是确定有穷自动机. 它通过 event 和当前的 state 得到下一个 state, 即 event + state = nextstate. 图 2-23 展示了其状态的转换.

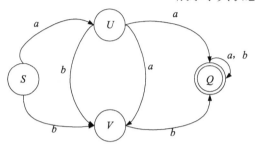

图 2-23 状态转换
Fig. 2-23 State transition

In this figure, the upper case letters S, U, V, Q are states, and the lower case letters a and b are actions.

In the algorithm for filtering sensitive words, we must reduce operations, while DFA has little calculation in DFA algorithm, only state conversion.

The key to the implementation of DFA in Java is the implementation of DFA. First, we analyze the above figure. In this process, we think the following structure will be clearer (Fig. 2-24).

图中, 大写字母 S, U, V, Q 都是状态, 小写字母 a 和 b 为动作.

在实现敏感词过滤的算法中, 必须要减少运算, 而 DFA 在 DFA 算法中几乎没有什么计算, 有的只是状态的转换.

在 Java 中实施敏感词过滤的关键就是 DFA 算法的实施. 首先对上图进行剖析. 在这个过程中考虑下面这种结构会更加清晰明了 (图 2-24).

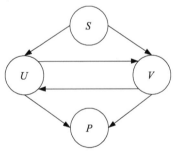

图 2-24 结构图
Fig. 2-24 Structure diagram

At the same time, there is no state transition, no action, and only Query (look up). We can think that through S query U and V, through U query V and P, and through V query U and P. Through this transformation, we can transform the transformation of state into lookup using Java set.

同时, 这里没有状态转换, 没有动作, 有的只是 Query (查找). 可以认为, 通过 S 查找 U, V, 通过 U 查找 V, P, 通过 V 查找 U, P. 通过这样的转换就可以将状态的转换转变为使用 Java 集合来查找.

Exercise 2

1. Fig. 2-25 shows the state diagram of two DFA M_1 and M_2. Please answer the following questions about these two machines.

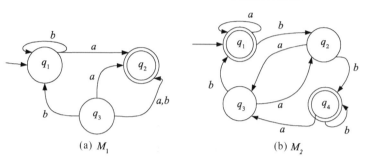

Fig. 2-25 Two DFAs

(a) What is the starting state of M_1?
(b) What is the acceptance state set of M_1?
(c) What is the starting state of M_2?
(d) What is the acceptance state set of M_2?
(e) For input $aabb$, what is the state sequence of M_1?
(f) Does M_1 accept string $aabb$?
(g) Does M_2 accept string ε?

2. Given a formal description of machines M_1 and M_2 drawn in question 1.

3. The formal description of DFA M is $(\{q_1, q_2, q_3, q_4, q_5\}, \{u, d\}, \delta, q_3, \{q_3\})$, where δ given in Tab. 2-5. Try to draw the state diagram of the machine.

Tab. 2-5 δ of M

	u	d
q_1	q_1	q_2
q_2	q_1	q_3
q_3	q_2	q_4
q_4	q_3	q_5
q_5	q_4	q_5

4. Draw the state diagram of the DFA that recognizes the following languages. Here, the alphabets are all $\{0,1\}$.

(a) $\{\omega|\omega$ starts from 1 and ends with $0\}$;

(b) $\{\omega|\omega$ contains at least three 1s$\}$;

(c) $\{\omega|\omega$ contains substring 0101, that is, for a certain x and y, $\omega=x0101y\}$;

(d) $\{\omega|\omega$ starts from 0 and has an odd length, or starts from 1 and has an even length$\}$;

(e) $\{\omega|\omega$ does not contain substring 110$\}$;

(f) $\{\omega|\omega$'s length does not exceed 5 $\}$;

(g) $\{\omega|\omega$ is any string except 11 and 111$\}$;

(h) $\{\omega|\omega$'s odd positions are all 1$\}$;

(i) $\{\omega|\omega$ contains at least two 0s and at most a 1$\}$;

(j) $\{\omega|\omega$ contains an even number of 0s or exactly two 1s$\}$;

(k) $\{\varepsilon, 0\}$;

(l) Empty set;

(m) All strings except the empty string.

5. Given an NFA that recognizes the following languages, and the required number of states must be met.

(a) Language $\{\omega|\omega$ ends with $00\}$, 3 states;

(b) Language $\{0\}$, 2 states;

(c) Language $0^*1^*0^*0$, 3 states;

(d) Language $\{\varepsilon\}$, 1 state;

(e) Language 0^*, 1 state.

6. Prove that every NFA can be converted into an equivalent NFA with only one accepting state.

7. (a) Prove: Assume M is a DFA that recognizes language B, and exchange the accepting state and non-accepting state of M to obtain a new DFA, then this new DFA recognizes the complement of B. Therefore, the regular language class is closed under the complement operation.

(b) Illustrate: Assume M is an NFA that recognizes language B. Exchange the acceptance and non-acceptance states of M to obtain a new NFA. This new

NFA does not necessarily recognize the complement of B. Are the language classes recognized by the NFA closed under the complement operation? Explain your answer.

8. Use the pumping theorem to prove that the following language is not regular.

(a) $A_1 = \{0^n 1^n 2^n \mid n \geqslant 0\}$.

(b) $A_2 = \{\omega\omega\omega \mid \omega \in \{a, b\}^*\}$.

新 NFA 不一定识别 B 的补集. NFA 识别的语言类在补运算下封闭吗？解释你的回答.

8. 利用泵引定理证明下述语言不是正则的.

(a) $A_1 = \{0^n 1^n 2^n \mid n \geqslant 0\}$.

(b) $A_2 = \{\omega\omega\omega \mid \omega \in \{a, b\}^*\}$.

习题 2 答案
Key to Exercise 2

Chapter 3　Context-Free Languages

3.1　Context-free grammar (CFG)

3.1.1　An overview and formalization of context-free grammar

When applying a production to a derivation, the part of the result that has been derived before and after is the context. Context-free means that as long as there is a production in the definition of the grammar, no matter what the string before and after a nonterminal symbol is, the corresponding production can be applied to the derivation.

For context-free grammar, we can denote them in the following ways:

Let $G=(V_N, V_T, P, S)$, if each of the production $\alpha \to \beta$ in P satisfies that α is a nonterminal symbol, $\beta \in (V_N, V_T)^*$, then the grammar becomes context-free. Context-free grammar is often used to describe the syntactic structures of today's programming languages, such as describing arithmetic expressions and describing various statements, etc.

Here is an example of a context-free grammar:

Example 3.1.1
$$A \to 0A1$$
$$A \to B$$
$$B \to {}^*$$

we call it G_1.

The derivation process of grammar G_1 generates the string 000#111 is

$A \Rightarrow 0A1 \Rightarrow 00A11 \Rightarrow 000A111 \Rightarrow 000B111 \Rightarrow 000{}^*111$

For context-free grammars, w_0 in the production $w_0 \to w_1$ must be a nonterminal symbol.

Example 3.1.2
$$S \to aSb$$
$$S \to ab$$

This grammar has two productions, each with only a nonterminal symbol S at the left, and this is a context-free grammar.

3.2 Parse trees

3.2.1 Definition of parse tree

Parsing tree is a graphical representation of language derivation. This representation reflects the essence of language and the derivation process of language.

Definition 3.2.1 For the sentence pattern of CFG G, an parse tree is defined as a tree with the following properties:

(1) Mark the root by the beginning symbol;

(2) Each leaf is marked by a terminal symbol, nonterminal symbol, or ε;

(3) Each internal node is a nonterminal symbol;

(4) If A is an internal marker of A node, and $X_1, X_2, \ldots, X_n$ are markers for all the children of the node from left to right. Then: $A \rightarrow X_1 X_2, \ldots, X_n$ is a production.

If $A \rightarrow \varepsilon$, then A node marked A may have only one child marked ε.

Take $E \Rightarrow -E \Rightarrow -(E) \Rightarrow -(E+E) \Rightarrow -(id+E) \Rightarrow -(id+id)$ as an example.

这个文法有两个产生式,每个产生式左边只有一个非终结符 S,这就是上下文无关文法.

3.2 语法分析树

3.2.1 语法分析树的定义

语法分析树是语言推导过程的图形化表示方法. 这种表示方法反映了语言的实质以及语言的推导过程.

定义 3.2.1 对于 CFG G 的句型, 分析树被定义为具有下述性质的一棵树:

(1) 根由开始符号所标记;

(2) 每片叶子由一个终结符、非终结符或 ε 标记;

(3) 每个内部结点都是非终结符;

(4) 若 A 是某结点的内部标记, 且 $X_1, X_2, \cdots, X_n$ 是该结点从左到右的所有孩子的标记, 则 $A \rightarrow X_1, X_2, \cdots, X_n$ 是一个产生式.

若 $A \rightarrow \varepsilon$, 则标记为 A 的结点可以仅有一个标记为 ε 的孩子.

以 $E \Rightarrow -E \Rightarrow -(E) \Rightarrow -(E+E) \Rightarrow -(id+E) \Rightarrow -(id+id)$ 为例.

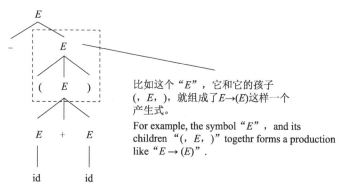

比如这个"E",它和它的孩子 $(, E,)$,就组成了 $E \rightarrow (E)$ 这样一个产生式。

For example, the symbol "E", and its children "$(, E,)$" togethr forms a production like "$E \rightarrow (E)$".

图 3-1 语法树的生成

Fig. 3-1 Syntax tree generation

Example 3.2.1 Syntax tree for $-(id+id)$ (Fig. 3-2).

例 3.2.1 $-(id+id)$ 的语法树 (图 3-2).

图 3-2 -(id+id) 的语法树
Fig. 3-2 Syntax tree for -(id+id)

3.2.2 Overview of Chomsky's paradigm

In computer science, a formal grammar is Chomsky normal if and only if all the generation rules have the following form: here A, B, and C are nonterminal symbols, a is a terminal symbol (a symbol for a constant value), S is a starting symbol, and ε is an empty string. Also, neither B nor C can be the starting sign. All of Chomsky's normal grammars are context-free, and in turn, all context-free grammars can be effectively transformed into equivalent grammars of Chomsky's normal grammars. In addition to (in grammars when including it might generate an empty string) optional rule $S \rightarrow \varepsilon$ is an exception, Chomsky's paradigm of all the rules of grammar are expansion, that is to say in the whole process of export string, the string of each terminal symbol and nonterminal symbol is either the same length or one more element than the previously exported string. The length n of the export of a string is always an exact $2n-1$ step.

Example 3.2.2 Let $G=(V_N, V_T, P, S)$, if the form of each production in P is $A \rightarrow aB$ or $A \rightarrow a$, where A and B are both nonterminal symbols and $a \in V_T^*$, then G is Chomsky's paradigm.

Example 3.2.3
$A \rightarrow BC$;
$A \rightarrow x$;
$A \rightarrow \varepsilon$

Where the uppercase letters are nonterminal symbols; lowercase letters are terminal symbols; ε is an empty

3.2.2 乔姆斯基范式概述

在计算机科学中,一个形式文法是乔姆斯基范式的,当且仅当所有产生规则都有如下形式:这里的 A,B 和 C 是非终结符,a 是终结符(表示常量值的符号),S 是起始符,而 ε 是空串. 另外,B 和 C 都不可以是起始符. 所有的乔姆斯基范式的文法都是上下文无关的,反过来,所有上下文无关的文法都可以有效地变换成等价的乔姆斯基范式的文法. 除了(在文法可能生成空串的时候包括的)可选规则 $S \rightarrow \varepsilon$,乔姆斯基范式的文法的所有规则都是扩张的. 就是说,在导出字符串的整个过程中,每个终结符和非终结符的字符串比起前面导出的字符串要么长度一样要么多出一个元素. 导出的字符串的长度 n 总是精确的 $2n-1$ 步长.

例 3.2.2 设 $G=(V_N, V_T, P, S)$,若 P 中的每一个产生式的形式都是 $A \rightarrow aB$ 或 $A \rightarrow a$,其中 A 和 B 都是非终结符,$a \in V_T^*$,则 G 是乔姆斯基范式.

例 3.2.3
$A \rightarrow BC$;
$A \rightarrow x$;
$A \rightarrow \varepsilon$

式中,大写字母为非终结符;小写字母为终结符;ε 为空串. 每个产生式右侧

string. The right side of each production begins with a terminal symbol and at most one terminal symbol.

3.2.3 Context-free grammar and the transformation of Chomsky's paradigm

1. Add start argument and rule: add a new start argument S_0 and a matching rule $S_0 \to S$, S is the old start argument.

(1) Purpose: the purpose of adding the start argument is that the start argument will always appear on the left;

(2) In Chomsky's paradigm, the starting argument is always on the left side of the rule, and the starting argument is not allowed on the right side of the rule;

(3) Corresponding to the Chomsky's paradigm rule: $A \to BC$ rule, A is an argument, B and C are also arguments, and B and C are not allowed to be starting arguments.

2. Eliminate all ε rules: eliminate all rules from arguments to null characters.

3. Eliminate all $A \to B$ rules: eliminate all single rules from a single argument to a single argument, those from a single argument to multiple arguments or constants are allowed. For example: $A \to B$ is to be deleted, $A \to BS$ can be retained.

4. Add argument: change $A \to BCD$ rule to $A \to ED$ rule, add the argument $E \to BC$.

Example 3.2.4 Transforming context-free grammar into Chomsky's paradigm:

$S \to ASA \mid aB$

$A \to B \mid S$

$B \to b \mid \varepsilon$

(1) Add the new start argument: S_0

$S_0 \to S$

$S \to ASA \mid aB$

$A \to B \mid S$

$B \to b \mid \varepsilon$

为终结符开头且最多仅有一个终结符.

3.2.3 上下文无关文法与乔姆斯基范式的转化

1. 添加开始变元及规则:添加一个新的开始变元 S_0 以及配套的规则 $S_0 \to S$,S 是旧的开始变元.

(1) 目的:添加开始变元的目的是使开始变元永远出现在左边;

(2) 乔姆斯基范式中开始变元始终在规则的左边,不允许开始变元在规则的右侧;

(3) 对应乔姆斯基范式规则:$A \to BC$ 规则,A 是变元,B 和 C 也是变元,并且 B 和 C 不允许是开始变元.

2. 消除所有的 ε 规则:消除所有从变元到空字符的规则.

3. 消除所有的 $A \to B$ 规则:消除所有从单个变元到单个变元的单条规则,允许从单个变元到多个变元或常元. 如:$A \to B$ 是需要删除的,$A \to BS$ 可以保留.

4. 添加变元:将 $A \to BCD$ 规则,转为 $A \to ED$ 规则,添加变元 $E \to BC$.

例 3.2.4 将上下文无关语法转化为乔姆斯基范式:

$S \to ASA \mid aB$

$A \to B \mid S$

$B \to b \mid \varepsilon$

(1) 添加新的开始变元:S_0

$S_0 \to S$

$S \to ASA \mid aB$

$A \to B \mid S$

$B \to b \mid \varepsilon$

(2) Eliminate $B \to \varepsilon$ rule: reconstruct the rule containing B according to the equivalence principle before and after elimination; eliminate $B \to \varepsilon$, that is, add the case that B is empty in the corresponding rule containing B. For aB, if B is empty, it is a; if B is not empty, it is ε.

$S_0 \to S$
$S \to ASA \mid aB \mid a$
$A \to B \mid \varepsilon \mid S$
$B \to b$

(3) Eliminate $A \to \varepsilon$ rule: reconstruct the rule containing A according to the equivalence principle before and after elimination; eliminate $A \to \varepsilon$, that is, add the case that A is empty in the corresponding rule containing A. For ASA, if A is empty, S, AS, SA will be generated (considering different cases where A is empty):

$S_0 \to S$
$S \to ASA \mid AS \mid SA \mid aB \mid a$
$A \to B \mid S$
$B \to b$

(4) Eliminate $A \to B$ rule: find the situation where B appears on the left, find the $B \to b$ rule, directly use $A \to b$ to replace $A \to B$ rule (Note: $B \to b$ rule is unchanged):

$S_0 \to S$
$S \to ASA \mid AS \mid SA \mid S \mid aB \mid a$
$A \to b \mid S$
$B \to b$

(5) Eliminate the rule of $S_0 \to S$: find the case where S appears on the left, find $S \to ASA \mid AS \mid SA \mid S \mid aB \mid a$, replace $S_0 \to S$ with $S_0 \to ASA \mid AS \mid SA \mid S \mid aB \mid A$ (Note: the rule of $S \to ASA \mid AS \mid SA \mid aB \mid a$ does not change):

$S_0 \to ASA \mid AS \mid SA \mid S \mid aB \mid a$
$S \to ASA \mid AS \mid SA \mid aB \mid a$
$A \to b \mid ASA \mid AS \mid SA \mid aB \mid a$
$B \to b$

(6) Add argument: add new rule $R \to SA$;

$S_0 \rightarrow AR \mid AS \mid SA \mid S \mid aB \mid a$

$S \rightarrow AR \mid AS \mid SA \mid aB \mid a$

$A \rightarrow b \mid AR \mid AS \mid SA \mid aB \mid a$

$R \rightarrow SA$

$B \rightarrow b$

Example 3.2.5 Converting context-free grammar to Chomsky's paradigm:

$A \rightarrow BAB \mid B \mid \varepsilon$

$B \rightarrow 00 \mid \varepsilon$

(1) Add a new start argument:

$S_0 \rightarrow A$

$A \rightarrow BAB \mid B \mid \varepsilon$

$B \rightarrow 00 \mid \varepsilon$

(2) Eliminate $B \rightarrow \varepsilon$ rule: according to the equivalence principle of the elimination before and after, rebuild the rule containing B, that is, adding and using ε to replace various cases of B, such as BAB replaces two cases of 1 B, and replaces one case of 2 Bs:

$S_0 \rightarrow A$

$A \rightarrow BAB \mid BA \mid AB \mid A \mid B \mid \varepsilon$

$B \rightarrow 00$

(3) Eliminate $A \rightarrow \varepsilon$ rule: according to the equivalence principle of the elimination before and after, rebuild the rule containing A, such as BAB, if A is empty, A is BB, if A is empty, AB has one more B:

$S_0 \rightarrow A$

$A \rightarrow BAB \mid BA \mid AB \mid A \mid B \mid BB$

$B \rightarrow 00$

(4) Eliminate $A \rightarrow B$ rule: find the situation where B appears on the left, find the $B \rightarrow 00$ rule, directly use $A \rightarrow 00$ rule to replace $A \rightarrow B$ rule (Note: $B \rightarrow 00$ rule is unchanged):

$S_0 \rightarrow A$

$A \rightarrow BAB \mid BA \mid AB \mid A \mid 00 \mid BB$

$B \rightarrow 00$

(5) Eliminate rule $S_0 \rightarrow A$: Find the case where A appears on the left, and find the rule $A \rightarrow BAB \mid BA \mid AB \mid A \mid 00 \mid BB$. Replace rule $S_0 \rightarrow A$ with rule $S_0 \rightarrow BAB$

例 3.2.5 将上下文无关文法转化为乔姆斯基范式：

$A \rightarrow BAB \mid B \mid \varepsilon$

$B \rightarrow 00 \mid \varepsilon$

（1）添加新的起始变元：

$S_0 \rightarrow A$

$A \rightarrow BAB \mid B \mid \varepsilon$

$B \rightarrow 00 \mid \varepsilon$

（2）消除 $B \rightarrow \varepsilon$ 规则：根据消除前后等价原则，重构含有 B 的规则，即添加使用 ε 替换 B 的各种情况。如 BAB 替换 1 个 B 的两种情况，替换 2 个 B 的一种情况：

$S_0 \rightarrow A$

$A \rightarrow BAB \mid BA \mid AB \mid A \mid B \mid \varepsilon$

$B \rightarrow 00$

（3）消除 $A \rightarrow \varepsilon$ 规则：根据消除前后等价原则，重构含有 A 的规则，如 BAB。如果 A 为空就是 BB，如果 A 为空，则 AB 多出一个 B：

$S_0 \rightarrow A$

$A \rightarrow BAB \mid BA \mid AB \mid A \mid B \mid BB$

$B \rightarrow 00$

（4）消除 $A \rightarrow B$ 规则：找 B 出现在左边的情况，发现有 $B \rightarrow 00$ 规则，直接使用 $A \rightarrow 00$ 规则替换 $A \rightarrow B$ 规则（注意：$B \rightarrow 00$ 规则不变）：

$S_0 \rightarrow A$

$A \rightarrow BAB \mid BA \mid AB \mid A \mid 00 \mid BB$

$B \rightarrow 00$

（5）消除 $S_0 \rightarrow A$ 规则：找 A 出现在左边的情况，发现有 $A \rightarrow BAB \mid BA \mid AB \mid A \mid 00 \mid BB$ 规则，直接使用 $S_0 \rightarrow BAB \mid BA \mid AB \mid$

|BA|AB|A|00|BB directly (Notice that $A \to BAB$ the rule of |BA|AB|A|00|BB doesn't change):

$S_0 \to BAB|BA|AB|A|00|BB$
$A \to BAB|BA|AB|A|00|BB$
$B \to 00$

(6) Add argument: add new rule $R \to BA$; the purpose is to replace a three-argument rule with a two-argument rule:

$S_0 \to BAB|BA|AB|A|00|BB$
$A \to RB|BA|AB|A|00|BB$
$B \to 00$
$R \to BA$

(7) Add argument: add new rule $C \to 0$; the purpose is to convert two terminal symbols in $B \to 00$ into two arguments:

$S_0 \to RB|BA|AB|A|CC|BB$
$A \to RB|BA|AB|A|CC|BB$
$B \to CC$
$R \to B$
$C \to 0$

3.3 Pushdown automata (PDA)

3.3.1 Overview of PDA

Pushdown automata is an abstract computing model defined in automata theory. Pushdown automata are more complex than finite-state automata; in addition to finite-state components, they also include a stack of unrestricted length; the state transition of the pushdown automata should not only refer to the finite state part, but also refer to the current state of the stack. State transition includes not only the transition of finite state, but also the process of a stack out or push (Fig. 3-3).

The definition of pushdown automata is as follows:

$$M = (\Sigma, Q, \Gamma, \delta, q_0, Z_0, F)$$

Σ is a finite set of input symbols;
Q is a finite set of states;
Γ is a finite set of pushdown memory symbols;

δ is a mapping from $Q\times(\Sigma\cup(\varepsilon))\times\Gamma$ to $Q\times\Gamma^*$ subset;

$q_0 \in Q$ is the initial state;

$Z_0 \in \Gamma$ is the symbol initially appearing at the top of the pushdown memory;

F is the set of termination states, and $F \subseteq Q$.

δ 是从 $Q\times(\Sigma\cup(\varepsilon))\times\Gamma$ 到 $Q\times\Gamma^*$ 子集的映射；

$q_0 \in Q$ 是初始状态；

$Z_0 \in \Gamma$ 是最初出现在下推存储器顶端的符号；

F 是终止状态集合，$F \subseteq Q$。

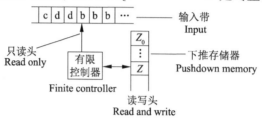

图 3-3 下推自动机示意图
Fig. 3-3 Schematic diagram of the pushdown automaton

Example 3.3.1 Fig. 3-4 shows the transition from state q_1 to state q_2 in a PDA, labeled $a, b \to c$, this means that in state q_1, if we encounter an input character string "a" and the top symbol of the stack is "b", we pop up "b", push "c" to the top of the stack and move it to state q_2.

例 3.3.1 图 3-4 表示了 PDA 中从状态 q_1 到状态 q_2 的转化，标记为 $a, b \to c$，这意味着在状态 q_1，如果我们遇到一个输入字符 "a"，并且堆栈的顶部符号是 "b"，那么我们弹出 "b"，将 "c" 推到堆栈顶部并移动到状态 q_2。

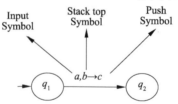

图 3-4 PDA 状态转化
Fig. 3-4 PDA state change

3.4 Context-free grammar and pushdown automata

3.4 上下文无关文法与下推自动机

The equivalence of PDA to CFG means that for any context-free grammar (CFG), there is correspondingly a PDA that accepts it. This proof of equivalence is, for most students, one of the absolute difficulties in formal languages that can be compared with Turing machine construction. The most difficult thing to understand is how to convert from a PDA to a corresponding CFG.

PDA 与 CFG 是等价的，意味着对任意上下文无关文法（CFG），都相应地存在一个 PDA 接受它。而这个等价性的证明对于大部分学生而言是形式语言中能与图灵机构造相提并论的绝对难点之一。而这其中最让人难以理解的便是如何由 PDA 转换为相应的 CFG。

Transformation steps:

(1) Convert a given CFG production to GNF;

(2) PDA has only one state $\{q\}$;

(3) CFG initial symbol will become the initial symbol in the PDA;

(4) For non-terminal symbols, add the following rule: $\delta(q,\varepsilon,A)=(q,\alpha)$, the production rule is $A\rightarrow\alpha$;

(5) For each terminal symbol, add the following rule: $\delta(q,a,a)=(q,\varepsilon)$.

Example 3.4.1 Convert the following syntax to a PDA that accepts the same language.

$S\rightarrow 0S1\mid A$

$A\rightarrow 1A0\mid S\mid\varepsilon$

(1) First, CFG can be simplified by eliminating unit production:

$S\rightarrow 0S1\mid 1S0\mid\varepsilon$

(2) Convert the CFG to GNF:

$S\rightarrow 0SX\mid 1SY\mid\varepsilon$

$X\rightarrow 1$

$Y\rightarrow 0$

(3) PDA can be

$R_1:\delta(q,\varepsilon,S)=\{(q,0SX)\mid(q,1SY)\mid(q,\varepsilon)\}$

$R_2:\delta(q,\varepsilon,X)=\{(q,1)\}$

$R_3:\delta(q,\varepsilon,Y)=\{(q,0)\}$

$R_4:\delta(q,0,0)=\{(q,\varepsilon)\}$

$R_5:\delta(q,1,1)=\{(q,\varepsilon)\}$

3.5 Context-free and non-context-free languages

3.5.1 Pumping Lemma for context-free languages

Assuming A is a context-free language (CFL), there must be a pumping length p, such that the string length in A is greater than or equal to p, and each string in A can be divided into five parts $S=uvxyz$, satisfying the following three conditions:

(1) $i \geq 0$, $uv^ixy^iz \in A$, where v^i denotes i vs in series, so v^3 is vvv;

(2) $|vy| \geq 0$;

(3) $|vxy| \leq p$.

This is a necessary condition that if a language is context-free, then the above requirements are met; Conversely, if the above requirements are not met, nothing is represented, and the language may or may not be CFL.

Example 3.5.1 Proof of pumping lemma using context free language (CFL)

$$C = \{ww \mid w \in \{0,1\}^*\}$$

it is not a context free language (CFL).

(1) Assumptions: C is a context-free language (CFL).

(2) Pumping length: according to the pump lemma, there is a pump length p.

(3) S is a string for language C: $S = 0^p1^p0^p1^p$.

(4) Pumping lemma:

① $i \geq 0$, $uv^ixy^iz \in A$, where v^i denotes i vs in series, so v^3 is vvv;

② $|vy| \geq 0$;

③ $|vxy| \leq p$.

(5) According to the pumping lemma, it is divided into 5 parts.

(6) If v and y are equal, discuss:

Where v and y cannot be 0 or 1 at the same time. If v and y are both the same number (0 or 1) at the same time, if v and y parts are repeated, the repeated number will increase, such as 0 increases, 1 does not increase, resulting in the string does not meet the requirements of the language.

So v and y have to be different characters, one is 0 and the other is 1.

(7) Discuss the cases where v and y are not equal and between 1 and 0 in the middle.

If v and y are between 1 and 0 in the middle, then the numbers of the first 0 and the third 0 are no longer equal, and the numbers of the second 1 and the fourth 1 are no longer equal after v and y increase at the same time, which does not meet the language requirements.

(8) Discuss the case where v and y have different values and are between 0 and 1 on the left:

If v and y are between 0 and 1 on the left, the number of the first 0 is no longer equal to the third 0, and the number of the second 1 is no longer equal to the fourth 1 after v and y increase at the same time, which does not meet the language requirements.

(9) Discuss the case where v and y have different values and are between 0 and 1 on the right:

If v and y are between 0 and 1 on the right, then the numbers of the first 0 and the third 0 are no longer equal, and the numbers of the second 1 and the fourth 1 are no longer equal, which does not meet the language requirements.

(10) Conclusion: the string does not satisfy the pumping lemma of CFL; The assumption is not true, so the language C is not a context-free language.

Extension Pushdown automata are unable to identify the C language, because pushdown automata are first-in-last-out stacks, the characters which are pushed first stack out later, therefore the string which is equal before and after can not be identified, specular reflection of a string can be identified. If the stack will be replaced by first-in-first-out queue, it can identify the language C.

第三个0个数不再相等,这第二个1与第四个1个数不再相等,这不符合语言要求。

(8)讨论v和y取值不等并处于左侧的0和1之间的情况:

如果v和y处于左侧的0和1之间,那么v和y同时增加后,第一个0与第三个0个数不再相等,第二个1与第四个1个数不再相等,这不符合语言要求。

(9)讨论v和y取值不等并处于右侧的0和1之间的情况:

如果v和y处于右侧的0和1之间,那么v和y同时增加后,第一个0与第三个0个数不再相等,第二个1与第四个1个数不再相等,这不符合语言要求。

(10)结论:该字符串不满足上下文无关语言(CFL)的泵引理,假设不成立,因此该语言C不是上下文无关语言。

引申 下推自动机之所以无法识别C这个语言,是因为下推自动机的栈是先进后出的,先入栈的字符后出来,这样就使得前后相等的字符串无法被识别,镜面反射的字符串可以被识别。如果将栈替换成先进先出的队列,那么就可以识别语言C了。

Exercise 3

1. Set the context-free grammar G:

$$R \to XRX \mid S$$
$$S \to aTb \mid bTa$$
$$T \to XTX \mid X \mid \varepsilon$$

Answer the following questions:

(a) What is the argument and terminal symbol of G? What's the starting argument?

(b) Give three strings that are in $L(G)$.

(c) Give three strings that are not in $L(G)$.

(d) $T \Rightarrow abc$ is true or false?

(e) $T \Rightarrow aba$ is true or false?

(f) Describe $L(G)$ in ordinary language.

2. Using the language $A = \{a^m b^n c^n \mid m, n \geq 0\}$ and $B = \{a^n b^n c^m \mid m, n \geq 0\}$ (language $B = \{a^n b^n c^m \mid m, n \geq 0\}$ is not context-free), prove context-free languages under the operation of exchange is not closed.

3. PDA $P = (\{p, q\}, \{0, 1\}, \{x, z\}, \delta, q, z)$ is converted to CFG, where δ is

(a) $\delta(q, 1, Z) = \{(q, XZ)\}$;

(b) $\delta(q, 1, X) = \{(q, XX)\}$;

(c) $\delta(q, 0, X) = \{(p, X)\}$;

(d) $\delta(q, \varepsilon, X) = \{(q, \varepsilon)\}$;

(e) $\delta(p, 1, X) = \{(p, \varepsilon)\}$;

(f) $\delta(p, 0, Z) = \{(q, Z)\}$.

4. True or false: Finite automata can recognize context-free languages.   (   )

5. Construct one each of the context-free grammars that produce the following languages:

(a) $\{a^n b^m c^{2m} \mid n, m \geq 0\}$;

(b) $\{wcw^R \mid w \in \{a, b\}^*\}$;

(c) $\{a^m b^n c^k \mid m = n \text{ or } n = k\}$.

Key to Exercise 3

Part III Theory of Computability

Chapter 4 Turing Machine

4.1 Definition of Turing machine (TM)

4.1.1 TM definition

The Turing machine, proposed by Alan Mathisom Turing in 1936, is a general-purpose computing model. Through researching Turing machine, we can study recursive enumerable sets and partial recursive functions. Research on algorithms and computability provides formal description tools.

The main content of this chapter: the basic definition of Turing machine, real-time description, Turing machine accepted language; the construction technology of Turing machine; Turing machine deformation; Church-Turing thesis; universal Turing machine; computable languages, undecidability, $P-NP$ problems.

4.1.2 Basic concepts

Turing proposed that Turing machine has the following two properties with finite description.

The process must consist of discrete, mechanically executable steps. The basic model includes: a finite controller, an input tape with an infinite number of squares, a reading and writing head. A move will complete the following three actions: change the state of the finite controller; print a symbol in the square where the currently read symbol is located; move the reading and writing head one space to the right or left.

4.2 Church-Turing thesis

● $M = (q, \Sigma, \Gamma, \delta, q_0, B, f)$

(1) q is a finite set of states, $\forall q \in Q$, q is a state of M.

第Ⅲ部分 可计算性理论

第4章 图灵机

4.1 图灵机(TM)定义

4.1.1 TM 定义

图灵机是由图灵在1936年提出的,它是一个通用的计算模型.通过研究图灵机,来研究递归可枚举集和部分递归函数.对算法和可计算性进行研究提供形式化描述工具.

本章的主要内容:图灵机的基本定义、即时描述、图灵机接受的语言;图灵机的构造技术;图灵机的变形;丘奇-图灵论题;通用图灵机;可计算语言、不可判定性、$P-NP$问题.

4.1.2 基本概念

图灵提出,图灵机具有以下两个有穷描述性质.

过程必须由离散的、可以机械执行的步骤组成.基本模型包括一个有穷控制器、一条含有无穷多个带方格的输入带、一个读写头.一个转移将完成以下三个动作:改变有穷控制器的状态;在当前读到的符号所在的带方格的输入带中打印一个符号;将读写头向右或向左移一格.

4.2 丘奇-图灵论题

● $M = (q, \Sigma, \Gamma, \delta, q_0, B, f)$

(1) q 为状态的有穷集合,$\forall q \in Q$,q 为 M 的一个状态.

(2) $q_0 \in Q$ is the starting state of M, for a given input string, M starts from state q_0, and the reading head is looking at the leftmost symbol of the input tape.

(3) $F \subseteq Q$ is the termination state set of M, and $\forall q \in f$, q is a termination state of M. Unlike FA and PDA, generally, once M enters the termination state, it stops running.

(4) Γ is a tape symbol table, $\forall X \in \Gamma$, X is a symbol of M, indicating that X can appear on the input tape a certain time during the operation of M.

(5) $B \in \Gamma$, it is called blank symbol, and the square with blank symbol is considered to be empty.

(6) $\Sigma \subseteq \Gamma - \{B\}$ for the input alphabet, $\forall a \in \Sigma$, a is an input symbol of M. Except for the blank symbol B, only the symbol in Σ can appear on the input band when M is started.

(7) $\delta: q \times \Gamma \rightarrow q \times \Gamma \times \{R, L\}$, is the transaction function of M.

(8) $\delta(q, X) = (p, Y, R)$ indicates that M reads the symbol X in state q, changes the state to p, prints the symbol Y in the square where X is located, and then moves the reading head one grid to the right.

(9) $\delta(q, X) = (p, Y, L)$ indicates that M reads the symbol X in state q, changes the state to p, prints the symbol Y in the square where X is located, and then moves the reading head one grid to the left.

Example 4.2.1 Let $M_1 = (\{q_0, q_1, q_2\}, \{0, 1\}, \{0, 1, B\}), \delta, q_0, B, \{q_2\})$, the definition of δ is as follows.

$$\delta(q_0, 0) = (q_0, 0, R)$$
$$\delta(q_0, 1) = (q_1, 1, R)$$
$$\delta(q_1, 0) = (q_1, 0, R)$$
$$\delta(q_1, B) = (q_2, B, R)$$

● $\alpha_1 \alpha_1 \in \Gamma^*$, $q \in Q$, $\alpha_1 q_1 \alpha_2$ is called the immediate description of M

(2) $q_0 \in Q$ 是 M 的开始状态,对于一个给定的输入串,M 从状态 q_0 启动,读写头正对着输入带最左端的符号.

(3) $F \subseteq Q$,是 M 的终止状态集,$\forall q \in f, q$ 为 M 的一个终止状态. 与 FA 和 PDA 不同,通常,一旦 M 进入终止状态,它就停止运行.

(4) Γ 为带符号表,$\forall X \in \Gamma, X$ 为 M 的一个带符号,表示在 M 的运行过程中,X 可以在某一时刻出现在输入带上.

(5) $B \in \Gamma$,被称为空白符,含有空白符的带方格被认为是空的.

(6) $\Sigma \subseteq \Gamma - \{B\}$ 为输入字母表,$\forall a \in \Sigma, a$ 为 M 的一个输入符号. 除了空白符号 B 之外,只有 Σ 中的符号才能在 M 启动时出现在输入带上.

(7) $\delta: q \times \Gamma \rightarrow q \times \Gamma \times \{R, L\}$,为 M 的转移函数.

(8) $\delta(q, X) = (p, Y, R)$ 表示 M 在状态 q 读入符号 X,将状态改为 p,并在这个 X 所在的带方格中印刷符号 Y,然后将读头向右移一格.

(9) $\delta(q, X) = (p, Y, L)$ 表示 M 在状态 q 读入符号 X,将状态改为 p,并在这个 X 所在的带方格中打印符号 Y,然后将读头向左移一格.

例 4.2.1 设 $M_1 = (\{q_0, q_1, q_2\}, \{0, 1\}, \{0, 1, B\}, \delta, q_0, B, \{q_2\})$,其中 δ 的定义如下.

$$\delta(q_0, 0) = (q_0, 0, R)$$
$$\delta(q_0, 1) = (q_1, 1, R)$$
$$\delta(q_1, 0) = (q_1, 0, R)$$
$$\delta(q_1, B) = (q_2, B, R)$$

● $\alpha_1 \alpha_1 \in \Gamma^*$, $q \in Q$, $\alpha_1 q_1 \alpha_2$ 称为 M 的即时描述

(1) q is the current state of M.

(2) $\alpha_1\alpha_2$ is the symbol string composed of the leftmost to rightmost nonblank symbols of the input tape of M, or the symbol string composed of the symbols in the grid watched by the reading head of the input tape of M from the leftmost to M.

(3) M is watching the leftmost symbol of α_2.

Example 4.2.2 let $X_1X_2\ldots X_{i-1}qX_iX_{i+1}\ldots X_n$ be an ID of M.

If $\delta(q, X_i) = (p, y, R)$, then the next ID of M is
$$X_1X_2\ldots X_{i-1}YpX_{i+1}\ldots X_n$$

Record as
$$X_1X_2\ldots X_{i-1}qX_iX_{i+1}\ldots X_n$$
$$\vdash_M X_1X_2\ldots X_{i-1}YpX_{i+1}\ldots X_n$$

Indicates that M changes the ID into $X_1X_2\ldots X_{i-1}qX_iX_{i+1}X_n$ after one movement under ID $X_1X_2\ldots X_{i-1}YpX_{i+1}\ldots X_n$.

If $\delta(q, X_i) = (p, Y, L)$, then when $i \neq 1$, the next ID of M is
$$X_1X_2\ldots pX_{i-1}YX_{i+1}\ldots X_n$$

Record as
$$X_1X_2\ldots X_{i-1}qX_iX_{i+1}\ldots X_n$$
$$\vdash_M X_1X_2\ldots pX_{i-1}YX_{i+1}\ldots X_n$$

Indicates that M changes the ID into $X_1X_2\ldots pX_{i-1}YX_{i+1}\ldots$ after one movement under ID $X_1X_2\ldots X_{i-1}qX_iX_{i+1}\ldots X_n$;

● $\vdash_M$ **is a binary relation on** $\boldsymbol{\Gamma}^*q\boldsymbol{\Gamma}^* \times \boldsymbol{\Gamma}^*q\boldsymbol{\Gamma}^*$

(1) $\vdash_M^n$ denotes the n-th power of $\vdash_M$: $\vdash_M^n = (\vdash_M)^n$;

(2) $\vdash_M^+$ denotes the positive closure of $\vdash_M$: $\vdash_M^+ = (\vdash_M)^+$;

(3) $\vdash_M^*$ denotes the kleene closure of $\vdash_M$: $\vdash_M^* = (\vdash_M)^*$.

When the meaning is clear, use $\vdash$、$\vdash^n$、$\vdash^+$、$\vdash^*$ denotes $\vdash_M$、$\vdash_M^n$、$\vdash_M^+$、$\vdash_M^*$ respectively.

Example 4.2.3 ID transformation sequence experienced by M_1 processing input string given in example 4.2.1.

(1) The ID transformation sequence experienced in processing the input string 000100 is as follows:

$q_0 000100 \vdash_M 0q_0 00100 \vdash_M 00q_0 0100$
$\vdash_M 000q_0 100 \vdash_M 0001q_1 00 \vdash_M 00010q_1 0$
$\vdash_M 000100\ q_1 \vdash_M 000100Bq_2$

(2) The ID transformation sequence experienced in processing the input string 0001 is as follows:

$q_0 0001 \vdash_M 0q_0 001 \vdash_M 00q_0 01$
$\vdash_M 000q_0 1 \vdash_M 0001q_1 \vdash_M 0001Bq_2$

(3) The ID transformation sequence experienced in processing the input string 000101 is as follows:

$q_0 000101 \vdash_M 0q_0 00101 \vdash_M 00q_0 0101$
$\vdash_M 000q_0 101 \vdash_M 0001q_1 01 \vdash_M 00010q_1 1$

(4) The ID transformation sequence experienced in processing input string 1 is as follows:

$q_0 1 \vdash_M 1q_1 \vdash_M 1Bq_2$

(5) The ID transformation sequence experienced in processing the input string 00000 is as follows:

$q_0 00000 \vdash_M 0q_0 0000 \vdash_M 00q_0 000$
$\vdash_M 000q_0 00 \vdash_M 0000\ q_0 0 \vdash_M 00000q_0 B$

● **Language accepted by Turing machine**

$L(M) = \{x \mid x \in \Sigma^* \ \& \ q_0 x \vdash_M^* \alpha_1 q_1 \alpha_2 \ \& \ q \in f \ \& \ \alpha_1, \alpha_2 \in \Gamma^*\}$

The language accepted by Turing machine is called recursive enumerable language.

If there is a Turing machine, $M = (q, \Sigma, \Gamma, \delta, q_0, B, f)$, $L = L(M)$, and for each input string x, M is stopped, then L is called recursive language.

Example 4.2.4 Let $M_2 = (\{q_0, q_1, q_2, q_3\}, \{0,1\}, \{0,1,B\}, \delta, q_0, B, \{q_3\})$, where δ is defined as follows:

$\delta(q_0, 0) = (q_0, 0, R)$
$\delta(q_0, 1) = (q_1, 1, R)$
$\delta(q_1, 0) = (q_1, 0, R)$
$\delta(q_1, 1) = (q_2, 1, R)$

$$\delta(q_2, 0) = (q_2, 0, R)$$
$$\delta(q_2, 1) = (q_3, 1, R)$$

In order to understand the language accepted by M_2, we need to analyze its working process.

(1) The ID transformation sequence experienced in processing the input string 00010101 is as follows:

$q_000010101 \vdash 0q_0010101 \vdash 00q_0010101$
$\vdash 000q_010101 \vdash 0001q_10101 \vdash 00010q_1101$
$\vdash 000101q_201 \vdash 0001010q_21 \vdash 00010101q_3$

M_2 is in the q_0 state. When it encounters 0, the state remains q_0. At the same time, move the reading head one grid to the right and point to the next symbol.

When the first 1 is encountered in the q_1 state, the state is changed to q_1, and continue to move the reading head to the right to find the next 1; when the second 1 is encountered, the action is similar, but the state is changed to q_2; when the third 1 is encountered, it enters the termination state q_3. At this time, it has just finished scanning the complete input symbol string, indicating that the symbol string is accepted by M_2.

(2) The ID transformation sequence experienced in processing the input string 1001100101100 is as follows:

$q_01001100101100 \vdash 1q_1001100101100$
$\vdash 10\ q_101100101100 \vdash 100q_11100101100$
$\vdash 1001\ q_2100101100 \vdash 10011q_300101100$

When M_2 encounters the third 1, it enters the termination state q_3, and the suffix 00101100 of the input string has not been processed. However, since M_2 has entered the termination state, the symbol string 1001100101100 is accepted by M_2.

(3) The ID transformation sequence experienced in processing the input string 000101000 is as follows:

$q_0000101000 \vdash 0q_000101000 \vdash 00q_00101000$
$\vdash 000q_0101000 \vdash 0001q_101000 \vdash 00010q_11000$
$\vdash 000101q_2000 \vdash 0001010\ q_200 \vdash 00010100$
$q_20 \vdash 000101000\ q_2B$

When the ID of M_2 changes to $000101000q_2B$, it stops because the next movement cannot be made, and the input string 000101000 is not accepted.

The languages accepted by M_2 are 0 and 1 symbol strings containing at least three 1s on the alphabet $\{0,1\}$. Please consider how to construct a TM that accepts the symbol string of exactly three 1s on the alphabet $\{0,1\}$.

Example 4.2.5 Construct TM M_3 so that $L(M) = \{0^n1^n2^n \mid n \geq 1\}$.

Analysis

(1) It cannot be checked by the number of "number" 0, 1, or 2.

(2) The most primitive method is to compare whether their numbers are is the same: eliminate a 0, then eliminate a 1, and finally eliminate a 2.

(3) An X is printed on the eliminated 0's square, a Y is printed on the eliminated 1's square, and a Z is printed on the eliminated 2's square.

(4) Normally, the general form of the symbol string on the input tape is

$$00\ldots0011\ldots1122\ldots22$$

(5) After TM is started, after a period of operation, the general condition of inputting the symbol string on the tape is

$$X\ldots X0\ldots 0Y\ldots Y1\ldots 1Z\ldots Z2\ldots 2BB$$

(6) Close attention needs to be paid to the border situation.

Boundary condition is

$X\ldots XX\ldots XY\ldots YY\ldots YZ\ldots Z2\ldots 2BB$
$X\ldots XX\ldots XY\ldots Y1\ldots 1Z\ldots Z2\ldots 2BB$
$X\ldots X0\ldots 0Y\ldots YY\ldots YZ\ldots Z2\ldots 2BB$
$X\ldots X0\ldots 0Y\ldots Y1\ldots 1Z\ldots ZZ\ldots ZBB$
$X\ldots X0\ldots 0Y\ldots YY\ldots YZ\ldots ZZ\ldots ZBB$

4.2.5 Structure of Turing machine

1. Utilization of finite storage function state

Example 4.2.6 Construct Turing machine M_4 so that $L(M_4) = \{x \mid x \in \{0,1\}^* \ \& \ x \text{ contains at}$

most three 1}.

Analysis M_4 only records the number of 1 that have been read.

$q[0]$ indicates that zero 1s have been read;
$q[1]$ indicates that one 1 has been read;
$q[2]$ indicates that two 1s have been read;
$q[3]$ indicates that three 1s have been read.

$M_4 = (\{q[0], q[1], q[2], q[3], q[f]\}, \{0,1\}, \{0,1,B\}, \delta, q[0], B, \{q[f]\})$

$\delta(q[0], 0) = (q[0], 0, R)$ $\delta(q[2], 0) = (q[2], 0, R)$
$\delta(q[0], 1) = (q[1], 1, R)$ $\delta(q[2], 1) = (q[3], 1, R)$
$\delta(q[0], B) = (q[f], B, R)$ $\delta(q[2], B) = (q[f], B, R)$
$\delta(q[1], 0) = (q[1], 0, R)$ $\delta(q[3], 0) = (q[3], 0, R)$
$\delta(q[1], 1) = (q[2], 1, R)$ $\delta(q[3], B) = (q[f], B, R)$
$\delta(q[1], B) = (q[f], B, R)$
$\delta(q[2], 0) = (q[2], 0, R)$
$\delta(q[2], 1) = (q[3], 1, R)$
$\delta(q[2], B) = (q[f], B, R)$
$\delta(q[3], 0) = (q[3], 0, R)$
$\delta(q[3], B) = (q[f], B, R)$

Turing machine is the one that accepts and only accepts 0 and 1 strings containing exactly three 1s, modify M_4 to obtain M_5.

$L(M_5) = \{x \mid x \in \{0,1\}^* \ \& \ x \text{ contains and only contains three 1s}\}$

$M_5 = (\{q[0], q[1], q[2], q[3], q[f]\}, \{0,1\}, \{0,1,B\}, \delta, q[0], B, \{q[f]\})$

$\delta(q[0], 0) = (q[0], 0, R)$
$\delta(q[0], 1) = (q[1], 1, R)$
$\delta(q[1], 0) = (q[1], 0, R)$
$\delta(q[1], 1) = (q[2], 1, R)$

$\delta(q[2], 0) = (q[2], 0, R)$
$\delta(q[2], 1) = (q[3], 1, R)$
$\delta(q[3], 0) = (q[3], 0, R)$
$\delta(q[3], B) = (q[f], B, R)$
$L(M_6) = \{x \mid x \in \{0,1\}^* \ \& \ x \text{ contains at least three 1s}\}$
$M_6 = (\{q[0], q[1], q[2], q[f]\}, \{0, 1\}, \{0, 1, B\}, \delta, q[0], B, \{q[f]\})$
$\delta(q[0], 0) = (q[0], 0, R)$
$\delta(q[0], 1) = (q[1], 1, R)$
$\delta(q[1], 0) = (q[1], 0, R)$
$\delta(q[1], 1) = (q[2], 1, R)$
$\delta(q[2], 0) = (q[2], 0, R)$
$\delta(q[2], 1) = (q[f], 1, R)$

Example 4.2.7 Construct Turing machine M_7. Its input alphabet is $\{0,1\}$. Now M_7 is required to add substring 101 at the end of its input symbol string.

Analysis

(1) The substring 101 to be added is stored in the finite controller.

(2) First find the end of the symbol string.

(3) The symbols among a given symbol string are sequentially printed on the input tape.

(4) Every time a symbol is printed, it is deleted from the "memory" of the finite controller. When the "memory" is empty, the Turing machine completes its work.

$M_7 = (\{q[101], q[01], q[1], q[\varepsilon]\}, \{0, 1\}, \{0, 1, B\}, \delta, q[101], B, \{q[\varepsilon]\})$

Where δ is defined as:
$\delta(q[101], 0) = (q[101], 0, R)$
$\delta(q[101], 1) = (q[101], 1, R)$
$\delta(q[101], B) = (q[01], 1, R)$
$\delta(q[01], B) = (q[1], 0, R)$
$\delta(q[1], B) = (q[\varepsilon], 1, R)$

Example 4.2.8 Construct Turning machine M_8. Its input alphabet is $\{0, 1\}$. M_8 is required to add substring 101 at the beginning of its input symbol string.

Divide the "memory" in the finite controller into two parts: the first part is used to store the substring to be added; the second part is used to store the substring with no square on the input tape that needs to be be moved due to the addition of symbol string.

(1) The general form is $q[x, y]$, where

x is the substring to be added;

y is the substring temporarily without square storage on the input tape that needs to be moved at present.

(2) $q[x, \varepsilon]$ is the start state.

(3) $q[\varepsilon, \varepsilon]$ is terminated.

(4) Set a and B as input symbols.

(5) $\delta(q[ax, y], B) = (q[x, yB], a, R)$.

(6) Indicates that the first character of the substring to be inserted should be printed on the square currently scanned of TM before the printing of the substring to be inserted is completed.

(7) $\delta(q[\varepsilon, ay], B) = (q[\varepsilon, yB], a, R)$.

(8) Indicates that after the insertion of the substring to be inserted is completed, the substring must be moved back in sequence.

(9) $\delta(q[\varepsilon, ay], B) = (q[\varepsilon, y], a, R)$.

(10) Indicates that the square band currently referred to by the reading head is blank. Now, the current first symbol a in the second part of the "memory" is printed on this square tape, and this symbol is deleted from the "memory".

2. Multi-track technology

Example 4.2.9 Construct M_{11} so that $L(M_{11}) = \{xcy | x, y \in \{0,1\}^+ \text{ and } x \neq y\}$.

Analysis

(1) Taking the symbol c as the dividing line, the symbols before c are compared with the symbols after c one by one.

(2) When the corresponding symbols are found to be different, it enters the termination state.

(3) When it is found that the lengths of x and y are different, it enters the termination state.

(4) Shut down when they are found to be the same.

(5) One track stores the checked symbol string and the other stores the marker.

Construction idea

$M_8 = (\{q[\varepsilon], q[0], q[1], p[0], p[1], q, p, s, f\}, \{[B,0], [B,1], [B,c]\}, \{[B,0], [B,1], [B,c], [\surd,0], [\surd,1], [B,B]\}, \delta, q[\varepsilon], [B,B], \{f\})$

$\delta(q[\varepsilon], [B,0]) = (q[0], [\surd,0], R)$
$\delta(q[\varepsilon], [B,1]) = (q[1], [\surd,1], R)$
$\delta(q[a], [B,d]) = (q[a], [B,d], R)$
$\delta(q[a], [B,c]) = (p[a], [B,c], R)$
$\delta(p[a], [\surd,B]) = (p[a], [\surd,B], R)$
$\delta(p[a], [B,a]) = (p, [\surd,a], L)$
$\delta(p, [\surd,B]) = (p, [\surd,B], L)$
$\delta(p, [B,c]) = (q, [B,c], L)$
$\delta(q, [B,a]) = (q, [B,a], L)$
$\delta(q, [\surd,a]) = (q[\varepsilon], [\surd,a], R)$
$\delta(p[a], [B,B]) = (f, [B,B], R)$
$\delta(p[a], [B,B]) = (p, [B,B], R)$
$\delta(s, [\surd,B]) = (s, [\surd,B], R)$
$\delta(s, [B,a]) = (f, [B,a], R)$

3. Subroutine technology

(1) The design of Turing machine is regarded as a special program design, and the concept of subroutine is introduced.

(2) A Turing machine M' completing a given function starts from a state q and ends at a fixed state f.

(3) Take these two states as two general states of another Turing machine M.

(4) When M enters state q, it is equivalent to starting M' (calling the subroutine corresponding to M'); when M' enters state f, it is equivalent to returning to state f of M.

Example 4.2.10 Construct M_9 to complete the multiplication of positive integers.

(4) 发现它们相同时停机.

(5) 一个道存放被检查的符号串,另一个道存放标记符.

构造思路

$M_8 = (\{q[\varepsilon], q[0], q[1], p[0], p[1], q, p, s, f\}, \{[B,0], [B,1], [B,c]\}, \{[B,0], [B,1], [B,c], [\surd,0], [\surd,1], [B,B]\}, \delta, q[\varepsilon], [B,B], \{f\})$

$\delta(q[\varepsilon], [B,0]) = (q[0], [\surd,0], R)$
$\delta(q[\varepsilon], [B,1]) = (q[1], [\surd,1], R)$
$\delta(q[a], [B,d]) = (q[a], [B,d], R)$
$\delta(q[a], [B,c]) = (p[a], [B,c], R)$
$\delta(p[a], [\surd,B]) = (p[a], [\surd,B], R)$
$\delta(p[a], [B,a]) = (p, [\surd,a], L)$
$\delta(p, [\surd,B]) = (p, [\surd,B], L)$
$\delta(p, [B,c]) = (q, [B,c], L)$
$\delta(q, [B,a]) = (q, [B,a], L)$
$\delta(q, [\surd,a]) = (q[\varepsilon], [\surd,a], R)$
$\delta(p[a], [B,B]) = (f, [B,B], R)$
$\delta(p[a], [B,B]) = (p, [B,B], R)$
$\delta(s, [\surd,B]) = (s, [\surd,B], R)$
$\delta(s, [B,a]) = (f, [B,a], R)$

3. 子程序技术

(1) 将图灵机的设计看成一种特殊的程序设计,将子程序的概念引进来.

(2) 一个完成某一个给定功能的图灵机 M' 从一个状态 q 开始,到达某一个固定的状态 f 结束.

(3) 将这两个状态作为另一个图灵机 M 的两个一般的状态.

(4) 当 M 进入状态 q 时,相当于启动 M'(调用 M' 对应的子程序);当 M' 进入状态 f 时,相当于返回到 M 的状态 f.

例 4.2.10 构造 M_9 完成正整数的乘法运算.

Analysis

(1) Let two positive integers be m and n respectively.

(2) The input string is $0^n 10^m$.

(3) The output should be 0^{n*m}.

(4) Algorithm idea: every time one of the n 0s is changed to B, m 0s are copied after the input string.

(5) During the operation of M_9, the content of the input tape is

$$B^h 0^{n-h} 10^m 10^{m*h} B$$

● **Multiplication of positive integers**

(1) Initialization. Finish changing the first 0 to B and write 1 after the last 0. We use q_0 to denote the startup state and q_1 to denote the state after initialization. First, eliminate the first 0 of the first n 0s,

$$q_0 0^n 10^m \vdash + B q_1 0^{n-1} 10^m 1$$

(2) Master control system. Starting from the state q_1, scanning the remaining 0 of the first n zeros and the first 1, and pointing the reading head to the first 0 of m 0s. At this time, the state is q_2. Its ID changes to

$$B^h q_1 0^{n-h} 10^m 10^{m*(h-1)} B$$
$$\vdash^+ B^h 0^{n-h} 1 \, q_2 0^m 10^{m*(h-1)} B$$

When the subroutine completes the replication of m 0s, it returns to q_3. This state is equivalent to the return (termination) state of the subroutine. Then, in q_3 state, move the reading head back to the first 0 of the remaining 0 of the first n 0s, change this 0 to B, enter q_1 state, and prepare for the next cycle

$$B^h 0^{n-h} 1 q_3 0^m 10^{m*h} B$$
$$\vdash^+ B^{h+1} q_1 0^{n-h-1} 10^m 10^{m*h} B$$

After completing the replication of $m*n$ 0s, clear the nonblank symbols other than these $m*n$ 0s on the input tape. q_4 is terminated state:

$$B^n q_1 10^m 10^{m*n} B$$
$$\vdash^+ B^{n+1+m+1} q_4 0^{m*n} B$$

(3) Subroutine. Complete the task of copying m

0s to the following. Start from q_2 to q_3 and return to the main control program.

$$B^{h+1}0^{n-h-1}1\ q_2 0^m 10^{m*h} B$$
$$\vdash^+ B^{h+1}0^{n-h-1}1\ q_3 0^m 10^{m*h+1} B$$

4.3 Variants of Turing machine

The Turing machine is expanded from different aspects.

(1) Two-way infinite tape Turing machine.

(2) Multi-tape Turing machine.

(3) Uncertain Turing machine.

(4) Multidimensional Turing machine, etc.

They are equivalent to the basic Turing machine.

4.3.1 Two-way infinite tape Turing machine

(1) Two-way infinite tape.

(2) Turing machine $M = (q, \Sigma, \Gamma, \delta, q_0, B, f)$.

(3) It is allowed that the reading head of M can still move to the left when it is at the leftmost end of the input string.

(4) Current ID of M: $X_1 X_2 \ldots X_{i-1} q X_i X_{i+1} \ldots X_n$.

(5) If $\delta(q, X_i) = (p, Y, R)$.

When $i \neq 1$, and $Y \neq B$, the next ID of M is

$$X_1 X_2 \ldots X_{i-1} Y p X_{i+1} \ldots X_n$$

record as

$$X_1 X_2 \ldots X_{i-1} q X_i X_{i+1} \ldots X_n$$
$$\vdash_M X_1 X_2 \ldots X_{i-1} Y p X_{i+1} \ldots X_n$$

indicating that M changes the ID into $X_1 X_2 \ldots X_{i-1} Y p X_{i+1} \ldots X_n$, after one movement under ID $X_1 X_2 \ldots X_{i-1} q X_i X_{i+1} \ldots X_n$.

When $i = 1$, and $Y = B$, the next ID of M is

$$p X_2 \ldots X_n$$

record as

$$q X_1 X_2 \ldots X_n \vdash_M p X_2 \ldots X_n$$

That is to say, just as when the basic Turing machine is all B on the right side of the reading head, these Bs do not appear in the ID. When it is all B on the left side of the reading head of the two-way infinite band Turing machine, these Bs do not appear in the ID of the Turing machine.

(6) If $\delta(q, X_i) = (p, Y, L)$.

When $i \neq 1$, the next ID of M is
$$X_1 X_2 \ldots p X_{i-1} Y X_{i+1} \ldots X_n$$
record as
$$X_1 X_2 \ldots X_{i-1} q X_i X_{i+1} \ldots X_n$$
$$\vdash_M X_1 X_2 \ldots p X_{i-1} Y X_{i+1} \ldots X_n$$
indicating that M changes the ID to $X_1 X_2 \ldots p X_{i-1} Y X_{i+1} \ldots X_n$ after one movement under ID $X_1 X_2 \ldots X_{i-1} q X_i X_{i+1} \ldots n$.

When $i = 1$, the next ID of M is
$$p B Y X_2 \ldots X_n$$
record as
$$q X_1 X_2 \ldots X_n \vdash_M p B Y X_2 \ldots X_n$$
indicating that M changes its ID into $p B Y X_2 \ldots X_n$ after one movement under ID $q X_1 X_2 \ldots X_n$.

Theorem 4.3.1 For any two-way infinite tape Turing machine M, there is an equivalent basic Turing machine M'.

Key points of proof

(1) Simulation of two-way infinite storage: a basic TM with two tracks is used to simulate: one track stores the contents stored in the square watched by the reading head when M starts and all the squares on the right; the other track stores the contents stored in all the squares to the left of the squares watched by the reading head at the beginning of startup in the reverse order.

(2) Simulation of two-way movement: on track 1, the moving direction is consistent with the original moving direction, and on track 2, the moving direction is opposite to the original moving direction.

4.3.2 Multi-tape Turing machine

(1) The Turing machine is allowed to have multiple two-way infinite tapes, each with an independent reading head.

(2) The k-tape Turing machine completes the following three actions during one movement:

① change the current status;

② each reading head prints a desired symbol on the square they are looking at;

③ each reading head moves a square towards its desired direction.

Theorem 4.3.2 The multi-band Turing machine is equivalent to the basic Turing machine.

Key points of proof

(1) For a k-tape Turing machine, a two-way infinite tape Turing machine M' with $2k$ tracks is used to simulate the k-tape Turing machine M.

(2) Corresponding to each tape of M, M' is simulated with two tracks. One track is used to store the contents of the corresponding tape, and the other track is specially used to mark the position of the reading head on the corresponding tape.

4.3.3 Uncertain Turing machine

(1) The difference between the uncertain Turing machine and the basic Turing machine is for any $(q, X) \in Q \times \Gamma$,
$\delta(q, X) = \{(q_1, Y_1, D_1), (q_2, Y_2, D_2), \ldots, (q_k, Y_k, D_k)\}$.

(2) D_j is the moving direction of the reading head, that is, $D_j \in \{R, L\}$.

(3) Indicating that when M is in the state of q and reads X, it can selectively enter state q_j, print character Y_j, and move the reading head accroding to D_j.

(4) $L(M) = \{w | w \in \Sigma^*$, and $ID_1 \vdash^* IDn$, and IDn contains the termination state of $M\}$.

Theorem 4.3.3 The uncertain Turing machine is equivalent to the basic Turing machine.

Key points of proof

(1) Let the equivalent basic Turing machine M' have three tapes.

(2) The first tape is used to store the input.

(3) Various possible moving sequences of M are systematically generated on the second tape.

（2）k 带图灵机在一次转移中完成如下 3 个动作：

① 改变当前状态；

② 每个读写头在它们所对的带方格上打印一个希望的符号；

③ 每个读写头向各自希望的方向转移一个带方格．

定理 4.3.2 多带图灵机与基本图灵机等价．

证明要点

（1）就一个 k 带图灵机而言，用一条具有 $2k$ 道的双向无穷带图灵机 M'，实现对这个 k 带图灵机 M 的模拟．

（2）对应 M 的每一条带，M' 用两个道来实现模拟．其中一条道用来存放对应的带的内容，另一条道专门用来标记对应带上的读写头所在的位置．

4.3.3 不确定的图灵机

（1）不确定图灵机与基本图灵机的区别是对于任意的 $(q, X) \in Q \times \Gamma$，
$\delta(q, X) = \{(q_1, Y_1, D_1), (q_2, Y_2, D_2), \ldots, (q_k, Y_k, D_k)\}$．

（2）D_j 为读写头的转移方向，即 $D_j \in \{R, L\}$．

（3）表示 M 在状态 q，读到 X 时，可以有选择地进入状态 q_j，打印字符 Y_j，按 D_j 转移读写头．

（4）$L(M) = \{w \mid w \in \Sigma^*$，且 $ID_1 \vdash^* IDn$，且 IDn 含 M 的终止状态$\}$．

定理 4.3.3 不确定的图灵机与基本的图灵机等价．

证明要点

（1）让等价的基本图灵机 M' 具有 3 条带．

（2）第 1 条带用来存放输入．

（3）M 的各种可能的移动序列系统地生成在第 2 条带上．

(4) M' processes the input string according to the moving sequences given on the second tape on the third band. If it is successful, it is accepted. If it is unsuccessful, the next possible moving sequences is generated on the second tape and a new round of "trial processing" is started.

4.3.4 Multi-dimensional Turing machine

(1) In multi-dimensional Turing machine, the reading head can move along multiple dimensions.

(2) Turing machine can move along k dimensions.

(3) The tape of k-dimensional Turing machine is composed of k-dimensional array and is infinite in all $2k$ directions. Its reading head can move towards any of $2k$ directions.

Theorem 4.3.4 Multi-dimensional Turing machine is equivalent to basic Turing machine.

(1) k-dimensional contents are denoted in one-dimensional form, just as multi-dimensional arrays are stored in one-dimensional form in computer memory.

(2) Segments are used to denote the contents in one-dimension.

(3) Use # as the segment separator.

(4) ¢ is used as the start flag of the string and $ is used as the end flag of the string.

● **Basic Turing machine simulates two-dimension Turing machine**

$$¢\ Ba1a2a3a4BBBBBB\ \#$$
$$Ba5Ba6a7a8a9a10BBB\ \#$$
$$Ba11BBBBa12Ba13Ba14\ \#$$
$$a15\ a16BBBBBBBB\ a16\#$$
$$BBB\ a17BBBBBa18B\ \#$$
$$a19a20BBBBBBBBB\ \#$$
$$BBBBBBBBBBa21\ \$\ BB\ \#$$
$$BBBBBBBBBBa21\ \$$$

(4) M'在第3条带上按照第2条带上给出的移动序列处理输入串,如果成功,则接受之,如果不成功,则在第2条带上生成下一个可能的移动序列,开始新一轮的"试处理".

4.3.4 多维图灵机

(1) 多维图灵机读头可以沿着多个维度转移.

(2) 图灵机可以沿着 k 个维度转移.

(3) k-维图灵机的带由 k 维阵列组成,而且在所有的 $2k$ 个方向上都是无穷的,它的读写头可以向着 $2k$ 个方向中的任一个转移.

定理4.3.4 多维图灵机与基本图灵机等价.

(1) 用一维的形式表示 k 维的内容,就像多维数组按照一维的形式存储在计算机的内存中一样.

(2) 段用来表示一维上的内容.

(3) 用#作为段的分割符.

(4) ¢ 用作该字符串的开始标志,$ 用作该字符串的结束标志.

● **基本图灵机模拟二维图灵机**

$$¢\ Ba1a2a3a4BBBBBB\#$$
$$Ba5Ba6a7a8a9a10BBB\ \#$$
$$Ba11BBBBa12Ba13Ba14\ \#$$
$$a15\ a16BBBBBBBB\ a16\#$$
$$BBB\ a17BBBBBa18B\ \#$$
$$a19a20BBBBBBBBB\ \#$$
$$BBBBBBBBBBa21\ \$\ BB\ \#$$
$$BBBBBBBBBBa21\ \$$$

4.3.5 Other Turing machines

1. Multi-head Turing machine

It means that there are multiple read heads on a tape, which are uniformly controlled by M's finite controller. M determines the movement to be performed according to the current state and the characters currently read by these heads. In each action of M, the characters printed by each reading head and the moving direction can be independent of each other.

Theorem 4.3.5 Multi-head Turing machine is equivalent to the basic Turing machine.

A basic Turing machine with $k+1$ tracks can be used to simulate a Turing machine with k heads (k-head Turing machine). One track is used to store the contents of the original input tape, and the other k tracks are used as the marks of k reading head positions respectively.

2. Offline Turing machine

(1) A multi-tape Turing machine with one input tape is read-only tape.

(2) The symbols ¢ and $ are used to define its input string storage area, ¢ on the left and $ on the right.

(3) The reading head on the tape is not allowed to move out of the area defined by ¢ and $, it is called an offline Turing machine.

(4) If only the reading head on the read-only tape is allowed to move from left to right, it is called an on-line Turing machine.

Theorem 4.3.6 Offline Turing machine is equivalent to basic Turing machine.

Proof point Let the offline Turing machine simulating M have one more tape than M, and use the extra tape to copy the input string of M. Then this tape is regarded as the input tape of M, and M is simulated for corresponding processing.

4.3.5 其他图灵机

1. 多头图灵机

指在一条带上有多个读写头,它们受 M 的有穷控制器的统一控制,M 根据当前的状态和这些多头当前读到的字符确定要执行的转移. 在 M 的每个动作中,各个读头所打印的字符和所转移的方向都可以是相互独立的.

定理 4.3.5 多头图灵机与基本的图灵机等价.

一条具有 $k+1$ 个道的基本图灵机可以用来模拟一个具有 k 个头的图灵机(k 头图灵机). 其中一条道用来存放原输入带上的内容,其余 k 条道分别用来作为 k 个读写头位置的标记.

2. 离线图灵机

(1) 有一条输入带的多带图灵机是只读带(read-only tape).

(2) 符号 ¢ 和 $ 用来限定其输入串存放区域,¢ 在左边,$ 在右边.

(3) 该带上的读头不允许移出由 ¢ 和 $ 限定的区域,称之为离线的图灵机.

(4) 若只允许只读带上的读头从左向右转移,则称之为在线图灵机.

定理 4.3.6 离线图灵机与基本的图灵机等价.

证明要点 让模拟 M 的离线图灵机比 M 多一条带,并且用这条多出来的带复制 M 的输入串,然后将这条带看作 M 的输入带,模拟 M 进行相应的处理.

3. Turing machine as enumerator

(1) One tape of multi-tape Turing machine is specially used as the output tape to record each sentence of the generated language.

(2) In an enumerator, once a character is written on the output tape, it cannot be changed. If the normal movement direction of the reading head on the tape is to move to the right, the reading head on the tape is not allowed to move to the left.

(3) If the language has infinite sentences, it will never stop. Every time it produces a sentence, it prints a separator "#" after it.

(4) The language generated by the enumerator is denoted as $G(M)$.

(5) Canonical order.

Theorem 4.3.7 A necessary and sufficient condition for L to be a recursively enumerable language is that there exists a Turing machine M such that $L = G(M)$.

Theorem 4.3.8 A necessary and sufficient condition for a language L to be a recursive language is that there is a Turing machine M such that $L = G(M)$, and L is generated by M in canonical order.

4. Multi-stack machine

(1) Multi-stack machine is an uncertain Turing machine with one read-only input tape and multiple storage tapes.

① The reading head on the read-only tape of the multi-stack machine cannot be moved to the left.

② The reading head on the tape can be moved left and right.

a. When moving to the right, a non-blank character is usually printed on the square with the current gaze.

b. When moving to the left, the blank character B must be printed on the square with the current gaze.

(2) A deterministic double stack machine is a deterministic Turing machine, which has a read-only

input tape and two storage tapes. When the reading head on the storage tape is moved to the left, only the blank symbol B can be printed.

(3) Pushdown automaton is a non-deterministic multi-tape Turing machine. It has a read-only input tape and a storage tape.

Theorem 4.3.9 An arbitrary single tape Turing machine can be simulated by a deterministic double stack machine.

5. Counter

(1) An offline Turing machine with a read-only input tape and several unidirectional infinite tapes for counting.

(2) A counter with n counting tapes is called an n counter.

(3) There are only two characters on the counting tape. One is Z, which is the symbol at the bottom of the stack. This character can also be regarded as the first symbol of the counting tape, which only appears at the leftmost end of the counting tape; the other is the blank symbol B. The number recorded on this tape is the number of B from Z to the current position of the reading head.

Theorem 4.3.10 Turing machine can be simulated by a double counting machine.

6. Church-Turing thesis and random access machine

(1) For any problem that can be solved by an effective algorithm, there is a Turing machine to solve this problem.

(2) Random access machine (RAM) contains an infinite number of storage units, which are numbered according to 0, 1, 2,..., and each storage unit can store an arbitrary integer, where is an arithmetic register that can hold any integer. These integers can be decoded into the usual types of computer instructions.

带和两条存储带. 存储带上的读写头左移时,只能打印空白符号 B.

(3) 下推自动机是一种非确定的多带图灵机. 它有一条只读的输入带和一条存储带.

定理 4.3.9 一个任意的单带图灵机可以被一个确定的双栈机模拟.

5. 计数机

(1) 有一条只读输入带和若干个用于计数的单向无穷带的离线图灵机.

(2) 拥有 n 个用于计数带的计数机被称为 n 计数机.

(3) 用于计数的带上仅有两个字符:一个为相当于栈底符号的 Z,该字符也可以看作计数带的首符号,它仅出现在计数带的最左端;另一个就是空白符 B. 这个带上所记的数就是从 Z 开始到读写头当前位置所含的 B 的个数.

定理 4.3.10 图灵机可以被一个双计数机模拟.

6. 丘奇-图灵论题与随机存取机

(1) 对于任何可以用有效算法解决的问题,都存在解决此问题的图灵机.

(2) 随机存取机(RAM)含有无穷多个存储单元,这些存储单元按照 0, 1, 2, … 进行编号,每个存储单元可以存放一个任意的整数;它是有穷个能够保存任意整数的算术寄存器. 这些整数可以被破解成通常的各类计算机指令.

Theorem 4.3.11 If the basic instructions of RAM can be realized by Turing machine, RAM can be realized by Turing machine.

4.3.6 Universal Turing machine

(1) Realize the simulation of all Turing machines.

(2) Coding system can not only realize the representation of Turing machine, but also realize the representation of sentences processed by TM.

These tape symbols other than black characters are encoded with 0 and 1. At the same time, the moving function of Turing machine can also be coded with 0 and 1.

(3) $M = (\{q_1, q_2, \ldots, q_n\}, \{0,1\}, \{0,1,B\}, \delta, q_1, B, \{q_2\})$.

(4) X_1, X_2 and X_3 are used to denote 0, 1 and B respectively, and D_1 and D_2 are used to denote R and L respectively.

(5) $\delta(q_i, X_j) = (q_k, X_l, D_m)$ can be denoted by $0^i 1 0^j 1 0^k 1 0^l 1 0^m$.

(6) Turing machine M can be denoted as

111 code$_1$ 11 code$_2$ 11 ... 11 code$_r$ 111

(7) The code$_t$ is an encoding of the form of $0^i 1 0^j 1 0^k 1 0^l 1 0^m$ of the action $\delta(q_i, X_j) = (q_k, X_l, D_m)$.

(8) The Turing machine M and its input string w can be denoted as 111 code$_1$ 11 code$_2$ 11 ... 11 code$_r$ 111w.

(9) The symbol lines representing Turing machine and the symbol lines representing input are sorted according to the standard order.

(10) $Ld = \{w \mid w$ is the j-th sentence, and the j-th Turing machine does not accept it$\}$ is not a recursive enumerable language.

(11) Universal language

$Lu = \{<M, w> \mid M \text{ accept } w\}$

$<M, w>$ is a string as the following form, denoting Turing machine $M = (\{q_1, q_2, \ldots, q_n\}, \{0,1\},$

$\{0,1,B\},\delta,q_1,B,\{q_2\})$ and its input string w.

111 code$_1$ 11 code$_2$ 11... 11 code$_r$ 111w

Example 4.3.1 Let Turing machine $M_{10} = (\{q_1,q_2,q_3,q_4\},\{0,1\},\{0,1,B\}),\delta,q_4,B,\{q_3\})$, where δ is defined as follows:

$$\delta(q_4, 0) = (q_4, 0, R)$$
$$\delta(q_4, 1) = (q_1, 1, R)$$
$$\delta(q_1, 0) = (q_1, 0, R)$$
$$\delta(q_1, 1) = (q_2, 1, R)$$
$$\delta(q_2, 0) = (q_2, 0, R)$$
$$\delta(q_2, 1) = (q_3, 1, R)$$

Code as

11100001010000101011000010010100101101010101011010010010010110010100101011
00100100010010111

The universal Turing machine checks whether M accepts the string:

0011011101110000101000010101100001010010110101010101101001001001011001010010101100100010010111001101110

4.4 Computing with Turing machine

Encoding nonnegative integers—unary numeral.

The nonnegative integer n is denoted by the symbol string $0n_1$.

Assume k-ary function $f(n_1, n_2, \ldots, n_k) = m$, TM $M = (q, \Sigma, \Gamma, \delta, q_0, B, f)$ accept input string.

Total recursive function: k-ary function $f(n_1, n_2, \ldots, n_k)$ is assumed. If any $n_1, n_2, \ldots, n_k$ and f are defined, that is, the Turing machine calculating f can always give a certain output, then f is called a total recursive function.

Partial recursive function: the function calculated by Turing machine is called partial recursive function.

Example 4.4.1 Construct TM M_{11}, and calculate $m+n$ for any nonnegative integer n, m, M_{11}.

Analysis The input of M_{11} is 0^n10^m and the output is 0^{n+m} symbol string. The case where n and m are 0 requires special consideration.

(1) When n is 0, the calculation is completed only by changing 1 into B. At this time, it is not necessary to check whether m is 0;

(2) When m is 0, it is necessary to scan the symbol 0 denoting n and change 1 to B;

(3) When neither n nor m is 0, we need to change the symbol 1 to 0 and the last 0 to B.

Construction idea

$M_{11} = (\{q_0, q_1, q_2, q_3\}, \{0,1\}, \{0,1,B\}, \delta, q_0, B, \{q_1\})$

$$\delta(q_0, 1) = (q_1, B, R)$$
$$\delta(q_0, 0) = (q_2, 0, R)$$
$$\delta(q_2, 0) = (q_2, 0, R)$$
$$\delta(q_2, 1) = (q_2, 0, R)$$
$$\delta(q_2, B) = (q_3, B, L)$$
$$\delta(q_3, 0) = (q_1, B, R)$$

Example 4.4.2 Construct Turing machine M_{12}. For any nonnegative integer n, m, M_{12}, calculate the following function:

$$n \dotdiv m = \begin{cases} n-m, & n \geqslant m \\ 0, & n < m \end{cases}$$

Construction idea

$M_{12} = (\{q_0, q_1, q_2, q_3, q_4, q_5, q_6,\}, \{0,1\}, \{0,1,X,B\}, \delta, q_0, B, \{q_6\})$

$$\delta(q_4, X) = (q_4, B, L)$$
$$\delta(q_4, 1) = (q_6, 0, R)$$
$$\delta(q_5, X) = (q_5, B, R)$$
$$\delta(q_5, 0) = (q_5, B, R)$$
$$\delta(q_5, B) = (q_6, B, R)$$

4.5 Definition of algorithm

Algorithm refers to the accurate and complete description of the problem-solving scheme. It is a series of clear instructions to solve the problem. The algorithm represents the strategy mechanism to solve the problem in a systematic way. In other words, the required output can be obtained in a limited time for a certain standard input. If an algorithm is defective or not suitable for a problem, executing the algorithm will not solve the problem. Different algorithms may use different time, space or efficiency to complete the same task. The advantages and disadvantages of an algorithm can be measured by space complexity and time complexity.

The instructions in the algorithm describe a calculation. When it runs, it can start from an initial state and (possibly empty) initial input, go through a series of limited and clearly defined states, and finally produce output and stop at a final state. The transition from one state to another is not necessarily deterministic. Some algorithms, including randomization algorithm, include some random inputs.

解决问题的策略机制. 也就是说, 能够对一定规范的输入, 在有限时间内获得所要求的输出. 如果一个算法有缺陷, 或不适合于某个问题, 那么执行这个算法将不会解决这个问题. 通过不同的算法可能用不同的时间、空间或效率来完成同样的任务. 一个算法的优劣可以用空间复杂度与时间复杂度来衡量.

算法中的指令描述的是一个计算, 当其运行时能从一个初始状态和(可能为空的)初始输入开始, 经过一系列有限而清晰定义的状态, 最终产生输出并停止于一个终态. 一个状态到另一个状态的转移不一定是确定的. 包含随机化算法在内的一些算法, 还包含了一些随机输入.

Exercise 4

1. Give a formal definition of enumerator (It can be regarded as a two-tape Turing machine, using its second tape as a printer), including the definitions of the languages it enumerates.

2. Check the formal definition of Turing machine, answer the following questions and explain your reasoning:

   (a) Can Turing machine write blank symbols on its tape?

   (b) Can the input alphabet > and the tape alphabet R be the same?

   (c) Can the read-write head of Turing machine be in the same position in two consecutive steps?

   (d) Can Turing machines contain only one state?

3. The following description is not a legal Turing machine, explain why.

   $MBad$ = on the input $<p>$, where p is a polynomial on the arguments $X_1, \ldots, X_k$: (a) Let $X_1, X_2, \ldots, X_k$ take all possible integer values; (b) Find the value of p for all these values; (c) As long as a value makes p be 0, it is accepted; otherwise, it is rejected.

4. The following languages are all languages on the alphabet $\{0, 1\}$, so as to realize the level description and give the Turing machine for deciding these languages:

   (a) $\{w | vw$ contains the same number of 0 and 1$\}$.

   (b) $\{w | w$ contains twice as many 0s as 1$\}$.

   (c) $\{w | ve$ contains more than twice as many 0s as 1$\}$.

5. Let k-PDA denote pushdown automata with k stacks. Therefore, 0-PDA is an nfA and 1-PDA is a common PDA. It is known that 1-PDA is stronger than 0-PDA (recognizing larger language classes).

   (a) It is proved that 2-PDA is stronger than 1-PDA.

   (b) It is proved that 3-PDA is not stronger than 2-

PDA. (Tip: use two stacks to simulate a Turing tape)

6. Turing machine with double infinite tape is similar to ordinary Turing machines, except that its tapes are infinite to the left and right. At the beginning, except for the input area, the tape is filled with blank characters. The calculation is also defined as usual, but it will not encounter the end of the tape when it moves to the left. It is proved that this type of Turing machine can recognize Turing recognizable language classes.

7. Write-once Turing machine is a single tape Turing machine. It can only change its content once (including the input area on the band) on each square at most. It is proved that this deformation of Turing machine model is equivalent to ordinary Turing machine model. (Tip: first consider the following: you can modify the square up to twice and use multiple tapes)

8. Turing machine with left reset is similar to ordinary Turing machines, except that its transfer function has the following form: $\delta: q \times \rightarrow q \times R \times \{R, RESET\}$ if $\delta(g,a) = (R, B, RESET)$, then when the machine is in state q and reads a, write B on the tape and enter state R, and the read-write head jumps to the left end of the tape. Note that such a machine does not have the ordinary ability to move its read-write head one symbol to the left. It is proved that the left reset Turing machine recognizes the Turing recognizable language class.

9. Turing machine with stay put instead of left is similar to ordinary Turing machine, but its transfer function has the following form: $\delta: q \times R \rightarrow q \times R \times \{R, S\}$, at any time, the machine can move the read-write head to the right or keep in place. It is proved that such a Turing machine is not equivalent to an ordinary Turing machine. What language class does such a Turing machine recognize?

6. 双无限带图灵机与普通图灵机相似,所不同的是它的带子向左和向右都是无限的. 此带在开始的时候,除了输入区域外,其他都填以空白符,计算也像通常一样定义,只是在它向左转移时不会遇到带的端点. 证明这种类型的图灵机能够识别图灵可识别语言类.

7. 只写一次图灵机是一个单带图灵机,它在每个带方格上最多只能改变其内容一次(包括带上的输入区). 证明图灵机模型的这个变形等价于普通的图灵机模型. (提示:首先考虑如下图灵机:可以修改带方格最多两次,使用多个带子)

8. 左复位图灵机与普通图灵机类似,只是它的转移函数具有下列形式: $\delta: q \times \rightarrow q \times R \times \{R, RESET\}$. 如果 $\delta(g,a) = (R, B, RESET)$,那么当机器处于状态 q 且读 a 时,在带上写下 B 并进入状态 R 后,读写头就跳到带的左端点. 注意:这样的机器没有将它的读写头向左转移一个符号的通常的能力. 证明左复位图灵机能够识别图灵可识别语言类.

9. 以停留代替左移的图灵机与普通图灵机类似,只是它的转移函数具有下列形式:$\delta: q \times R \rightarrow q \times R \times \{R, S\}$. 在任何时候,机器可以将读写头向右移,或让其停留在原地不动. 证明这样的图灵机与普通图灵机并不等价. 这样的图灵机能够识别什么语言类?

习题4答案
Key to Exercise 4

Chapter 5 Decidability

Algorithm is a strategy for computer to solve problems, but what kind of problem can the algorithm solve? Is there a problem that the algorithm cannot solve? With these questions, this chapter will begin to study the ability of algorithms to solve problems. Computers can solve many problems at present, but there are still some unsolvable problems in the algorithm. By studying the unsolvable problems, you can think about how to modify or simplify the problem, so as to find an algorithm that can solve the problem. The computer is only a tool to solve the problem, only by knowing the scope of its capabilities can it be used efficiently. In addition, studying undecidability helps to improve logical thinking, and can deepen the knowledge and understanding of the concept of calculation.

5.1 Decidable languages

For a judgment question, if a program can be compiled with any element in the domain as input, the execution of the program can give the answer to the corresponding individual question, and the judgment question is called decidable. For a given problem, encode its instance into a string and convert it into a language. In this section, we will introduce two languages related to the issue of decidability: decidable language related to regular language and decidable language related to context-free language.

5.1.1 Decidable problem related to regular language

Regarding the decidable problem of regular language, the following three questions are first raised:

(1) Does a finite computer accept a string?

(2) Is the language of a finite automaton empty?

(3) Are two finite automata equivalent?

第 5 章 可判定性

算法是计算机解决问题的策略,但是算法能解决哪种问题?是否存在算法无法解决的问题?伴随着这些疑问,本章将开始研究算法解决问题的能力.计算机目前可以解决许多问题,但是仍然存在着一些在算法上不可解的问题,通过研究不可解问题,可以思考如何去修改或者简化问题,以便找到可以解决该问题的算法.计算机只是解决问题的工具,只有了解其能力范围才能高效利用它.此外,研究不可解性有助于提升逻辑思维,并且能够加深对计算这一概念的认识与理解.

5.1 可判定性语言

对于一个判定问题,如果能够以域中任意元素作为输入编出一个程序,执行该程序就能给出相应的个别问题的答案,称该判定问题为可判定的.对于给定的问题,对其实例进行编码成字符串,将其转换成一个语言.本节将介绍两个与可判定性问题有关的语言:与正则语言相关的可判定性语言和与上下文无关语言相关的可判定性语言.

5.1.1 与正则语言相关的可判定性语言

关于正则语言的可判定问题,首先提出以下三个问题:

(1) 一个有穷计算机是否接受一个串?

(2) 一个有穷自动机的语言是否为空?

(3) 两个有穷自动机是否等价?

First, discuss the acceptance problem of deterministic finite computers, that is, the **DFA acceptance problem**. Give a finite automaton, then a string is given and ask whether the finite automaton accepts this string. The problem is converted into language through coding, and it is recorded as A_{DFA}, and it has the following expressions:

$A_{DFA} = \{<B,w> \mid B \text{ is DFA}, w \text{ is string}, B \text{ accepts } w\}$

Use string w as input to this automaton, the result of the operation is acceptance, and automaton B accepts w, if and only if $<B,w>$ belongs to A_{DFA}. So the decidable problem that we study is to ask whether an input is in the language, which is equivalent to solving the DFA acceptability problem. Through coding, from the problem to the language, so as to solve the problem by proving the language.

Theorem 5.1.1 A_{DFA} is a decidable language.

Proof idea Design a Turing machine M used to determine A_{DFA}, and it is known that this Turing machine M is shutting down everywhere. The working principle of Turing machine M is to simulate the calculation process of finite automata B on string w. The key point is that the deterministic finite automaton always stops and there is no dead loop problem, so the Turing machine M must be able to complete this simulation process.

Proof To design a Turing machine M to determine A_{DFA}, there are:

$M = $ "For input $<B,w>$, B is DFA and w is string:

(1) Simulate B on input w.

(2) If the simulation ends in an accepting state, then accept it; if the simulation ends in an unacceptable state, then reject it."

Next, the above representation will be explained in detail. The first step is to check whether B and w are legal inputs. For all Turing machine programs, we need to check the legality of the input. How to check?

首先讨论确定型有穷计算机的接受性问题，即 **DFA 接受问题**. 给定一个有穷自动机，再给定一个串，问这个有穷自动机是否接受这个串. 将该问题编码转化成语言，记为 A_{DFA}，有下列表达形式：

$A_{DFA} = \{<B,w> \mid B \text{ 是 DFA}, w \text{ 是串}, B \text{ 接受 } w\}$

将串 w 作为输入给这个自动机，运行结果为接受，自动机 B 接受了 w，当且仅当 $<B,w>$ 属于 A_{DFA}. 所以我们研究判定问题就是问一个输入是否在该语言里，这也与求解 DFA 的接受性问题等价. 通过编码，从问题转换到语言，以便通过证明语言来解决问题.

定理 5.1.1 A_{DFA} 是一个可判定性语言.

证明思路 设计一个用于判定 A_{DFA} 的图灵机 M，能知道这个图灵机 M 是处处停机的. 图灵机 M 的工作原理是模拟有穷自动机 B 在串 w 上的计算过程. 关键的一点在于，确定型有穷自动机永远是停机的，不存在死循环问题，所以图灵机 M 一定能完成这个模拟的过程.

证明 设计一个判定 A_{DFA} 的图灵机 M，有

$M = $ "对于输入 $<B,w>$，B 是 DFA，w 是串：

（1）在输入 w 上模拟 B.

（2）若模拟以接受状态结束，则接受；若模拟以不接受状态结束，则拒绝."

接下来对上述表示进行详细解释. 第一步检查 B 和 w 是不是合法输入. 对所有的图灵机程序，我们都有必要检查输入的合法性，如何做检查呢？

For finite automata B, according to the definition, finite automation is a five-tuple, including the state machine Q, the alphabet Σ, the transfer function δ, the initial state q_0, and the final state set F. It is judged whether it is a finite automaton by checking whether B is composed of these five parts. If the check is passed, the second step is executed, and if the check is not passed, it is directly rejected.

In the second step, after the detection of B and w is reasonable, the Turing machine M starts to perform the simulation. Since the finite automata will stop and give an acceptance or rejection result, the Turing machine will accept or reject accordingly, so the Turing machine always can give the correct answer. Turing machine M tracks the current state and position of B when running on w by writing information on the tape. The status and location update is determined by the transfer function of B. When M has processed the last symbol of w, if B is in the accepting state, M will accept it, otherwise M will reject it.

Next, prove the acceptability of non-deterministic finite automata, which is to detect whether a given non-deterministic finite automaton accepts a given string. Translating the question into language, there are:

$A_{\text{NFA}} = \{<B,w> | B \text{ is NFA}, w \text{ is string}, B \text{ accepts } w\}$

Theorem 5.1.2 A_{NFA} is a decidable language.

There are two ways to prove the theorem.

Proof idea 1 Set a non-deterministic Turing machine N to simulate the calculation of the non-deterministic finite automaton B on the string w, similar to the proof process of Theorem 5.1.1.

Proof idea 2 The deterministic finite automaton has just been proved to be decidable, so you can first transform the non-deterministic finite automaton B into the deterministic finite automaton C, and then use the

对于有穷自动机 B，根据定义可知，有穷自动机是一个五元组，包括状态机 Q、字母表 Σ、转移函数 δ、初始状态 q_0、终态集合 F. 通过检查 B 是否由这五个部分组成从而判断其是否为有穷自动机. 若检查通过，则执行第二步，若没有通过检查，则直接拒绝.

第二步，B 和 w 检测合理后，图灵机 M 开始执行模拟，由于有穷自动机将会停机，给出一个接受或拒绝的结果，图灵机也会相应地接受或拒绝，所以图灵机能给出正确答案. 图灵机 M 通过在带上写下信息来跟踪 B 在 w 上运行时的当前状态和当前位置. 状态和位置的更新由 B 的转移函数来确定，当 M 处理完 w 的最后一个符号时，如果 B 处于接受状态，则 M 将接受，否则 M 将拒绝.

接下来证明非确定型有穷自动机的接受性问题，即检测一个给定的非确定型有穷自动机是否接受一个给定的串. 将问题转换成语言，有

$A_{\text{NFA}} = \{<B,w> | B$ 是 NFA, w 是串, B 接受 $w\}$

定理 5.1.2 A_{NFA} 是一个可判定性语言.

该定理有两个证明思路.

证明思路 1 设一个非确定图灵机 N 来模拟非确定型有穷自动机 B 在串 w 上的计算，类似于定理 5.1.1 的证明过程.

证明思路 2 刚才证明了确定型有穷自动机是可判定的，所以可以先把非确定型有穷自动机 B 转化成确定型有穷自动机 C，再使用图灵机来模拟

Turing machine to simulate the calculation of C on the string w.

Next, use the second proof idea to prove the theorem.

Proof Construct a Turing machine N for determining A_{NFA}, with:

N = "For input $<B, w>$, B is NFA, and w is a string:

(1) Automata transformation: transform the non-deterministic finite automata into an equivalent deterministic finite automata C.

(2) Reduction process: Calculate the transformed deterministic finite automaton C on the input string w:

① Imitation: Construct a Turing machine M, given the input string w, imitate the deterministic finite automaton C to perform calculations on the string w. ② Accept/Reject: If the above calculation is in the accepting state, let the Turing machine M accept it, otherwise let the Turing machine M reject it.

(3) Turing machine N result: If the abovementioned Turing machine M accepts, the result of the Turing machine N constructed this time is also acceptance; if the Turing machine M rejects, the result of Turing machine N is also rejection."

In this proof, special attention is needed to distinguish between Turing machine N and Turing machine M. M is a subroutine of N, and M only accept DFAs as input.

Next, we begin to prove the derivation problem of regular expressions, that is, give a regular expression, then a string is given, and ask whether this regular expression can derive the given string. Converting the question into language:

$A_{REX} = \{<R, w> |$ Regular expression R, derived string $w\}$

Theorem 5.1.3 A_{REX} is a decidable language.

C 在串 w 上的计算.

接下来使用第二种证明思路证明该定理.

证明 构造用于判定 A_{NFA} 的图灵机 N,有

N = "对于输入 $<B,w>$,B 是 NFA,w 是串:

(1) 自动机转化:把非确定型有穷自动机 B 转化成等价的确定型有穷自动机 C.

(2) 归约过程:将转化后的确定型有穷自动机 C,在输入字符串 w 上计算:

① 模仿:构造图灵机 M,给定输入字符串 w 之后,模仿确定性有限自动机 C 在串 w 上进行计算.② 接受/拒绝:如果上述计算处于接受状态,就让图灵机 M 接受,否则让图灵机 M 拒绝.

(3) 图灵机 N 的结果:若上述图灵机 M 接受,则本次构造的图灵机 N 的结果也是接受;若图灵机 M 拒绝,则图灵机 N 的结果也是拒绝."

该证明中需要特别注意的是区分图灵机 N 和图灵机 M,其中 M 是 N 的子程序,M 只接受 DFA 作为输入.

接下来证明正则表达式的派生性问题,即给定一个正则表达式,再给定一个串,问这个正则表达式是否能够派生出这个给定的串.将该问题转换成语言:

$A_{REX} = \{<R,w> |$ 正则表达式 R,派生串 $w\}$

定理 5.1.3 A_{REX} 是一个可判定性语言.

Proof Give a Turing machine P to determine the A_{REX} language, there are:

$P=$ "For input $<R,w>$, first perform a reasonableness check. If R is a regular expression and w is a string, then perform the following steps:

(1) Conversion: It is known that regular expressions and finite automata are equivalent, so first convert the regular expression R into an equivalent deterministic finite automata A.

(2) Run: Use the algorithm given in Theorem 5.1.1 to run the Turing machine M on the input $<A,w>$. Note that in this step, A is a deterministic finite automaton.

(3) Result: If M accepts, then accept it; otherwise the result is rejection."

The process of these proofs has already included the idea of the reduction, after solving the first problem, to solve the second and third problems, as long as these problems to be solved are transformed into the first problem. For decidable problems, since deterministic finite automata, non-deterministic finite automata, and regular expressions are equivalent to each other, if any one of these expressions is provided to the Turing machine, the Turing machine can recognize it.

The fourth problem is the emptiness of finite automata. Given a finite automaton, ask whether the finite automaton does not accept any string at all. The corresponding language is

$E_{DFA} = \{<A> | A \text{ is DFA and } L(A) \text{ is empty}\}$

where $L(A)$ is the language recognized by A.

Theorem 5.1.4 E_{DFA} is a decidable language.

Proof idea: First check all the strings one by one, and hand the strings to the automaton to run. But there are an infinite number of strings, so the DFA state diagram is used to determine whether to accept. DFA accepts a string if and only if the state diagram can reach the accepting state from the initial state, and there is a path in this process. If there is no path

证明 给出一个图灵机 P 来判定 A_{REX} 语言,有

$P=$ "对于输入$<R,w>$,首先进行合理性检测,若 R 是一个正则表达式并且 w 是一个字符串,则执行以下步骤:

(1) 转化:已知正则表达式和有穷自动机是等价的,所以首先把正则表达式 R 转化成等价的确定型有穷自动机 A.

(2) 运行:在输入$<A,w>$上使用定理 5.1.1 中给定的算法运行图灵机 M.注意在这一步中,A 是一个确定型有穷自动机.

(3) 结果:如果 M 接受,则结果接受;否则结果为拒绝."

这些证明的过程中都已经包含了归约的思想,解决了第一个问题,要解决第二个问题和第三个问题,只要将这些要解决的问题都转化为第一个问题即可.对于可判定问题,由于确定型有穷自动机、非确定型有穷自动机和正则表达式是相互等价的,将这其中任何一个表达提供给图灵机,图灵机都能够识别.

第四个问题,有穷自动机的空性问题.给定一个有穷自动机,问该有穷自动机是否根本不接受任何串,相应的语言为

$E_{DFA} = \{<A> | A \text{ 是 DFA 且 } L(A) \text{ 为空}\}$

其中 $L(A)$ 为 A 所识别的语言.

定理 5.1.4 E_{DFA} 是一个可判定语言.

证明思路:首先逐个检查所有的字符串,依次把串交给自动机去运行.但是存在无穷多个串,所以通过 DFA 的状态图来判断是否接受.当且仅当状态图可以从初始状态到达接受状态,并且这一过程存在一个路径,DFA

from the source point to the end point in this state diagram, it can be known that the automaton will not accept any strings.

Proof Let the Turing machine T have
$T=$ "For input $<A>$, A is DFA:
(1) Mark the initial state.
(2) Repeat the following steps until all states are marked.
(3) For a state, if a transition to it starts from a marked state, mark it.
(4) If no accept state is marked, it means that all states are unreachable, then accept it, otherwise reject it."

5.1.2 Decidable problems related to context-free languages

Four problems will be discussed in this section, including the derivation of the context-free grammar, the emptiness of the context-free grammar, the equivalence of the context-free grammar, and the determination of context-free language.

First study the first question. Given a context-free grammar G, ask whether this G can derive a specific string. This question is called a **CFG derivative question**. The corresponding language is as follows:
$A_{CFG} = \{<G, W> | \text{CFG } G \text{ derives string } w\}$

Theorem 5.1.5 A_{CFG} is a decidable language.

Proof idea: Let G traverse all the derivations to determine which one is the derived string w. If G produces string w, the process will be terminated; if G does not produce string w, then this process will be continued. The disadvantage of this method is that the Turing machine may sometimes not stop, which can only prove that A_{CFG} is a language recognized by Turing machine.

In order to solve this problem, CNF (the Chomsky's paradigm) is used here. According to CNF, it takes exactly $2n-1$ steps to derive a string of length n. In this case, first convert the CFG to the equivalent CNF, and then enumerate all the derivations with the length of $2n-1$ in the CNF.

Proof Turing machine S = "For input $<G,w>$, G is CFG:

(1) Convert G into equivalent CNF;

(2) List all the derivations of step $\max(1,2|w|-1)$;

(3) If one of these derivations produces a string w, accept it, otherwise reject it."

Check whether a context-free grammar does not derive any strings. This problem is called the emptiness problem of the context-free grammar, that is, the **CFG emptiness problem**. The corresponding language expression is as follows:

$E_{CFG} = \{<G> | G$ is CFG and $L(G)$ is empty$\}$

Theorem 5.1.6 E_{CFG} is a decidable language.

Proof idea: In the previous section, it is proved that the emptiness problem of finite automata is proved by state transition diagram. For the context-independent grammar, we let a Turing machine S check all possible strings w one by one, because there are infinitely many w, so this checking process will not stop. Another method is to check whether the initial argument produces a terminal symbol string. This is the ultimate goal of detection. If the initial argument can produce a terminal symbol string, it means that $L(G)$ is non-empty.

The method of checking the initial argument is recursion, that is, to check whether each argument can produce a terminal symbol string, similar to the marking algorithm of finite automata.

Proof Turing machine R = For input $<G>$, G is a context-free grammar:

(1) Mark all the terminal symbols in G;

(2) Repeat the following steps until no variables that can be marked are found;

(3) If G has a rule: $A \to U_1U_2U_3\ldots U_k$, and $U_1U_2U_3\ldots U_k$ has been marked, then the argument A will be marked;

(4) If the initial argument is not marked, accept it, otherwise reject it."

To determine whether two given context-free grammars derive the same language, this problem is called the equivalence problem of context-free grammars. The corresponding language is

$EQ_{CFG} = \{<G,H>|G$ and H are both CFG, and $L(G)=L(H)\}$

Note that this problem is undecidable, and the proof method for this problem will be introduced in Chapter 6.

Give a context-free language, then give a string, ask whether the string is in the context-free language, and refer to the problem **the decidable problem of the context-free language**.

Theorem 5.1.7 Every CFL is decidable.

Proof idea: Assuming that A is a context-free language (i.e., CFL), then A is either represented by a context-independent language, or recognized by a push down automaton, and A's push down automaton (PDA) is transformed into an equivalent non-deterministic Turing machine (NTM), some calculation branches of PDA may not stop, then the calculation branch of the corresponding NTM may not stop, so this Turing machine is not a decider. Next, we use the Turing machine S to determine the A_{CFG} to prove.

Proof Assume A is a CFL and G is the CFG of A. Then we will design the Turing machine M_G which can determine A. There are

$M_G = $ "For input w:

(1) Run Turing machine S on input $<G,w>$ (Theorem 5.1.5);

(2) If the Turing machine S accepts it, accept it, otherwise reject it."

So far, we have learned four language classes:

(3) 如果 G 有规则:$A \to U_1U_2U_3\cdots U_k$,且 $U_1U_2U_3\cdots U_k$ 都已经做过标记,则将变元 A 做上标记;

(4) 如果初始变元没有被标记,则接受,否则拒绝."

判定两个给定的上下文无关文法是不是派生同一语言,这个问题称为上下无关文法的等价性问题,相应的语言为

$EQ_{CFG} = \{<G,H> \mid G$ 和 H 都是 CFG,且 $L(G)=L(H)\}$

注意:该问题是不可判定的,该问题的证明方法将会在第六章介绍.

给定一个上下文无关语言,再给定一个串,问这个串在不在这个上下文无关语言里面,称该问题为**上下文无关语言的判定问题**.

定理 5.1.7 每个 CFL 是可判定的.

证明思路:假设 A 是一个上下文无关语言(即 CFL),那么 A 要么是由上下无关文语言来表示的,要么是由下推自动机来识别的,将 A 的下推自动机 PDA 转化成一个等价的非确定型图灵机 NTM,PDA 某些计算分支可能不停机,那么相应 NTM 的计算分支也可能不停机,所以这个图灵机不是判定器.接下来我们使用判定 A_{CFL} 的图灵机 S 来进行证明.

证明 设 A 是一个 CFL,G 是 A 的 CFG,下面设计判定 A 的图灵机 M_G,有

$M_G = $ "对于输入 w:

(1) 在输入 $<G,w>$ 上运行图灵机 S(定理 5.1.5);

(2) 如果图灵机 S 接受,则接受,否则就拒绝."

到此为止,我们已经学习了四个语

regular, context-free, decidable, and Turing recognizable languages. The relationship between the four languages is shown in Fig. 5-1.

言类:正则的、上下文无关的、可判定的和图灵可识别的.四个语言之间的关系如图 5-1 所示.

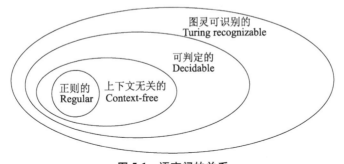

Fig. 5-1 The relationship between languages

图 5-1 语言间的关系

5.2 Halting problem

A decisive problem is a type of problem in which an answer of yes or no can be obtained based on input values selected from an infinite set. Therefore, the decisive problem is equally defined as a set of input sets whose return value is "yes". These input values can be natural numbers or other types of values. Using some encoding methods, such as Gödel numbers, these strings can be encoded as natural numbers. In order to maintain the simplicity of the formal definition, the decisive problem is expressed using a subset of natural numbers.

If the decisive problem A is a recursive set, then the problem can be called determinable or effectively solvable problem. If A is a recursively enumerable set, then the problem can be called partially decidable, or provable problem. This means that there is an algorithm that will eventually stop running when the answer to the question is "yes", but if the answer is "no", the algorithm may run forever. Partially decidable problems and any other undecidable problems are called "undecidable".

Alan Turing presented the first example of an undecidable problem in 1936: the halting problem. The halting problem refers to the input of a piece of pro-

5.2 停机问题

决定性问题是一类根据从一个无限集合中选取的输入值,得出是或否的回答的问题.因此,将决定性问题等同地定义为问题返回值为"是"的一组输入集.这些输入值可以是自然数,也可以是其他类型的值.使用一些编码方法,例如哥德尔编号,这些字符串可以编码为自然数.为了保持形式定义的简单性,决定性问题使用自然数的子集来表达.

如果决定性问题 A 是一个递归集,那么称该问题为可决定的或者有效可解的.如果 A 是一个递归可枚举集,那么该问题可以称为部分可判定、可制定或可证明的.这意味着存在一种算法,当问题的答案为"是"时,该算法最终会停止运行,但如果答案为"否",则该算法可能会永远运行.部分可判定的问题和任何其他不可判定的问题称为"不可解的".

艾伦·图灵在 1936 年提出了第一个不可制定问题的实例:停机问题.停机问题是指输入一段程序代码和针

gram code and the input for this program, whether it can be programmed to judge whether the program terminates after running this program.

First, give a Turing machine M and an input string w, the Turing machine receives the string w, which has:

$A_{TM} = \{<M,w>|M \text{ is a TM, and } w \in L(M)\}$

Theorem 5.2.1 A_{TM} is undecidable.

First, prove that A_{TM} is Turing recognizable. In this way, the above formula can show that the recognizer is more powerful than the decider.

Proof Assume there is language A, if a Turing machine M can be designed, for any string w, if $w \in A$, then after M reads w, the running result is that the Turing machine stops and enters the accepting state, then A is Turing recognizable. If the Turing machine can determine that it does not stop after accepting w before accepting the string w, it will reject w, but the Turing machine has no way to know the result in advance, so A_{TM} is sometimes called a **halting problem**.

Note that this proof does not impose restrictions on the situation where w does not belong to A, so if M reads w that does not belong to A, it may rejects, or it may be looped. The case where w does not belong to M is not considered.

5.2.1 Diagonalization method

Before proving the undecidability of the halting problem, we first introduce a classic proof method-Cantor's diagonal argument.

In the philosophy of mathematics, there is a view of "actual infinity" in the discussion of the question of infinity. The idea of "actual infinity" refers to: "all natural numbers" exist. Although they are inexhaustible, since they all exist, why can't we regard them as a "totality"? This view was first put forward by the ancient Greek philosopher Plato, called the "actual infinity" view. In other words, the infinite object is regarded as a self-fulfilling process or an infinite whole.

对此程序的输入,能否编程判断运行这个程序后程序是否终止.

首先给定一个图灵机 M 和一个输入字符串 w,图灵机接受字符串 w,有:

$A_{TM} = \{<M,w>|M \text{ 是一个 TM},且 w \in L(M)\}$

定理 5.2.1 A_{TM} 是不可判定的.

首先证明 A_{TM} 是图灵可识别的,这样,通过上述式子可以表明识别器比判定器更强大.

证明 假设有语言 A,如果能设计出一个图灵机 M,对于任意字符串 w,如果 $w \in A$,那么 M 读取 w 后运行结果为停机并进入接受状态,那么 A 就是图灵可识别的.如果图灵机在接受 w 串之前能判断出接受 w 后不停机,则会拒绝 w,但是图灵机没有办法提前知道结果,所以 A_{TM} 有时候被称为**停机问题**.

注意,该证明对 w 不属于 A 的情况没有做出限制,所以若 M 读取到不属于 A 的 w,则有可能会拒绝,也有可能循环.对于 w 不属于 M 的情况不予考虑.

5.2.1 对角化方法

在证明停机问题的不可判定性之前,首先介绍一种经典的证明方法——康托对角线论证法.

在数学哲学中,对无穷问题的探讨有一种"实无穷"的观点."实无穷"思想是指:"全体自然数"是存在的,虽然数不完,但既然都存在,我们为什么不可以把他看作一个"全体"?这最早是古希腊的哲学家柏拉图提出的观点,叫作"实无穷"观点.换言之,可以把无限的对象看成可以自我完成的过程或无穷整体.

Galileo (1564—1642) was the first scientist to seriously consider the "actual infinity". He had already considered comparing the correspondence between the two sets. He considered two real infinities, **N** is the set of all natural numbers $\{1, 2, 3, 4, 5,...\}$, and ε is the set of all perfect squares $\{1, 4, 9, 16, 25,...\}$ all constitute actual infinity. To compare the sizes of these two sets, intuitively, there are of course more natural numbers. Among the first 10 natural numbers, only 1, 4, and 9 are perfect squares; among the first 100 natural numbers, there are only 10 perfect squares, accounting for 10%; among the first 10,000 natural numbers, there are 100 perfect squares, only accounts for 1%; among the first 100 million natural numbers, the perfect square number is even more insignificant, accounting for only 0.01%. It seems that natural numbers are much more than perfect square numbers. However, from another perspective, if there is a natural number, there is a perfect square number. They have a one-to-one correspondence as shown in Fig 5-2:

伽利略(1564—1642)是第一个认真思考过"实无穷"的科学家. 他当时就已经考虑到比较两个集合的对应关系了. 他考虑两个实无穷, **N** 是全体自然数集合$\{1,2,3,4,5,\cdots\}$, ε 是全体完全平方数集合$\{1,4,9,16,25,\cdots\}$, 它们都构成了实无穷. 要比较这两个集合的大小, 直观上看, 当然是自然数多. 前 10 个自然数里, 完全平方数只有 1,4,9 三个; 在前 100 个自然数里, 完全平方数只有 10 个, 占 10%; 在前 1 万个自然数里, 完全平方数有 100 个, 只占 1%; 在前 1 亿个自然数里, 完全平方数更微不足道了, 只占 0.01%. 看起来, 自然数比完全平方数多得多. 但是, 另一个角度看, 有一个自然数, 便有一个完全平方数, 他们有如图 5-2 所示的一一对应关系:

$$\begin{array}{c|c} n & f(n) \\ \hline 1 & 1^2 \\ 2 & 2^2 \\ 3 & 3^2 \\ \vdots & \vdots \end{array}$$

图 5-2 数及其平方的对应关系
Fig. 5-2 Correspondence between number and its square

Regardless of whether it is a finite set or an infinite set, if a one-to-one correspondence can be established between the elements of the two sets A and B, it should be recognized that there are as many elements as A and B. This is the view put forward by Cantor, and it is also recognized by modern mathematics. Therefore, the size of the set is expressed in mathematical language: it is called **cardinality**. If the elements of two sets can establish a one-to-one correspondence, it is called equipotential. For infinite sets, what we have to do is to study the correspondence between elements.

不管是有穷集还是无穷集, 如果能在 A 与 B 两个集合的元素之间建立一一对应的关系, 就应该承认 A 与 B 的元素一样多. 这是康托提出的观点, 也是现代数学承认的观点. 所以集合的大小用数学语言表达就是**势**. 如果两个集合间的元素能建立起一一对应的关系, 就称其为等势. 对于无穷集合来说, 就是要研究它们之间元素的对应关系.

The diagonal argument method is a proof method proposed by George Cantor to show that the set of real numbers is an uncountable set.

The diagonal method was not Cantor's first proof method of the uncountable real numbers, it was published three years after his first proof. His first proof used neither decimal expansion nor any other number system. Since this technique was first used, similar proof construction methods have been used in a wide range of proofs.

Cantor's original proof shows that the number of points in the interval $[0,1]$ is not countable infinity. The proof is completed by contradiction, and the proof steps are as follows:

(1) Assume (derived from the original question) that the number of points in the interval $[0,1]$ is countable infinite.

(2) So we can arrange all the numbers in this interval into a series $(r_1, r_2, r_3, \dots)$.

(3) It is known that every such number can be expressed in decimal form.

(4) We arrange these numbers into series (these numbers do not need to be arranged in order; in fact, some countable sets, such as rational numbers, cannot be sorted according to the size of the numbers, but it is not a problem to just form a series of numbers). Some numbers have multiple expressions, such as $0.499\dots = 0.500\dots$, we choose the former.

(5) For example, if the decimal form of the sequence is as follows:

$r_1 = 0.5105110\dots$
$r_2 = 0.4132043\dots$
$r_3 = 0.8245026\dots$
$r_4 = 0.2330126\dots$
$r_5 = 0.4107246\dots$
$r_6 = 0.9937838\dots$
$r_7 = 0.0105135\dots$
…………

对角线论证法是乔治·康托提出的用于说明实数集合是不可数集的证明方法.

对角线法并非康托关于实数不可数的第一个证明法,而是在他的第一个证明法三年后发表.他的第一个证明既未用到十进制展开也未用到任何其他数字系统.自从该方法第一次使用以来,在很大范围内的证明中都用到了类似的证明构造方法.

康托的原始证明表明,区间$[0,1]$中的点数不是可数无穷大.该证明是用反证法完成的,证明步骤如下:

(1)假设(从原题中得出)区间$[0,1]$中的点数是可数无穷大的.

(2)于是乎我们可以把所有在这区间内的数字排成数列$(r_1, r_2, r_3, \dots)$.

(3)已知每一个这类的数字都能以小数形式表达.

(4)我们把这些数字排成数列(这些数字不需按序排列;事实上,有些可数集,例如有理数也不能按照数字的大小把他们全数排序,但只要排成数列就没有问题).某些有多种表达形式的数字,例如$0.499\dots = 0.500\dots$,我们选择前者.

(5)举例,如果该数列小数形式表示如下:

$r_1 = 0.5105110\dots$
$r_2 = 0.4132043\dots$
$r_3 = 0.8245026\dots$
$r_4 = 0.2330126\dots$
$r_5 = 0.4107246\dots$
$r_6 = 0.9937838\dots$
$r_7 = 0.0105135\dots$
…………

(6) Consider the k-th place after the decimal point of r_k. For convenience, add an underline and bold these numbers here, as shown below:

$r_1 = 0.\underline{\mathbf{5}}105110\ldots$
$r_2 = 0.4\underline{\mathbf{1}}32043\ldots$
$r_3 = 0.82\underline{\mathbf{4}}5026\ldots$
$r_4 = 0.233\underline{\mathbf{0}}126\ldots$
$r_5 = 0.4107\underline{\mathbf{2}}46\ldots$
$r_6 = 0.99378\underline{\mathbf{3}}8\ldots$
$r_7 = 0.010513\underline{\mathbf{5}}\ldots$
$\ldots\ldots\ldots$

(7) Let us set a real number $x \in [0,1]$, where x is defined in the following ways:

① If the k-th decimal place of r_k is equal to 5, then the k-th decimal place of x is 4.

② If the k-th decimal place of r_k is not equal to 5, then the k-th decimal place of x is 5.

(8) Obviously x is a real number in the interval $[0,1]$. Taking the previous number as an example, the corresponding x should be $0.4555554\ldots$.

(9) Since we assume that $(r_1, r_2, r_3, \ldots)$ includes all real numbers in the interval $[0,1]$, there must be a $r_n = x$.

(10) But the special definition of x makes the n-th decimal place of x and r_n different, so $x \notin (r_1, r_2, r_3, \ldots)$.

(11) So $(r_1, r_2, r_3, \ldots)$ cannot list all the real numbers in the interval $[0,1]$, which is a contradiction. It is concluded that the hypothesis "the number of points in the interval $[0,1]$ is countable infinite" proposed in the first point is not valid.

5.2.2 Undecidability proof of the halting problem

Next, use the diagonalization method to prove Theorem 5.2.1, which proves the undecidability of the following languages:

$A_{TM} = \{<M, w> | M \text{ is a TM, and } w \in L(M)\}$

Proof idea Construct a $D_{TM} = \{<M> | \text{TM}, M$ accepts the string $<M>\}$. D_{TM} is a special case of A_{TM}.

The difference between D_{TM} and A_{TM} is that A_{TM} has two inputs M and w, while D_{TM} has only one input M, and M itself has another input, the content is the encoded font of M, so D_{TM} is a special case of A_{TM}. Then use the diagonalization method to prove the undecidability of D_{TM}.

Proof(**Proof by contradiction**) Assuming that Turing machine H can determine A_{TM}, then

$$H(<M,w>)\begin{cases} \text{accept, if } M \text{ accepts } w \\ \text{reject, if } M \text{ rejects } w \end{cases}$$

Input the contents of M and w to H, H can give the correct answer, if M accepts w, H accepts the result, if M does not accept w, H rejects the result, both cases will cause H to stop without the problem of non-stop caused by infinite loops.

Then use H to construct a new Turing machine D_{TM}, the construction idea is as follows:

D = "For input $<M>$, where M is a Turing machine:

(1) Run H on the input $<M>$, since H is to solve a special acceptability problem, input the digital code of the Turing machine M itself to M, judge whether M accepts it, and H will give the answer.

(2) If H accepts, D rejects; if H rejects, D accepts."

The second step is to set the output result of D to be opposite to the output result of H, which is denoted as follows:

$$D(<M>)\begin{cases} \text{accept, if } M \text{ rejects } <M> \\ \text{reject, if } M \text{ accepts } <M> \end{cases}$$

So whether it is Turing machine H or Turing machine D, there will eventually be a shutdown result. But then there will be a question, what kind of result will be output when input $<D>$ on the Turing machine D? Denoted as follows:

$$D(<D>)\begin{cases} \text{accept, if } D \text{ rejects } <D> \\ \text{reject, if } D \text{ accepts } <D> \end{cases}$$

殊情况,不同的是,A_{TM} 有 M 和 w 两个输入,而 D_{TM} 只有一个输入 M,而 M 本身又有一个输入,内容为 M 的编码字体,所以 D_{TM} 是 A_{TM} 的特例.然后使用对角化方法证明 D_{TM} 的不可判定性.

证明(反证法) 假设图灵机 H 能够判定 A_{TM},则有

$$H(<M,w>)=\begin{cases} 接受,M 接受 w \\ 拒绝,M 不接受 w \end{cases}$$

将 M 与 w 的内容输入给 H,H 能够给出正确答案,若 M 接受 w,H 就接受该结果,若 M 不接受 w,H 就拒绝该结果,两种情况都会导致 H 停机,而不会出现死循环导致的不停机问题.

然后利用 H 来构造新的图灵机 D_{TM},构造思路如下:

D = "对于输入 $<M>$,其中 M 是图灵机:

(1)在输入 $<M>$ 上运行 H,其中由于 H 解决特殊的接受性问题,因此将图灵机 M 本身的数字编码输给 M,判断 M 是否接受,由 H 给出答案.

(2)如果 H 接受,D 就拒绝;如果 H 拒绝,D 就接受."

第二步是将 D 的输出结果设置为与 H 的输出结果相反,表示如下:

$$D(<M>)=\begin{cases} 接受,M 拒绝<M> \\ 拒绝,M 接受<M> \end{cases}$$

所以无论是图灵机 H 还是图灵机 D,最终都会有一个停机的结果.但是接下来就会出现一个问题,在图灵机 D 上输入 $<D>$ 会输出什么样的结果?表示如下:

$$D(<D>)=\begin{cases} 接受,D 不接受<D> \\ 拒绝,D 接受<D> \end{cases}$$

It can be seen that if D does not accept $<D>$, the result of D is acceptance, and if D accepts $<D>$, the result of D is rejection. The following conclusions can be drawn:

$$D(<D>) = \text{accept} \Leftrightarrow D(<D>) = \text{reject}$$

From the previous hypothesis on D, it can be seen that D is decidable. When D accepts $<D>$, the result should be acceptance, and when D rejects $<D>$, the result should be rejection, which contradicts the previous conclusion. It can be seen that the Turing machine H assumed in this proof does not exist, so there is no H to judge A_{TM}, and the conclusion is that A_{TM} is undecidable.

There is a Turing machine U, so that for any input $<M, w>$ (where M is a description of a Turing machine, w is an input string), U can simulate the situation what happens to M when the input is w, and U is called a **universal Turing machine**. Next, explain the essence of the proof process. The proof process uses the diagonalization method and uses a general Turing machine to simulate the operation. Using the one-to-one correspondence between Turing machines and natural numbers, list the Turing machines one by one, and treat each input as a special Turing machine to form a result as shown in Fig. 5-3 below.

可知如果 D 不接受 $<D>$, D 的结果为接受, 如果 D 接受 $<D>$, D 的结果为拒绝. 可以得出以下结论:

$$D(<D>) = 接受 \Leftrightarrow D(<D>) = 拒绝$$

由前面对 D 的假设可知, D 是可判定的, D 在接受 $<D>$ 时结果应该为接受, D 在拒绝 $<D>$ 时结果应该为拒绝, 这就与前面的结论相矛盾. 由此可知, 本证明所假设的图灵机 H 是不存在的, 所以不存在 H 判定 A_{TM}, 从而得出结论: A_{TM} 是不可判定的.

存在一个图灵机 U, 使得对于任意输入 $<M,w>$ (其中 M 是对一个图灵机的描述, w 是一个输入字符串), U 能够模拟出 M 在输入为 w 时的情况, 称 U 为**通用图灵机**. 接下来解释一下该证明过程的实质. 该证明过程使用了对角化方法, 使用通用图灵机来模拟运行. 利用图灵机与自然数的一一对应关系, 将图灵机一一列出来, 并将每一个输入当作一个特殊的图灵机, 构成一个如图 5-3 所示的结果.

图 5-3 U 的输出结果

Fig. 5-3 Output result of U

The above figure is the result of inputting content in the general Turing machine. If the Turing machine accepts it, it will be marked, and the result of stopping or not stopping is treated as blank.

Using Turing machine $H(<M, w>)$ to determine the acceptability problem, it is different from the general Turing machine U, which may be stop or non-stop. Turing machine H must be stop everywhere. The output results are shown in Fig 5-4.

	$<M_1>$	$<M_2>$	$<M_3>$	$<M_4>$	$<M_5>$	$<M_6>$	...
M_1	接受 Accept	拒绝 Refuse	接受 Accept	拒绝 Refuse	接受 Accept	接受 Accept	...
M_2	拒绝 Refuse	接受 Accept	拒绝 Refuse	拒绝 Refuse	接受 Accept	拒绝 Refuse	...
M_3	拒绝 Refuse	拒绝 Refuse	拒绝 Refuse	拒绝 Refuse	拒绝 Refuse	拒绝 Refuse	...
M_4	接受 Accept	拒绝 Refuse	接受 Accept	拒绝 Refuse	接受 Accept	拒绝 Refuse	...
M_5	接受 Accept	接受 Accept	接受 Accept	接受 Accept	接受 Accept	接受 Accept	...
M_6	拒绝 Refuse	接受 Accept	拒绝 Refuse	拒绝 Refuse	拒绝 Refuse	接受 Accept	...
⋮	⋮	⋮	⋮	⋮	⋮	⋮	

图 5-4 H 的输出结果
Fig. 5-4 Output result of H

Replace w in Turing machine H with $<M>$, D takes the opposite of the diagonal result of H in the running results of Fig 5-4, the expression is $D(<M>) = \neg H(M, <M>)$. At this time, if D is determinable, then D is calculated using one of the Turing machines listed in Fig. 5-3 above, and then $D(<D>)$ must be generated in a certain line of the running result. The contradiction of the results is shown in Fig. 5-5.

图 5-5 结果的矛盾
Fig. 5-5 The contradiction of the results

5.2.3 A non-Turing recognizable language

In this section, a non-Turing recognizable language is introduced. By proving the definition below, we can know that if a language is decidable, we can infer that the language and its complement are both recognizable.

Theorem 5.2.2 Language A is decidable $\Leftrightarrow$ Language A and A^C are recognizable ($A^C = \Sigma^* - A$, that is, A^C is the complement of A).

Before proving the theorem, let's first look at an example:

Example 5.2.1 Prove that the decidable language class is closed in the following operations.

a. Union.

Proof Let M_1 and M_2 be the deciders that recognize the decidable languages A_1 and A_2. Construct Turing machine M:

$M=$ "Input w,

(1) Run M_1 and M_2 on w respectively and run a step M_2 for every step of M_1.

(2) If one of M_1 and M_2 accepts, then accept, if both reject, then reject. "

M is the decider that recognizes $A_1 \cup A_2$. Therefore, it can be determined that the decidable language class is closed for union operation.

b. Connection.

Proof Let M_1 and M_2 be the deciders that recognize the decidable languages A_1 and A_2. Construct Turing machine M:

$M=$ "Input w,

(1) List all the ways ($|w|+1$) that divide w into two sections.

(2) For each segmentation method, run M_1 on the first segment and M_2 on the second segment. If both accept, then accept.

(3) Reject if none of the segmentation methods is accepted.

M is the decider of $A_1 \circ A_2$. Therefore, it can be determined that the decidable language class is closed

5.2.3 一个非图灵可识别语言

该小节介绍一个非图灵可识别语言,通过证明下面的定义可以知道,一个语言若是可判定的,可以推断出该语言和其补都是可识别的.

定理 5.2.2 语言 A 是可判定的 $\Leftrightarrow$ 语言 A 和 A^C 是可识别的 ($A^C = \Sigma^* - A$, 即 A^C 是 A 的补).

在证明该定理之前,首先看一个例子:

例 5.2.1 证明在下列运算中可判定语言类是封闭的.

a. 并.

证明 设 M_1, M_2 为识别可判定语言 A_1, A_2 的判定器. 构造图灵机 M:

$M=$ "输入 w,

(1) 分别在 w 上运行 M_1 和 M_2, 每运行一步 M_1 就运行一步 M_2.

(2) 若 M_1 和 M_2 中有一个接受,则接受;若都拒绝,则拒绝."

M 为识别 $A_1 \cup A_2$ 的判定器. 所以可判定语言类对并运算封闭.

b. 连接.

证明 设 M_1, M_2 为识别可判定语言 A_1, A_2 的判定器. 构造图灵机 M:

$M=$ "输入 w,

(1) 列出所有将 w 分成两段的方式 ($|w|+1$ 种).

(2) 对于每种分段方式,在第一段上运行 M_1, 第二段上运行 M_2. 若都接受,则接受.

(3) 若没有一种分段方式被接受,则拒绝.

M 为 $A_1 \circ A_2$ 的判定器. 所以,可判定语言类对连接运算封闭.

to the connection operation.

From this example, it can be seen that the decidable language is closed to Boolean operations, and then we start to prove Theorem 5.2.2.

Proof Assume language A is decidable, and it can be known that the decidable language is closed to Boolean operations, so the complement language A^C is decidable, and then the decidable languages are all Turing identifiable, so the following conclusions are drawn: A is decidable $\Rightarrow A$ and A^C are Turing recognizable.

Assume that both languages A and A^C are Turing recognizable. Assume the Turing machine M_1 recognizes A, and the Turing machine M_2 recognizes A^C. Each time a w is given, the recognizers of A and A^C are run at the same time to determine whether w is in language A, and whether w is in A or A^C, it can be known that either M_1 or M_2 must accept strings w, which means that either Turing machines M_1 or M_2 will stop and accept, so M will always stop. M will accept all strings in A and reject strings other than A, so an M can be constructed to determine A.

Next, a Turing machine M will be constructed to determine A.

M = "For input w:

(1) Run M_1 and M_2 in parallel on the input w. M has two tapes, one simulates M_1, one simulates M_2, and M simulates one step of two machines alternately, until one of them stops. (Note that you cannot simulate M_1 first and then simulate M_2, because M_1 may not stop and simulate forever.)

(2) If M_1 accepts, then accept; if M_2 accepts, then reject."

M can determine A, indicating that A is decidable, so it can be concluded that if language A and A^C are both Turing recognizable, A is decidable.

由这个例子可知可判定语言对布尔运算封闭,接下来开始证明定理5.2.2。

证明 假设语言 A 是可判定的,又可知可判定语言对布尔运算封闭,所以补语言 A^C 是可判定的,从而可判定语言都是图灵可识别的.得出以下结论:如果 A 是可判定的,那么 A 和 A^C 是图灵可识别的.

假设语言 A 和 A^C 都是图灵可识别的.设图灵机 M_1 识别 A,图灵机 M_2 识别 A^C,每给一个 w,同时运行 A 和 A^C 的识别器,判断 w 是否在语言 A 里面,无论 w 是在 A 还是在 A^C 里,可以知道 M_1 和 M_2 中必定会有一个接受串 w,这说明图灵机 M_1 和 M_2 会有一个停机并接受,所以 M 总会停机. M 会接受 A 中所有的串,拒绝除 A 以外的串,所以可以构造一个 M 用于判断 A.

下面构造图灵机 M 判定 A.

M = "对输入 w:

(1)在输入 w 上并行运行 M_1 和 M_2,M 有两条带,一条模拟 M_1,一条模拟 M_2,M 交替地模拟两个机器中的一步,直到其中一个停机为止.(注意,不能先模拟完 M_1 再模拟 M_2,因为 M_1 可能不停机并永远模拟下去)

(2)若 M_1 接受,就接受;若 M_2 接受,就拒绝."

M 能够判断 A,说明 A 是可判断的,所以可以得出结论:若语言 A 和 A^C 都是图灵可识别的,则 A 是可判定的.

Exercise 5

1. We can find specific examples of non-recursive language without the diagonalization argument. **Busy beaver function** $\beta: \mathbf{N} \mapsto \mathbf{N}$ is defined as follows: For each integer n, $\beta(n)$ is the largest number m such that there are Turing machines with alphabet $\{\triangleright, \sqcup, a, b\}$ and exactly n states. When it starts on the space tape, it finally stops in the configuration $(h, \triangleright, \sqcup, a^m)$.

(a) Prove that if f is any recursive function, there is an integer k_f such that $\beta(n+k_f) \geq f(n)$ (k_f is the number of states in the Turing machine M_f, if M_f is started on input a^m, then M_f will have $a^{f(m)}$ on the tape when it stops)

(b) Prove that β is not recursive. [Assuming it is, then so is $f(n) = \beta(2n)$. Apply the result in (a) above]

2. Consider the question of whether a DFA and a regular expression are equivalent, formulate the problem as a language and prove that it is decidable.

3. Let $ALL_{DFA} = \{<A> | A$ is a DFA that identifies $\Sigma^*\}$. Prove that ALL_{DFA} is decidable.

4. Let $A = \{<M> | M$ is DFA, it accepts any string containing an odd number of 1s$\}$.

Prove that A is decidable.

5. A useless state of the pushdown automata is a state that it will not enter on any input. Consider checking whether a pushdown automaton has a useless state, formalizing the problem into a language, and proving that it is decidable.

6. Let A and B be two disjoint languages. Say language C separates A and B, if $A \subseteq C$ and $B \subseteq \overline{C}$. Prove that any two disjoint-complementary Turing recognizable languages can be separated by a certain decidable language.

7. Check the operation of two DFAs on all strings when the scale is less than or equal to a certain number, and use this method to prove that the EQ_{DFA} is decidable. Calculate such a number.

7. 检查两个 DFA 在规模小于或者等于某个数时在所有串上的运行，并以此方法证明 EQ_{DFA} 是可判定的. 计算这样的一个数.

习题 5 答案
Key to Exercise 5

Chapter 6 Undecidability

Some languages can be determined by a decider (a Turing machine that always stops). For example, give a string w, this decider concludes whether w belongs to this language. These languages are **decidable languages**. So are there undecidable languages? That is, is there a language that does not have a decider that decides it? The answer is "yes".

Imagine the following language:

$A = \{(<M>, w) | M$ denotes Turing machine, w is a character string, M accepts $w\}$

That is, the string in language A consists of two parts: the first part is the Turing machine M; the second part is the string w. And M accepts w.

Assume the language A is decidable, that is, there exists a decider H. H accepts $(<M>, w)$ when M accepts w, and rejects $(<M>, w)$ when M does not accept w (Note that H is a decider that always stops, accepts or rejects $(<M>, w)$). Then we modify H slightly by reversing its result: when M accepts w, H rejects $(<M>, w)$; when M does not accept w, H accepts $(<M>, w)$.

Then a change is made to H. Its input string is simply a serialized string $<M>$ of a Turing machine M. The input given to this M is no longer some string w, but the string $<M>$ itself. Then the behavior of H becomes: the input to it is a string $<M>$ denoting some Turing machine. Taking $<M>$ as the input to the Turing machine M, if M accepts $<M>$, H rejects $<M>$; if M does not accept $<M>$, then H accepts $<M>$.

Here comes the highlight. What happens if H's own serialized string $<H>$ is provided to H? When H accepts $<H>$, H rejects $<H>$; when H does not accept $<H>$, H accepts $<H>$. This contradicts each other. The only conclusion can only be that H cannot exist. That is, it is impossible to construct a decider for language A. A is a **undecidable language**. Undecidable languages exist.

Reducibility is the transformation of a problem into another problem, so that the solution of the second problem is used to solve the first problem. This idea is similar to that of mathematical proofs, where if a proof is difficult to cut from the original proposition, converting the original proposition into a converse proposition based on the equivalence of the original proposition and its converse proposition will lead to an easy solution or entry point to the problem. For example, if you have a map, the road recognition problem is reduced to the map problem.

6.1 Undecidable problems from language theory

A_{TM} is undecidable, that is, the problem of determining whether a Turing machine accepts a given input is undecidable. Consider a related problem: $HALT_{TM}$, that is, the problem of determining whether a Turing machine stops (by accepting or rejecting) for a given input. If A_{TM} is reduced to $HALT_{TM}$, the undecidability of A_{TM} can be used to prove the undecidability of $HALT_{TM}$. Let

$HALT_{TM} = \{<M, w> \mid M$ is a TM and stops for input $w\}$

Theorem 6.1.1 $HALT_{TM}$ is undecidable.

Proof idea Use the proof by contradiction (to reduce the A_{TM} to $HALT_{TM}$). Assume that TM R determines $HALT_{TM}$, and using R, one can construct a TM S that determines A_{TM}. Using R, one can check whether M is down for w. If M is not down for w, S rejects it because $<M, w>$ is not in A_{TM}. If M is down for w, S simulates it and there is no risk of a dead loop. This way if TM R exists, it can determine A_{TM}, but A_{TM} is known to be undecidable. Thus contradicting and thus $HALT_{TM}$ is undecidable.

Proof Assume TM R determines $HALT_{TM}$, from

造一个判定器去判定它. A 是**不可判定语言**. 不可判定语言是存在的.

可归约性就是将一个问题转化为另一个问题,使用第二个问题的解来解第一个问题. 这种思想类似于数学证明中,如果一个证明很难从原命题切入,此时根据原命题与其逆否命题是等价的,将原命题转换成逆否命题求解,可得到极其简便的解法或者解题的切入点. 例如,假设要在一个城市中认路,如果有一张地图,就将认路问题归约为地图问题.

6.1 语言理论中的不可判定问题

A_{TM} 是不可判定的,即确定一个图灵机是否接受一个给定的输入问题是不可判定的. 考虑一个相关的问题:$HALT_{TM}$,即确定一个图灵机对给定的输入是否停机(通过接受或拒绝)问题. 若将 A_{TM} 归约到 $HALT_{TM}$,就可以利用 A_{TM} 的不可判定性证明 $HALT_{TM}$ 的不可判定性. 设

$HALT_{TM} = \{<M, w> \mid M$ 是一个 TM,且对输入 w 停机$\}$

定理 6.1.1 $HALT_{TM}$ 是不可判定的.

证明思路 使用反证法(把 A_{TM} 归约为 $HALT_{TM}$). 假设把 TM R 判定 $HALT_{TM}$,利用 R 可以构造一个判定 A_{TM} 的 TM S. 使用 R 可以检查 M 对 w 是否停机. 如果 M 对 w 不停机,S 就拒绝,因为$<M,w>$不在 A_{TM} 中. 如果 M 对 w 停机,S 就模拟它,而且不会有死循环的危险. 这样如果 TM R 存在,就能判定 A_{TM},但是 A_{TM} 已知是不可判定的. 因此矛盾,从而 $HALT_{TM}$ 是不可判定的.

证明:假设 TM R 判定 $HALT_{TM}$,由

which TM S can be constructed to determine A_{TM}, which is constructed as follows:

S = "On output $<M, w>$:

(1) Run TM R on input $<M, w>$.

(2) If R rejects, reject.

(3) If R accepts, simulate M on w until it stops.

(4) If M has accepted, accept; if M has rejected, reject."

Obviously, if R determines $HALT_{TM}$, S determines A_{TM}, and since A_{TM} is undecidable, $HALT_{TM}$ must also be undecidable.

That is, there is **no program that "can determine whether the program will stop or not"**.

To gain more insight into the proof using this method, additional theorems and proofs are presented below. First, the definition of Turing recognizable language is introduced: Let M be a Turing machine, and M is said to accept a string w if it can enter the accepted state and stop after M runs on the input string w. The set of all strings accepted by M is called the language recognized by M, or the language of M for short, and is denoted as $L(M)$. Let

$E_{TM} = \{<M> \mid M \text{ is a TM and } L(M) = \varnothing\}$

Theorem 6.1.2 E_{TM} is undecidable.

Proof idea Assume that E_{TM} is decidable, and use this to prove that A_{TM} is decidable. Let R be a TM that determines the E_{TM}, and consider using R to construct S that determines the A_{TM}. How should S run when it receives input $<M, w>$? One idea for constructing S is to run R on input $<M>$ and see if it accepts it. If yes, it is known that $L(M)$ is the empty set, so M does not accept w. If R rejects w, it is only known that $L(M)$ is not the empty set, i.e., M accepts some string, but does not know if it accepts this particular w. So we change the goal, i.e., modify $<M>$, so that M rejects all strings except w.

Proof Let's start by writing that modified machine M_1 described in the proof idea in standard terms.

$M_1 =$ "On input x:

(1) If $x \neq w$, reject.

(2) If $x = w$, then run M on x and accept when M accepts."

This machine takes w as a part of its description. The obvious way to check if $x = w$ holds is to scan the input and compare it to w character by character to determine if they are the same.

Assume again that TM R determines E_{TM}. Construct TM S for determining A_{TM} as follows.

$S =$ "On the input $<M, w>$, where $<M, w>$ is the encoding of TM M and the string w.

(1) Construct the above TM M_1 with the description of M and w.

(2) Run R on input $<M_1>$.

(3) If R accepts, reject; if R rejects, accept."

If R is the decider of E_{TM}, then S is the decider of A_{TM}. Since the decider of A_{TM} does not exist, E_{TM} must be undecidable.

That is, the problem of "**determining whether the language recognized by the Turing machine is the empty set**" is undecidable.

It is also interesting to mention a computational problem related to Turing machines: give a Turing machine and a language that can be recognized by some simpler computational model, determine whether the Turing machine recognizes this language.

For example, if $REGULAR_{TM}$ is the problem of determining whether a given Turing machine has an equivalent infinite automaton, then this problem is the same as the problem of **determining whether a given Turing machine recognizes a regular language**.

$REGULAR_{TM} = \{<M> \mid M \text{ is a TM and } L(M) \text{ is a regular language}\}$

Theorem 6.1.3 $REGULAR_{TM}$ is undecidable.

The proof of Theorem 6.1.3 is similar to the above proof method and will not be repeated here. Practically, there is a more general result about this problem, called **Lace's theorem**. It states that it is undecidable to check whether any property about a language is recognizable by a Turing machine.

Up to this point, all the above methods of proof are reductive from A_{TM}. However, in proving that some languages are undecidable, it may be simpler to start from another undecidable language (e.g. E_{TM}). A reduction from E_{TM} is given below as an example of another type of proof of undecidability, which can of course also be proved using a reduction from A_{TM}.

Let $EQ_{TM} = \{<M_1, M_2> \mid M_1 \text{ and } M_2 \text{ are both TM and } L(M_1) = L(M_2)\}$

Theorem 6.1.4 EQ_{TM} is undecidable.

Proof Reduce E_{TM} to EQ_{TM}. Let TM R determine EQ_{TM}. Construct TM S for determining E_{TM} as follows.

$S =$ "For input $<M>$, where M is TM.

(1) Run R on input $<M, M_1>$, where M_1 is the Turing machine that rejects all inputs.

(2) If R accepts, accept. If R rejects, then reject."

If R determines EQ_{TM}, then S determines E_{TM}, but from theorem 6.1.2, E_{TM} is undecidable, so EQ_{TM} is also undecidable. In other words, the above problem of "**determining whether two Turing machines are equivalent to each other**" is undecidable.

Next, we introduce the use of historical computing reduction, to compute history method is an important technology for proving that A_{TM} is reducible to a certain language. It is more convenient to use this method when proving the undecidability of a problem, if the problem involves checking the existence of something.

The computational history of a Turing machine on an input is the sequence of configurations that the Turing machine goes through when processing this input. It is a complete record of the computations experienced by the machine.

Definition 6.1.1 Let M be a Turing machine and w be a string. An acceptance computation history of M on w is a sequence of configurations $C_1, C_2, \ldots C_l$, where C_1 is the starting configuration of M on w, C_l is an accepting configuration of M, and each C_i is a legal result of C_{i-1}, that is, it conforms to the rules of M. A rejecting computation history of M on w can be defined similarly, except that C_l should be a rejecting configuration.

The computational histories are all finite sequences. If M does not stop on w, the two do not exist. Deterministic Turing machines have at most one computation history. Non-deterministic Turing machines may have multiple computational histories. But now we continue to concentrate on deterministic automata.

Definition 6.1.2 A linear bounding automaton (LBA) is a restricted Turing machine that does not allow its read-write head to leave the tape region containing the inputs. If this machine attempts to move its read-write head out of the two endpoints of the input, the read-write head stays in place. This is the same way that the read-write head of an ordinary Turing machine does not leave the left endpoint of the tape.

Let A_{LBA} be the question of whether the LBA accepts its input. Although A_{LBA} is the same as the undecidable problem A_{TM} when the Turing machine is restricted to LBA, we can still prove that A_{LBA} is decidable. Let

$A_{\text{LBA}} = \{<M, w> \mid M$ is an LBA that accepts string $w\}$

Before proving its decidability, it is necessary to prove the following theorem. It states that when the input of an LBA is a string of length n, there can only be a finite number of configurations.

Theorem 6.1.5 Let M be an LBA with q states and g tape symbols. For length n, M has exactly qng^n different configurations.

Proof The configuration of M is like a snapshot in the middle of a computation. The configuration consists of control states, read-write head positions, and tape contents. Here, M has q states. Its tape length is n, so the read-write head may be in one of the n positions and g^n tape symbol strings may appear on the tape. The product of these three quantities is the total number of configurations of M with tape length n.

Theorem 6.1.6 A_{LBA} is decidable.

Proof The algorithm for determining A_{LBA} is as follows.

L = "For input $<M, w>$, where M is LBA, w is the string:

(1) Simulate M qng^n steps on w, or until it stops.

(2) If M is down, accept when it accepts and reject when it rejects. If it is not yet down, reject."

If M is running qng^n steps on w before it stops, it must be repeating some configuration, i.e., it is stuck in a loop, according to Theorem 6.1.5. This is the reason why the algorithm rejects in this case.

Two other theorems for the decidable problem of linear bounded automata are given below, and their proofs are not repeated.

Theorem 6.1.7 $E_{\text{LBA}} = \{<M> \mid M$ is an LBA and $L(M) = \varnothing\}$ is undecidable.

Theorem 6.1.8 $ALL_{\text{CFG}} = \{<G> \mid G$ is a CFG and $L(G) = \Sigma^*\}$ is undecidable.

6.2 Undecidable problems about grammars

Undecidable problems appear not only in the domain related to Turing machines, but in fact in all mathematical domains. For example, there exist multiple undecidable problems related to grammars, as summarized below.

Theorem 6.2.1 Each of the following problems is undecidable:

(a) For a given grammar G and string w, determine whether $w \in L(G)$.

(b) For a given grammar G and input e, determine whether $e \in L(G)$.

(c) For two given grammars G_1 and G_2, determine whether $L(G_1) = L(G_2)$.

(d) For any grammar G, determine whether $L(G) = \varnothing$.

(e) In addition, there exists some fixed grammar G_0 such that deciding whether any given string w belongs to $L(G_0)$ is undecidable.

For all the above problems, a similar proof can be made from the halting problem to the reduction of (a). Most of the proofs about the above have been proved in the previous section. Of course, undecidable problems do not include discerning whether $w \in L(G)$, or whether $L(G) = \varnothing$. These problems can be solved algorithmically, and in fact these algorithms are effective. Many other problems, however, are not undecidable.

Theorem 6.2.2 Each of the following problems is undecidable:

(a) Given the context-free grammar G, does $L(G) = \Sigma^*$?

(b) Given two context-free grammars G_1 and G_2, does $L(G_1) = L(G_2)$?

(c) Given two pushdown automata M_1 and M_2, do they accept exactly the same language?

(d) Given a pushdown automaton M, find the equivalent pushdown automaton with the least number of states.

For the proof of (a), the proved undecidable problem in part (d) of Theorem 6.2.1 can be reduced to part (a), i.e., the problem of deciding whether a given general grammar generates any string or not.

For (b), if we can discern whether two context-free grammars generate Σ^*, let the second grammar indeed generate the trival grammar of Σ^*.

For (c), if we can discern whether two pushdown automata are equivalent, then we can seek to discern whether two context-free grammars are equivalent by transforming them into two pushdown automata that accept the same language, and then discerning the equivalence of these two pushdown automata.

For (d), if there exists an algorithm for minimizing the number of states of an arbitrary pushdown automaton, as there is for an infinite automaton, then we can try to discern whether a given pushdown automaton accepts Σ^*; it accepts when and only when the optimized pushdown automaton has only one state and accepts Σ^*. It is decidable whether a single-state pushdown automaton accepts Σ^* or not.

6.3 An undecidable tiling problem

Given a finite set of bricks, each brick is a unit square. We were asked to cover the first quadrant of the plane with replicas of these bricks, as shown in Fig. 6-1. We have an infinite number of replicas of each brick.

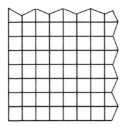

Fig. 6-1 Tiling problem

The only restrictions are that special bricks must be placed in the lower left corner (the origin bricks), that only specific pairs of bricks can be placed horizontally next to each other, and that only specific pairs of bricks can be placed vertically next to each other (bricks cannot be rotated or turned over). Given a finite set of bricks, origin bricks, and adjacent rules, is there an algorithm to determine whether the first quadrant can be covered?

This problem can be formalized as follows. The system of tiling is a quadruple $D = (D, d_0, H, V)$, where D is a finite set of bricks tiles, $d_0 \in D$, and $H, V \subseteq D \times D$. The tiling according to D is a function $f: \mathbf{N} \times \mathbf{N} \to D$ such that the following relations hold:
$$f(0,0) = d_0,$$
$$(f(m,n), f(m+1,n)) \in \mathbf{N}, m, n \in \mathbf{N},$$
$$(f(m,n), f(m,n+1)) \in V, m, n \in \mathbf{N}$$

Theorem 6.3.1 Given a tiling system, determine whether there exists a tiling according to this system, and this problem is undecidable.

Proof We are given a Turing machine M and the problem of deciding whether M does not stop on input e is reduced to the tiling problem. This problem is the complement of the stopping problem, so it is a undecidable problem. If this problem is reduced to the tiling problem, then the tiling problem must be undecidable.

The basic idea is to construct the tiling system D from an arbitrary Turing machine M such that the tiling according to D (if such tiling exists) denote the infinite computation of M starting from a tape of spaces. The configuration of M is denoted horizontally in the tiling; the succeeding configuration appears on top of the previous one. That is, the horizontal dimension denotes the tape of M and the vertical dimension denotes the time. If M never stops at the empty input, the succeeding rows are paved to infinity; but if M stops after k steps, paving to more than k rows is not possible.

It is helpful to consider that the edges of bricks are marked with some kind of information when constructing the relations H and V; we allow bricks to be adjacent to bricks horizontally or vertically only if the markings on adjacent edges are the same. On horizontal edges, these markings are either symbols from the alphabet of M or a combination of states and symbols. The tiling system is arranged so that if tiling is feasible, then by looking at the horizontal edge markers between rows n and $n+1$, we can read the configuration of M after the $n-1$ step calculation. Thus only one edge along such a boundary is marked with an ordered pair of states and symbols, all other edges are marked with a single symbol.

The marker on the vertical edge of a brick is either vacant (it can only match vertical edges that are also unmarked) or contains the state of M and the "direction" pointer that we indicated with an arrow. These markers on the vertical edges are used to indicate that the tape head moves from one brick to the next to the left or to the right.

6.4 Formal definition of mapping reducibility

Now define formal reducibility to denote computational problems in terms of languages.

Definition 6.4.1 Language A is mapped reducible to language B if there exists a computable function $f: \Sigma^* \to \Sigma^*$ such that for each w,
$$w \in A \Leftrightarrow f(w) \in B$$
denote by $A \leq_m B$. Call the function f the reductive contract from A to B.

The A to B mapping reduction provides a way to transform the membership test problem of A into a membership test problem of B. To check if $w \in A$ is present, this reduction f can be used to map w to $f(w)$ and then check if $f(w) \in B$ is present. The term mapping reduction comes from the function or mapping that provides the means for reduction.

If a problem mapping is reducible to a second problem, and the second problem has been previously solved, then the solution to the original problem is obtained. The following theorem illustrates this idea.

Theorem 6.4.1 If $A \leq_m B$ and B is decidable, then A is also decidable.

Proof Let M be the decider of B and f be the reduction from A to B. The decider N of A is described as follows:

$N=$ "For the input w:

(1) Compute $f(w)$.

(2) Run M on $f(w)$ and output the output of M."

Clearly, if $w \in A$, then $f(w) \in B$, since f is the reduction from A to B. Therefore, M accepts $f(w)$ as long as $w \in A$. Therefore, N runs as desired.

Theorem 6.4.2 If $A \leq_m B$ and A is undecidable, then B is also undecidable.

Theorem 6.4.3 If $A \leq_m B$ and B is Turing-recognizable, then A is also Turing-recognizable.

The proof of this theorem is similar to the proof of Theorem 6.4.1, except that M and N are changed to recognizers instead of deciders.

Theorem 6.4.4 If $A \leq_m B$, and A is not Turing-recognizable, then B is not Turing-recognizable either.

As a typical application of this theorem, let A be the complement of A_{TM}, $\overline{A_{\mathrm{TM}}}$. It is known that $\overline{A_{\mathrm{TM}}}$ is not Turing-recognizable. It is easy to see from the definition of mapping reducibility, $A \leq_m B$ has the same meaning as $\overline{A} \leq_m \overline{B}$. To prove that B is not recognizable, one can prove $A_{\mathrm{TM}} \leq_m \overline{B}$; one can also use mapping reducibility to prove that certain problems are neither Turing-recognizable nor complementary Turing-recognizable, as in the following theorem.

定理6.4.1 如果$A \leq_m B$且B是可判定的,那么A也是可判定的.

证明 设M是B的判定器,f是从A到B的归约.A的判定器N的描述如下：

$N=$"对于输入w:

(1) 计算$f(w)$.

(2) 在$f(w)$上运行M,输出M的输出."

显然,如果$w \in A$,则$f(w) \in B$,因为f是从A到B的归约.因此,只要$w \in A$,M就接受$f(w)$.故N的运行正如所求.

定理6.4.2 如果$A \leq_m B$且A是不可判定的,则B也是不可判定的.

定理6.4.3 如果$A \leq_m B$且B是图灵可识别的,则A也是图灵可识别的.

此定理的证明与定理6.4.1的证明类似,只是将M和N改成识别器而非判定器.

定理6.4.4 如果$A \leq_m B$,且A不是图灵可识别的,那么B也不是图灵可识别的.

作为这个定理的一个典型应用,设A是A_{TM}的补集,$\overline{A_{\mathrm{TM}}}$.可知$\overline{A_{\mathrm{TM}}}$不是图灵可识别的.由映射可归约性的定义不难看出,$A \leq_m B$与$\overline{A} \leq_m \overline{B}$有相同的含义.为证明$B$不是可识别的,可以证明$A_{\mathrm{TM}} \leq_m \overline{B}$;还可以使用映射可归约性来证明某些问题既不是图灵可识别的,也不是补图灵可识别的,就像下面的定理那样.

Theorem 6.4.5 EQ_{TM} is neither Turing-recognizable nor complementary Turing-recognizable.

Proof First prove that EQ_{TM} is not Turing-recognizable. To do so, it is sufficient to prove that A_{TM} is reducible to $\overline{EQ_{TM}}$. The reduction function is as follows.

$F = $ "On input $<M,w>$, where M is TM, w is the string.

(1) Construct the following two machines M_1 and M_2.

$M_1 = $ "For any input:

a. Reject."

$M_2 = $ "For any input:

a. Run M on w. If it accepts, accept."

(2) Output $<M_1, M_2>$."

Here, M_1 accepts nothing, and if M accepts w, then M_2 accepts every input, so the two machines are not equivalent. Conversely, if M does not accept w, then M_2 accepts nothing, so they are equivalent. In this way f reduces A_{TM} to $\overline{EQ_{TM}}$, which is exactly what we expect.

To prove that $\overline{EQ_{TM}}$ is not Turing recognizable, it is sufficient to give a reduction from A_{TM} to the complement of $\overline{EQ_{TM}}$ (i.e., EQ_{TM}). Thus to prove $A_{TM} \leq_m EQ_{TM}$. The following TM G computes the reductive function g.

$G = $ "On the input $<M, w>$, where M is a TM and w is a string.

(1) Construct the following two machines M_1 and M_2.

$M_1 = $ "For any input.

a. Accept."

$M_2 = $ "For any input.

a. Run M on w.

b. If it accepts, accept."

(2) Output $<M_1, M_2>$."

The only difference between f and g is on machine M_1. In f, machine M_1 is always rejected. While in g, it always accepts. In both f and g, M accepts w when and only when M_2 accepts all strings. In g, M accepts w when and only when M_1 and M_2 are equivalent. This is the reason why g is reduced from A_{TM} to EQ_{TM}.

f 和 g 之间的唯一区别在机器 M_1 上. 在 f 中, 机器 M_1 总是被拒绝. 而在 g 中, 它总是接受. 在 f 和 g 中, 当且仅当 M_2 接受所有串时, M 接受 w. 在 g 中, 当且仅当 M_1 和 M_2 等价时, M 接受 w. 这就是 g 是从 A_{TM} 到 EQ_{TM} 的归约的原因.

Exercise 6

1. If $A \leq_m B$ and B is a regular language, does this imply that A is also a regular language? Why?

2. Prove that all Turing identifiable problems map to be reducible to A_{TM}.

3. A useless state of a Turing machine is a state that it does not enter for any input. Consider the problem of checking whether a Turing machine has a useless state. Formalize this problem into a language and prove that it is undecidable.

4. Check that the Turing machine is on output w. Has it ever tried to move the read-write head to the left somewhere in the computation process? Formalize this problem into a language and prove that it is decidable.

5. Rice's theorem. Let P be a nontrivial property of the language of any Turing machine. Prove that the problem of determining whether the language of a particular Turing machine has property P is undecidable. More formally, let P be a language that consists of descriptions of Turing machines and P satisfies the following two conditions: first, P is nontrivial, it contains some TM descriptions but not all; second, P is a property of the language of TM. Whenever $L(M_1) = L(M_2)$, there is $<M_1> \in P$ when and only when $<M_2> \in P$, and in addition M_1 and M_2 are arbitrary TM. Prove that P is an undecidable language.

6. Assume that the rules of the tiling game become instead of specifying special tiles to be placed at the origin, but specifying a special set of tiles, and specifying that only those tiles can be used to lay the first row. Prove that the tiling problem is still undecidable.

Key to Exercise 6

Part IV Theory of Computational Complexity

Chapter 7 Time Complexity

This chapter aims to introduce the basics of time complexity theory. First, we will introduce a method of measuring the time required to solve the problem, then introduce how to classify the problem according to the time required, and finally discuss some decidable problems that require a lot of time, and how to recognize them when encountering such problems.

7.1 Measuring complexity

Consider the following examples. Language $A = \{0^k1^k \mid k \geqslant 0\}$, obviously A is a decidable language. Let's analyze the time that single-tape Turing machine spends in determining A. Consider the single-tape TM M_1 that determines A first. Give a low-level description of the Turing machine, including the actual movement of the read-write head on the tape, so as to calculate the number of steps taken by M_1 when it runs.

$M_1 = $ "For the input string w:

(1) Scan the tape, and if a 0 is found to the right of 1, it will be rejected.

(2) If there are both 0 and 1 on the tape, repeat the next step.

(3) Scan the tape, delete a 0 and a 1.

(4) If all 1s are deleted and there are still 0s, or all 0s are deleted and there are still 1s, reject it. Otherwise, if neither 0 nor 1 remains on the tape, accept it."

Now consider the time required for the algorithm to determine the A's Turing machine M_1.

On a particular input, the number of steps used by the algorithm may be related to several parameters. For example, if the input is a graph, the number of steps may depends on the number of nodes, the number of edges, and the maximum degree of the graph, or a combination of these factors, or some combination of them and other factors. For the sake of simplicity, the running time of the algorithm is calculated purely as a function of the length of the input string, regardless of other parameters. In the worst-case analysis, consider the longest running time on all inputs of a certain length. In the average case analysis, consider the average of the running time on all inputs of a certain length.

Definition 7.1.1 Let M be a deterministic Turing machine that stops on all inputs. The running time or **time complexity** of M is a function $f: \mathbf{N} \to \mathbf{N}$, where $\mathbf{N}$ is a set of non-negative integers, $f(n)$ is the maximum number of steps that M takes when running on all inputs of length n. If $f(n)$ is the running time of M, it is said that M runs in time $f(n)$, and M is the time $f(n)$ Turing machine. Usually, n denotes the length of the input.

7.1.1 Big O and Small o notation

Since the precise running time of an algorithm is usually a complex expression, it is generally only an estimate of its trend and level. For this reason, only the highest order term of the expression of the algorithm running time is considered, and the coefficient of this term and other secondary terms are ignored.

For example, function $f(n) = 6n^3 + 2n^2 + 20n + 45$ has four terms, and the highest order term is $6n^3$. Ignore the coefficient 6, and say that f is gradually not greater than n^3. The asymptotic notation or big O notation for expressing this relationship is $f(n) = O(n^3)$.

Definition 7.1.2 Let f and $g: \mathbf{N} \to \mathbf{R}_+$. Call $f(n) = O(g(n))$, if there are positive integers c and n_0, so that for all $n \geq n_0$, there is
$$f(n) \leq cg(n)$$
When $f(n) = O(g(n))$, it is said that $g(n)$ is the **asymptotic upper bound** of $f(n)$.

Intuitively, $f(n) = O(g(n))$ means that if the difference of a constant factor is ignored, then f will be less than or equal to g. You can think of O as a hidden constant. In practice, most of the functions f that may be encountered have an obvious highest order term h. In this case, write it as $f(n) = O(g(n))$, where g is h without coefficients.

Example 7.1.1 Let $f_1(n)$ be the function $5n^3 + 2n^2 + 22n + 6$. Keep the highest order term $5n^3$, and discard its coefficient 5 to get $f_1(n) = O(n^3)$.

Verify that the result meets the above formal definition. Let c equal 6 and n_0 equal 10, then for all $n \geqslant 10$, there is $5n^3 + 2n^2 + 22n + 6 \leqslant 6n^3$.

In addition, there is $f_1(n) = O(n^4)$, because n^4 is greater than n^3, so it is also an asymptotic upper bound of f_1.

Example 7.1.2 The Big O notation interacts with the logarithm in a special way. Usually when writing a logarithm, you must specify the base (or called the base of the logarithm), such as $x = \log_2 n$. Here the cardinality 2 shows that this equation is equivalent to equation $2^x = n$. The value of $\log_b n$ is multiplied by the corresponding constant as the base b changes, because there is identity $\log_b n = \log_2 n / \log_2 b$. Therefore, there is no need to specify the base when writing $f(n) = O(\log n)$, because the constant factor will eventually be ignored.

Let $f_2(n)$ be function $3n\log_2 n + 5n\log_2 \log_2 n + 2$. There is $f_2(n) = O(n\log n)$, at this time, because $\log n$ is more dominant than $\log \log n$.

Big O notation can also appear in arithmetic expressions, such as the expression $f(n) = O(n^2) + O(n)$. At this time, each occurrence of the symbol O represents a different hidden constant. Because $O(n^2)$ is more dominant than $O(n)$, this expression is equivalent to $f(n) = O(n^2)$. When the symbol O appears on an exponent as in the expression $f(n) = 2^{O(n)}$, the meaning is the same. This expression represents an upper bound of 2^{cn}, where c is a certain constant.

Expression $f(n) = 2^{\log n}$ will appear in some analyses. From the identity equation $n = 2^{\log_2 n}$ to get $n^c = 2^{c\log_2 n}$, it can be seen that $2^{O(\log n)}$ represents an upper bound of n^c, where c is a constant. The expression $n^{O(1)}$ represents the same bounds in another way, because the expression $O(1)$ represents a value that does not exceed a certain fixed constant.

We often derive a bound of the form n^c, where c is a constant greater than 0. Such bounds are called **polynomial bounds**. A boundary shaped like $2^{(n^\delta)}$ when δ is a real number greater than 0, it is called **an exponential boundary.**

Along with the big O notation is the small o notation. The big O notation means that one function is asymptotically not greater than another function. To say that one function is asymptotically smaller than another function, use the small o notation. The difference between big O and small o notation is similar to the difference between $\leq$ and $<$.

Definition 7.1.3 Let f and g be two functions $f, g : \mathbf{N} \to \mathbf{R}_+$. If
$$\lim_{n \to \infty} \frac{f(n)}{g(n)} = 0$$
then we say that $f(n) = o(g(n))$. In other words, $f(n) = o(g(n))$ means that for any real number $c > 0$, there is a number n_0, for all $n \geq n_0$, there is $f(n) < cg(n)$.

Example 7.1.3 It is easy to verify the following equation.
(1) $\sqrt{n} = o(n)$
(2) $n = o(n\log(\log n))$
(3) $n\log(\log n) = o(n\log n)$
(4) $n\log n = o(n^2)$
(5) $n^2 = o(n^3)$

However, $f(n)$ will not equal $o(f(n))$.

7.1.2 Analysis algorithm

This section analyzes the Turing machine algorithm corresponding to language $A = \{0^k 1^k | k \geq 0\}$. For readability, this algorithm is repeated here.

M_1 = "For the input string w:

(1) Scan the tape, if you find 0 to the right of 1, reject it.

(2) If there are both 0 and 1 on the tape, repeat the next step.

(3) Scan the tape and delete a 0 and a 1.

(4) If there are still 0s after deleting all 1s, or there are still 1s after deleting all 0s, reject it. Otherwise, if neither 0 nor 1 remains on the tape, accept it."

In order to analyze M_1, consider its four steps separately. In step 1, the machine scans the tape to verify that the input form is 0^*1^*. It takes n steps to perform this scan. As previously agreed, n is usually used to denote the length of the input. Repositioning the read-write head on the left end of the tape requires an additional n steps. So this step requires a total of $2n$ steps. Using big O notation, it is said that this stage requires $O(n)$ steps. Note that there is no mention of the operation of repositioning the read-write head in the machine description. The asymptotic notation allows to omit operating details that do not affect runtime by more than a constant factor in the machine description.

In steps (2) and (3), the machine scans the tape repeatedly, deleting a 0 and a 1 in each scan. Each scan requires $O(n)$ steps. Because two symbols are deleted in each scan, scan at most $n/2$ times. So steps (2) and (3) require $(n/2)O(n) = O(n^2)$ step.

In step (4), the machine scans once to decide whether to accept or reject. The time required for this step is at most $O(n)$. Therefore, the total time M spent on the input of length n is $O(n)+O(n^2)+O(n)$, or $O(n^2)$.

7.1.2 分析算法

本小节分析与语言 $A = \{0^k 1^k | k \geq 0\}$ 对应的图灵机算法. 为了便于阅读, 这里重述一遍此算法.

M_1 = "对输入串 w:

(1) 扫描带子, 如果发现 0 在 1 的右边, 就拒绝.

(2) 如果带子上既有 0 也有 1, 就重复下一步.

(3) 扫描带子, 删除一个 0 和一个 1.

(4) 如果删除所有 1 后还有 0, 或者删除所有 0 以后还有 1, 就拒绝. 否则, 如果在带子上既没有剩下 0 也没有剩下 1, 就接受."

为了分析 M_1, 将其四个步骤分开来考虑. 步骤 1 中, 机器扫描带子以验证输入的形式是 0^*1^*. 执行这次扫描需要 n 步. 如前所述, 通常用 n 表示输入的长度. 将读写头重新放置在带子的左端另外需要 n 步. 所以这一步骤总共需要 $2n$ 步. 用大 O 记法, 称这一阶段需要 $O(n)$ 步. 注意, 在机器描述中没有提及重新放置读写头的操作. 渐近记法允许在机器描述中忽略那些对运行时间的影响不超过常数倍的操作细节.

在步骤(2)和(3)中, 机器反复扫描带子, 在每一次扫描中删除一个 0 和一个 1. 每一次扫描需要 $O(n)$ 步. 因为每一次扫描删除两个符号, 所以最多扫描 $n/2$ 次. 于是步骤(2)和(3)需要 $(n/2)O(n) = O(n^2)$ 步.

在步骤(4)中, 机器扫描一次决定是接受还是拒绝. 这一步需要的时间最多是 $O(n)$. 所以, M 在长度为 n 的输入上总共耗时为 $O(n)+O(n^2)+O(n)$, 或

In other words, its running time is $O(n^2)$. This completes the time analysis of the machine.

In order to classify languages according to time requirements, some notations are defined below.

Definition 7.1.4 Let $t: \mathbf{N} \to \mathbf{R}_+$ be a function. Define the **time complexity class** $TIME(t(n))$ as the set of all languages determined by the Turing machine in $O(t(n))$ time.

The previous analysis of language $A = \{0^k 1^k \mid k \geq 0\}$ shows that because M determines A at time $O(n^2)$, and $TIME(n^2)$ includes all languages that can be determined at time $O(n^2)$, so $A \in TIME(n^2)$.

Is there a machine that determines A asymptotically faster? In other words, does A belong to $TIME(t(n))$ for a certain $t(n) = o(n^2)$? Delete two 0s and two 1s in each scan, instead of just one each, it can reduce running time, because doing so cuts the number of scans in half, but does not affect the asymptotic running time. The following machine M uses a different method to determine A asymptotically faster. It shows $A \in TIME(n\log n)$.

$M_2 =$ "For the input string w:

(1) Scan the tape, if you find 0 to the right of 1, reject it.

(2) As long as there are 0 and 1 on the tape, repeat the following steps.

(3) Scan the tape to check whether the total number of remaining 0 and 1 is even or odd. If it is odd, reject it.

(4) Scan the tape again, starting with the first 0, and deleting one 0 every other; then starting from the first 1, and deleting one 1 every other.

(5) If there are no more 0s and 1s on the tape, accept it. Otherwise, reject it."

Before analyzing M_2, let's verify that it can indeed determine A. In step (4), every time a scan is performed, the total number of remaining 0s is reduced by half, and the other 0s are deleted. Therefore, if there are 13 0s at the beginning, then only 6 0s will remain after step (4) is executed once. There are 3, 1 and 0 0s left for each subsequent execution. This step has the same effect on the number of 1.

Now let's check the parity of the numbers of 0 and 1 each time step (3) is executed. Assume again that there are 13 0s and 13 1s at the beginning. When step (3) is executed for the first time, there will be an odd number of 0s (because 13 is an odd number) and an odd number of 1s. After each execution, there are only even numbers (6), then odd numbers (3), and again, odd number (1) 0s and 1s. Due to the loop condition specified in step (2), when there are 0 0s or 0 1s left, step (3) is no longer executed. For the obtained parity sequence (odd, even, odd, odd), if you replace even with 0 and odd with 1, and reverse this sequence, you get 1101, which is 13, which is the binary representation of the number of 0s and 1s at the beginning. The parity sequence always gives a binary representation in reverse.

When step (3) checks that the total number of remaining 0 and 1 is even, it is actually checking whether the parity of the number of 0 is consistent with the parity of the number of 1. If the two parities are always the same, then the binary representations of the numbers of 0 and 1 are the same, so that the two numbers are equal.

In order to analyze the running time of M_2, first notice that each step consumes $O(n)$ time, and then determine the number of times each step needs to be executed. Steps (1) and (5) are executed once, and a total of $O(n)$ time is required. Step (4) deletes at least half of the 0s and 1s each time it is executed, so at most $1+\log_2 n$ cycles can delete all characters. So steps (2)~(4) consume a total of time $(1+\log_2 n)O(n)$, which is $O(n\log n)$. The running time of M_2 is $O(n)+O(n\log n)=O(n\log n)$.

$A \in TIME(n^2)$ has been proved before, and now there is a better bound, namely $A \in TIME(n\log n)$. This result cannot be further improved on a single-tape Turing machine. In fact, the languages determined by the single-tape Turing machine in time $O(n\log n)$ are all regular languages.

If the Turing machine has a second tape, language A can be determined in $O(n)$ time (also called **linear time**). The following dual-tape Turing machine M_3 determines A in linear time. The operating mode of machine M_3 is different from the above machines that determines A. It simply copies all 0s to the second tape, and then uses them to match with 1.

$M_3 =$ "For the input string w:

(1) Scan tape 1, and if a 0 is found to the right of 1, reject it.

(2) Scan the 0 on tape 1 and stop until the first 1, and copy the 0 to tape 2 at the same time.

(3) Scan the 1 on tape 1 until the end of the input. Every time a 1 is read from tape 1, a 0 is deleted from tape 2. If all 0s are deleted before the 1 is read, it is rejected.

(4) If all 0s are deleted, accept it. If there are still 0s left, refuse."

The analysis of this machine is very simple. Each of the four steps obviously requires $O(n)$ steps, so the total running time is $O(n)$, and it is linear. Note that this is the best possible running time, because just reading the input requires n steps.

Summarize the results about the time complexity of A, that is, the time required to determine A. Given a single-tape Turing machine M_1, A can be determined within time $O(n^2)$, and a faster single-tape Turing machine M_2 can determine A within time $O(n\log n)$. Later, a dual-tape Turing machine M_3 will be given, which can determine A within time $O(n)$. Therefore, the time complexity of A on a single-tape Turing machine is $O(n\log n)$, and it is $O(n)$

前面已经证明 $A \in TIME(n^2)$, 而现在有更好的界, 即 $A \in TIME(n\log n)$. 这个结果在单带图灵机上不可能进一步改进. 实际上, 单带图灵机在 $O(n\log n)$ 时间内判定的语言都是正则语言.

如果图灵机有第二条带子, 就可以在 $O(n)$ 时间 (也称为**线性时间**) 内判定语言 A. 下面的双带图灵机 M_3 在线性时间内判定 A. 机器 M_3 的运行方式和上面那些判定 A 的机器不同, 它只是简单地将所有 0 复制到第二条带子上, 然后用来和 1 进行匹配.

$M_3 =$ "对输入串 w:

(1) 扫描带子 1, 如果发现 0 在 1 的右边, 就拒绝.

(2) 扫描带 1 上的 0, 直到第一个 1 时停止, 同时把 0 复制到带 2 上.

(3) 扫描带 1 上的 1 直到输入结束. 每次从带 1 上读到一个 1, 就在带 2 上删除一个 0, 如果在读完 1 之前所有的 0 都被删除, 就拒绝.

(4) 如果所有的 0 都被删除, 就接受. 如果还有 0 剩下, 就拒绝."

这个机器分析起来很简单. 显然, 四个步骤的每一步需要 $O(n)$ 步, 所以总的运行时间是 $O(n)$, 从而是线性的. 注意, 这是可能的最好运行时间, 因为只是读输入就需要 n 步.

总结一下关于 A 的时间复杂度的结果, 即判定 A 所需要的时间. 给出一个单带图灵机 M_1, 能够在 $O(n^2)$ 时间内判定 A, 而一个更快的单带图灵机 M_2, 能够在 $O(n\log n)$ 时间内判定 A. 随后将给出一个双带图灵机 M_3, 能够在 $O(n)$ 时间内判定 A. 因此在单带图灵机上 A 的时间复杂度是 $O(n\log n)$, 在双带图灵机上是 $O(n)$. 注意, A 的

on a dual-tape Turing machine. Note that the complexity of A is related to the selected calculation model.

The above discussion highlights a major difference between complexity theory and computability theory. In the theory of computability, the Church-Turing thesis asserts that all reasonable calculation models are equivalent, that is, the language classes they determine are the same. In complexity theory, the choice of model affects the time complexity of the language. For example, a language that is decidable in linear time on one model is not necessarily decidable in linear time on another model.

In complexity theory, problems are classified according to their time complexity. But which model is used to measure time? The same language may take different time on different models.

Fortunately, for a typical deterministic model, the time requirements are not very different. Therefore, as long as the classification system is not very sensitive to relatively small differences in complexity, the choice of deterministic model is not relevant. In the following sections, this idea will be discussed further.

7.1.3 Complexity relationship between models

Now examine how the choice of calculation model affects the time complexity of the language. Investigate three models: single-tape Turing machine, multi-tape Turing machine and non-deterministic Turing machine.

Theorem 7.1.1 Let $t(n)$ be a function, $t(n) \geq n$. Then each multi-tape Turing machine at $t(n)$ time is equivalent to a single-tape Turing machine at $O(t^2(n))$ time.

Proof idea The proof idea of this theorem is very simple. First, it is clear how to transform a multi-tape Turing machine into a single-tape Turing machine that simulates it. Analyze this simulation to determine how much extra time it will take. Prove that

复杂度与选择的计算模型有关.

上面的讨论突出了复杂性理论与可计算性理论之间的一个重大区别. 在可计算性理论中,丘奇—图灵论题断言,所有合理的计算模型都是等价的,即它们所判定的语言类都是相同的. 在复杂性理论中,模型的选择影响语言的时间复杂度,如在一个模型上线性时间内可判定的语言在另一个模型上就不一定是线性时间内可判定的.

在复杂性理论中,根据计算问题的时间复杂度对问题分类. 但是用哪种模型来度量时间呢? 同一个语言在不同的模型上可能需要不同的时间.

幸运的是,对于典型的确定型模型,所需要的时间差别不是太大. 所以,只要分类体系对复杂性上相对较小的差异不是很敏感,那么对确定型模型的选择就关系不大. 在下面几节中,将进一步讨论这一想法.

7.1.3 模型间的复杂性关系

现在考察计算模型的选择是怎样影响语言的时间复杂度的. 考察三种模型:单带图灵机、多带图灵机和非确定型图灵机.

定理7.1.1 设$t(n)$是一个函数,$t(n) \geq n$,则每一个$t(n)$时间的多带图灵机都和某一个$O(t^2(n))$时间的单带图灵机等价.

证明思路 此定理的证明思路非常简单. 首先清楚怎样把一个多带图灵机转变为一个模拟它的单带图灵机. 分析这种模拟,来确定它需要多少额外的时间. 证明模拟多带机的每一步

each step of the simulation of a multi-tape machine requires at most $O(t(n))$ steps of a single-tape machine. Therefore, the total time required is $O(t^2(n))$ step.

Proof Let M be a k-tape Turing machine running in time $t(n)$. Construct a single-tape Turing machine S running in time $O(t^2(n))$.

The machine S simulates the operation of M, and S uses one of its tapes to represent the contents of all k tapes of M. These tapes are stored continuously, and the positions of M's read-write heads are marked on the appropriate squares.

At the beginning, S lets its tape form a format that denotes all tapes of M, and then simulates the steps of M. In order to simulate the step of M, S scans all the information on the tape to determine the symbol under M's read-write head. Then S scans the tape again, updating the tape content and the position of the read-write head. If M's read-write head moves to the right to a position on the tape that has not been read before, then S must increase the storage space allocated to this tape. To do this, it moves part of its tape one space to the right.

Now let's analyze this simulation. For each step of M, machine S scans the active part of the tape twice. The first time to obtain the information necessary to determine the next action; the second time to complete this step, the length of the active part of the S tape determines how long it takes for S to scan one time, so the upper bound of this length must be determined. To this end, take the sum of the lengths of the active parts on the k tapes of M. Because in step $t(n)$, if M's read-write head moves to the right in each step, M uses up $t(n)$ squares. If it still moves to the left, it does not need that many squares, so the length of each active part is at most $t(n)$. So S scans its active part once and requires $O(t(n))$ steps.

证明 设 M 是一个在时间 $t(n)$ 内运行的 k 带图灵机. 构造一个在时间 $O(t^2(n))$ 内运行的单带图灵机 S.

机器 S 模拟 M 运行, S 用它的一条带子表示 M 的所有 k 条带子的内容. 这些带子连续存放, M 读写头的位置都标在恰当的方格上.

开始时, S 让其带子形成表示 M 的所有带子的格式, 然后模拟 M 的步骤. 为了模拟 M 的步骤, S 扫描带子上的所有信息, 确定在 M 的读写头下的符号. 然后 S 再次扫描带子, 更新带上内容和读写头位置. 如果 M 的读写头向右移动到带子上还没有读到的位置, 那么 S 必须增加分配给这条带子的存储空间. 为此, 它把其带子的一部分向右移动一格.

现在来分析这种模拟. 对于 M 的每一步, 机器 S 两次扫描带子上活跃的部分. 第一次获取决定下一步动作所必需的信息; 第二次完成这一步的动作, S 带上活跃部分的长度决定了 S 扫描一次需要多长时间, 所以必须确定这个长度的上限. 为此, 取 M 的 k 条带子上活跃部分的长度之和. 因为在 $t(n)$ 步中, 如果 M 的读写头每一步都向右移动, 则 M 用掉 $t(n)$ 个方格. 若它还向左移, 则不用那么多方格, 所以每一个活跃部分的长度最多是 $t(n)$. 于是 S 扫描一次它的活跃部分需要 $O(t(n))$ 步.

Simulate each step of M, S performs two scans, it is also possible to move to the right k times at most. Each time it takes $O(t(n))$, so simulating the one-step operation of M, S takes a total of time $O(t(n))$.

Now let's define the total time required for the simulation. In the initial stage, S makes its tape form the proper format, which requires $O(n)$ step. Subsequently, S simulates the $t(n)$ step operation of M, and each simulation step requires $O(t(n))$ step, so the simulation part requires $t(n) \times O(t(n)) = O(t^2(n))$ step. Therefore, the entire simulation process of M requires $O(n) + O(t^2(n))$ steps.

Assuming $t(n) \geq n$ (this is a reasonable assumption, because if there is less time, M can't even finish reading the input), then the running time of S is $O(t^2(n))$, and the proof is complete.

Let's consider the similar theorem for non-deterministic single-tape Turing machines. Prove that the language determined by this machine is also decidable on a deterministic single-tape Turing machine, but it takes more time. Before that, the running time of the non-deterministic Turing machine must be defined. Recall that a non-deterministic Turing machine becomes a decision machine when all its computational branches stop on all inputs.

Definition 7.1.5 Let N be a non-deterministic Turing machine and a decider. The **running time** of N is the function $f: N \to N$, where $f(n)$ is the maximum number of steps in all calculation of branches on any input of length n, as shown in Fig. 7-1.

The definition of non-deterministic Turing machine running time is not used to correspond to any actual computing device. On the contrary, it will be explained later that it is a useful mathematical definition that helps characterize the complexity of a class of important computational problems.

模拟 M 的每一步，S 执行两次扫描，还可能最多向右移动 k 次。每一次用时 $O(t(n))$，所以模拟 M 的一步操作，S 总共耗时 $O(t(n))$。

现在来界定模拟所需要的全部时间。在开始阶段，S 让它的带子形成恰当的格式，这需要 $O(n)$ 步。随后，S 模拟 M 的 $t(n)$ 步操作，每模拟一步需要 $O(t(n))$ 步，所以模拟部分一共需要 $t(n) \times O(t(n)) = O(t^2(n))$ 步。因此，整个 M 的模拟过程需要 $O(n) + O(t^2(n))$ 步。

假定 $t(n) \geq n$（这是合理的假定，因为如果时间更少，M 连输入都读不完），则 S 的运行时间是 $O(t^2(n))$。证毕。

下面探讨对非确定型单带图灵机的类似定理。证明这种机器所判定的语言在确定型单带图灵机上也是可判定的，但需要更多的时间。在此之前，必须定义非确定型图灵机的运行时间。回忆一下，当一个非确定型图灵机的所有计算分支在所有输入上都停机时，它便成为一个判定机。

定义7.1.5 设 N 是一个非确定型图灵机，并且是一个判定机。N 的**运行时间**是函数 $f: N \to N$，其中 $f(n)$ 是在任何输入长度为 n 上所有计算分支中的最大步数，如图7-1 所示。

非确定型图灵机运行时间的定义不是用来对应任何实际的计算设备的。相反，稍后将说明，它是一个有用的数学定义，有助于刻画一类重要的计算问题的复杂性。

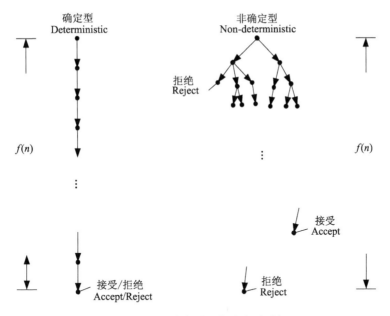

Fig. 7-1　Measuring deterministic and non-deterministic time

Theorem 7.1.2　Let $t(n)$ be a function, $t(n) \geqslant n$. Then every non-deterministic single-tape Turing machine at $t(n)$ time is equivalent to a certain deterministic single-tape Turing machine at $2^{O(t(n))}$ time.

Proof　Let N be a non-deterministic Turing machine running in time $t(n)$. Construct a deterministic Turing machine D. D simulates N by searching for N's non-deterministic calculation tree. Now analyze this simulation.

On the input of length n, the length of each branch of the non-deterministic calculation tree of N is at most $t(n)$, and each node of the tree has at most b children, where b is the maximum value of the legal choice determined by the transfer function of N. Therefore, the total number of leaves is $b^{t(n)}$ at most.

The simulation process uses the breadth-first method to explore this tree. In other words, before visiting a node with a depth of $d+1$, first visit all nodes with a depth of d. When visiting a certain node, start from the root and descend to that node. This approach is very inefficient, but even improving this inefficiency will not change the current theorem, so there is no need to improve it. The total number of nodes in the tree is less than twice the maximum number of leaves, so use $O(b^{t(n)})$ as its upper bound. The time to descend from the root to a node is $O(t(n))$. Therefore, the running time of D is $O(b^{t(n)})$.

Turing machine D has three tapes. According to Theorem 7.1.1, turning it into a single-tape Turing machine at most powers the running time. In this way, the running time of the single-tape simulator is $(2^{O(t(n))})^2 = 2^{O(2t(n))} = 2^{O(t(n))}$, and the theorem is proved.

7.2 Class P

Theorem 7.1.1 and 7.1.2 show an important difference. On the one hand, the time complexity of the problem is at most the difference of square or polynomial on deterministic single-tape and multi-tape Turing machines; on the other hand, on the deterministic and non-deterministic Turing machines, the time complexity of the problem is at most the difference of exponential.

7.2.1 Polynomial time

The polynomial difference in running time can be considered small, while the exponential difference is considered large. Take a look at why we choose to distinguish polynomials and exponents, rather than the other two types of functions.

First, notice that there is a huge difference in growth rate between a typical polynomial (such as n^3) and a typical exponent (such as 2^n). For example, let n be 1000, which is a reasonable size for the input of an algorithm. In this case, n^3 is 1 billion, although it is a large number, it can still be processed. However, 2^n is a number much larger than the number of atoms in the universe. The polynomial time algorithm is fast enough for many purposes, while the exponential time algorithm is rarely used.

生改变,所以无须改进它. 树上结点的总数小于最大叶数的两倍,因此用 $O(b^{t(n)})$ 作为它的上界. 从根出发下行到一个结点的时间是 $O(t(n))$. 因此,D 的运行时间是 $O(b^{t(n)})$.

图灵机 D 有三条带子. 按照定理 7.1.1,把它转变为单带图灵机最多把运行时间乘方. 这样,单带模拟机的运行时间是 $(2^{O(t(n))})^2 = 2^{O(2t(n))} = 2^{O(t(n))}$,定理得证.

7.2 P 类

定理 7.1.1 和 7.1.2 有一个重要的差别. 一方面,问题的时间复杂度在确定型单带和多带图灵机上最多是平方或多项式的差异;另一方面,在确定型和非确定型图灵机上,问题的时间复杂度最多是指数的差异.

7.2.1 多项式时间

运行时间的多项式差异可以认为是较小的,而指数差异则被认为是较大的. 看一下为什么要选择区分多项式和指数,而不是选择别的某两类函数.

首先,注意到典型的多项式(如 n^3)与典型的指数(如 2^n)在增长率上存在巨大的差异. 例如,令 n 是 1000,这是一个算法输入的合理规模. 在这种情况下,n^3 是 10 亿,虽然是大数,但还可以处理. 然而,2^n 是一个比宇宙中的原子数还大得多的数. 多项式时间算法就很多目标而言是足够快了,而指数时间算法则很少使用.

The typical exponential time algorithm comes from searching the solution space to solve the problem, which is called **brute force search**. For example, one way to decompose a number into prime factors is to search through all possible factors. The size of the search space is exponential, so this search requires exponential time. Sometimes, by understanding the problem more deeply, brute force search can be avoided and a more practical polynomial time algorithm may be found.

All reasonable deterministic calculation models are **polynomial equivalent**, that is, any one of them can simulate the other, and the running time only increases by polynomial times. When it is said that all reasonable deterministic models are polynomial equivalent, we are not trying to define what is reasonable. But there is a concept in our mind, it is broad enough to accommodate those models that are similar to the actual computer running time. For example, Theorem 7.1.1 shows that deterministic single-tape and multi-tape Turing machine models are polynomial equivalent.

From now on, we will focus on the aspects of time complexity theory that are not affected by only polynomial differences in running time. Ignoring such differences allows us to study theories without relying on the choice of specific calculation models. Remember, our goal is to give the basic properties of calculations, not the properties of Turing machines or other special models.

You may find it absurd to ignore the polynomial difference in running time. In fact, programmers of course care about this difference, and they are desperate to make the program run faster. However, when introducing the asymptotic notation above, the constant factor was ignored. It is now recommended to ignore polynomial differences much larger than this, such as differences between time n and n^3.

The decision to ignore polynomial differences does not mean that such differences are not important, on the contrary, the difference between time n and n^3 is important. But whether certain problems (such as factorization problems) are polynomial or non-polynomial really have nothing to do with polynomial differences, and these problems are also very important. We only focus on this type of problem here. Looking at the forest apart from the trees does not mean that one thing is more important than another, it simply provides a different perspective.

Definition 7.2.1 P is the decidable language class of the deterministic single-tape Turing machine in polynomial time. In other words,
$$P = \bigcup_k TIME(n^k)$$

In theory, class P plays a central role, and its importance lies in:

(1) For all calculation models equivalent to deterministic single-tape Turing machine polynomials, P is invariant.

(2) P roughly corresponds to the type of problem that is actually solvable on the computer.

Article (1) shows that in mathematics, P is a robust class, and it is not affected by the specific calculation model used.

Article (2) shows that from a practical point of view, P is appropriate. When a problem is in P, there is a way to solve it in time n^k (k is a constant). Whether it is practical for such a long time depends on k and the actual application. Of course, the runtime of n^{100} is unlikely to have any practical applications. Regardless, polynomial time has been proven useful as a criterion for actual solvability. Once you find a polynomial time algorithm for a problem that originally seemed to require exponential time, you must understand some of its key aspects, and you can usually further reduce its complexity to a practical level.

忽略多项式差异的决定并不是说这样的差异不重要,相反,时间 n 和 n^3 之间的差异是重要的. 但是某些问题(如因数分解问题)是多项式的还是非多项式的确与多项式差异无关,而且这些问题也很重要. 我们在这里仅关注这种类型的问题. 撇开树看森林并不意味着一件事比另一件事更重要,而只是提供一种不同的视角.

定义 7.2.1 P 是确定型单带图灵机在多项式时间内可判定的语言类. 换言之,
$$P = \bigcup_k TIME(n^k)$$

在理论中,P 类扮演核心的角色,它的重要性在于:

(1) 对于所有与确定型单带图灵机多项式等价的计算模型来说,P 是不变的.

(2) P 大致对应于在计算机上实际可解的那一类问题.

第(1)条表明,在数学上,P 是一个稳健的类,它不受所采用的具体计算模型的影响.

第(2)条表明,从实用的观点看,P 是恰当的. 当一个问题在 P 中的时候,就有办法在时间 n^k(k 是常数)内求解. 至于这么长时间是否实用就取决于 k 和实际的应用情况. 当然,n^{100} 的运行时间不太可能有任何实际应用. 不管怎样,把多项式时间作为实际可解性的标准已经被证明是有用的. 一旦为某个原先似乎需要指数级时间的问题找到了多项式时间的算法,就一定是了解了它的某些关键方面,通常能进一步降低它的复杂性,达到实用的程度.

7.2.2 Examples of problems in P

When the polynomial time algorithm is given, it is a high-level description of the algorithm, without mentioning the characteristics of the specific calculation model. This avoids the cumbersome details of the movement of the tape and the read-write head. When describing the algorithm, it is necessary to follow certain habits so that its polynomial can be analyzed.

We continue to describe the algorithm as numbered steps. Implementing one step of an algorithm on a Turing machine usually requires many steps of the Turing machine. Therefore, we must be sensitive to the number of steps in each step of the Turing machine to implement the algorithm and the total number of steps in the algorithm.

When analyzing an algorithm and proving that it runs in polynomial time, two things need to be done. First, we must give a polynomial upper bound for the number of steps required by the algorithm to run on an input of length n (usually a big O notation). Second, every step in the algorithm description must be examined to ensure that they can all be realized by a reasonable deterministic model in polynomial time. When describing the algorithm, carefully determine its steps to make the second part of the analysis easy. When both parts of the work are completed, we can conclude that the algorithm runs in polynomial time. Since it has been proven that it requires polynomial steps, each step can be completed in polynomial time, and the combination of polynomials is still polynomial.

What needs attention is the encoding method used for the problem. We continue to use bracket notation $<\cdot>$ to point out a reasonable encoding to turn one or more objects into a string, without specifying any specific encoding method. Now, a reasonable approach is to allow encoding/decoding of objects into natural internal representations or other reasonable encodings in polynomial time. The well-known encoding methods for graphs, automata, and similar things are all reasonable. But please note that the unary notation of the coded number (such as the number 17 encoded as a unary string 11111111111111111) is unreasonable, because it is exponentially larger than a truly reasonable encoding (such as any notation based on $k \geq 2$).

Many of the computational problems encountered in this chapter involve the encoding of graphs. A reasonable encoding of a graph is a list of its nodes and edges, and the other is an **adjacency matrix**, where if there is an edge from node i to node j, the (i, j) item is 1, otherwise it is 0. When analyzing the algorithm on the graph, the running time may be calculated based on the number of nodes instead of the size of the graph. In a reasonable graph representation method, the size denoted by the polynomial of the number of nodes. Therefore, if you analyze an algorithm and prove that its running time is a polynomial (or exponent) of the number of nodes, then you know that it is a polynomial (or exponent) of the input length.

The first problem is related to directed graphs. The directed graph G contains nodes s and t, as shown in Fig. 7-2. The PATH problem is to determine whether there is a directed path from s to t.

Let PATH = $\{<G,s,t>|G$ is a directed graph with a directed path from s to $t\}$

码方法都是合理的. 但请注意, 编码数字的一进制记法(如数字 17 编码为一进制字符串 11111111111111111)是不合理的, 因为它比真正合理的编码(如以任何 $k \geq 2$ 为基的记法)大指数倍.

本章碰到的许多计算问题都包含图的编码. 图的一种合理编码是它的结点和边的列表, 另一种是**相邻矩阵**, 其中若从结点 i 到结点 j 有边, 则这个 (i, j) 项为 1, 否则为 0. 当分析图上的算法时, 运行时间可能会根据图的结点数而不是大小来计算. 在合理的图表示方法中, 根据结点数的多项式表示大小. 因此, 如果分析某个算法, 并证明其运行时间是结点数的多项式(或指数), 那么就知道它是输入长度的多项式(或指数)了.

第一个问题与有向图有关. 有向图 G 包含结点 s 和 t, 如图 7-2 所示. PATH 问题就是要确定是否存在从 s 到 t 的有向路径.

令 PATH = $\{<G, s, t>|G$ 是具有从 s 到 t 的有向路径的有向图$\}$

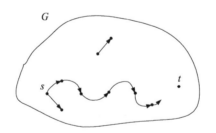

图 7-2 PATH 问题: 是否存在从 s 到 t 的路径

Fig. 7-2 PATH problem: Is there a path from s to t

Theorem 7.2.1 $PATH \in P$.

Proof idea Prove the theorem by giving a polynomial time algorithm to determine $PATH$. Before describing the algorithm, it should be noted that the brute force algorithm for this problem is not fast enough.

The brute force algorithm of $PATH$ examines all possible paths in G to determine whether there is a directed path from s to t. A possible path is the sequence of nodes in G with a length of at most m, where m is the number of nodes in G. (If there is a directed path from s to t, then there is a directed path no longer than m, because there is no need to repeat nodes on the path.)

But the number of these possible paths is m^m, which is an exponential multiple of the number of nodes in G. Therefore, the brute force algorithm consumes exponential time.

In order to obtain the polynomial time algorithm, we must try to avoid brute force search. One way is to use graph search methods, such as breadth-first search. Consecutively mark all nodes in G that start from s and have lengths 1, 2, 3, and, until all nodes can be reached by the directional paths of m. A polynomial can be used to easily define the running time of the strategy.

Proof A polynomial time algorithm M of $PATH$ runs as follows:

$M=$ "For input $<G, s, t>$, G is a directed graph containing nodes s and t:

(1) Mark the node s.

(2) Scan all sides of G. If an edge (a, b) is found, a is marked but b is not marked, then the mark b.

(3) Repeat step (2) until no more nodes are marked.

(4) If t is marked, accept; otherwise, reject."

定理 7.2.1 $PATH \in P$.

证明思路 通过给出判定 $PATH$ 的多项式时间算法来证明该定理. 在描述算法之前,要注意该问题的蛮力算法是不够快的.

$PATH$ 的蛮力算法通过检察 G 中所有可能的路径来确定是否存在从 s 到 t 的有向路径. 一条可能的路径就是 G 中长度最多为 m 的结点序列, m 是 G 中的结点数. (如果从 s 到 t 存在有向路径, 那么就存在长度不超过 m 的有向路径, 因为路径上无须重复结点.)

但是这些可能的路径数是 m^m, 这是 G 中结点数的指数倍. 因此, 该蛮力算法消耗指数级时间.

为了获得多项式时间算法, 必须设法避免蛮力搜索. 一种办法是采用图搜索方法, 如宽度优先搜索. 连续标记 G 中从 s 出发, 长度为 $1, 2, 3, \cdots, m$ 的有向路径可达到所有结点. 用多项式可以容易地界定该策略的运行时间.

证明 $PATH$ 的一个多项式时间算法 M 运行如下:

$M=$ "对输入 $<G, s, t>$, G 是包含结点 s 和 t 的有向图:

(1) 在结点 s 上做标记.

(2) 扫描 G 的所有边. 如果找到一条边 (a, b), a 被标记而 b 没有被标记, 那么标记 b.

(3) 复重步骤(2), 直到不再有结点被标记.

(4) 若 t 被标记, 则接受; 否则, 拒绝."

Analyze the algorithm and prove that it runs in polynomial time. Obviously, steps (1) and (4) are executed only once. Step (3) is executed at most m times, because every time except the last time, an unmarked node in G must be marked. So the total number of steps used is at most $1+1+m$, which is a polynomial of the scale of G.

Steps (1) and (4) of M are easily implemented in polynomial time using any reasonable deterministic model. Step (3) needs to scan the input to check whether some nodes are marked, which is also easy to achieve in polynomial time. So M is the polynomial time algorithm of $PATH$.

Look at another example of a polynomial time algorithm. It is said that two numbers are relatively prime, if 1 is the largest integer that can divide them at the same time. For example, 10 and 21 are relatively prime, although neither of them are prime. But 10 and 22 are not relatively prime, because they can be divisible by 2. Let $RELPRIME$ represent the problem of checking whether two numbers are relatively prime, namely

$RELPRIME = \{<x, y> | x$ and y are relatively prime$\}$

Theorem 7.2.2 $RELPRIME \in P$.

Proof idea   An algorithm to solve this problem is to search through all possible common factors of these two numbers, and if no common factor greater than 1 is found, accept it. However, the size of a number denoted in binary or any other notation based on $k(k \geq 2)$ is an exponential multiple of its length. Therefore, the brute force algorithm needs to be searched multiple possible factors of the exponent, and consume the running time of the exponent.

We use an ancient numerical process to solve the problem, called **Euclidean algorithm**, to calculate the greatest common factor. The **greatest common factor** of two natural numbers x and y is recorded as

分析该算法,证明它在多项式时间内运行. 显然,步骤(1)和(4)只执行一次. 步骤(3)最多执行 m 次,因为除最后一次外,每一次执行都要标记 G 中的一个未标记的结点. 所以用到的总步骤数最多是 $1+1+m$,这是 G 的规模的多项式.

M 的步骤(1)和(4)很容易用任何合理的确定型模型在多项式时间内实现. 步骤(3)需要扫描输入以检查某些结点是否被标记,这也容易在多项式时间内实现. 所以 M 是 $PATH$ 的多项式时间算法.

来看另一个多项式时间算法的例子. 称两个数是互为素数的,若 1 是能同时整除它们的最大整数. 例如,10 和 21 是互素的,虽然它们自己都不是素数. 但是 10 和 22 不是互素的,因为它们都能被 2 整除. 令 $RELPRIME$ 代表检查两个数是否互素的问题,即

$RELPRIME = \{<x, y> | x$ 与 y 互素$\}$

定理 7.2.2 $RELPRIME \in P$.

证明思路   解决该问题的一种算法是搜遍这两个数的所有可能的公因子,如果没有发现大于 1 的公因子,那么就接受. 然而,用二进制或其他任何以 k 为基的记法($k \geq 2$)表示的数字的大小是它表示长度的指数倍. 因此该蛮力算法需要搜遍指数多个可能的因子,消耗指数的运行时间.

使用一种古老的数值过程来求解该问题的方法,称为**欧几里得算法**,以此可计算最大公因子. 两个自然数 x 和 y 的**最大公因子**记为 $\gcd(x,y)$,它是能

$\gcd(x,y)$, which is the largest integer that can divide x and y at the same time. For example, $\gcd(18,24) = 6$. Obviously, the necessary and sufficient condition for x and y to be relatively prime is $\gcd(x,y) = 1$. In the proof, Euclidean algorithm is described as algorithm E, which uses the function mod, $x \bmod y$ is equal to the remainder obtained by dividing x by y.

Proof Euclidean algorithm E is as follows:

$E = $ "For input $<x,y>$, x and y are natural numbers in binary representation.

(1) Repeat the following operations until $y = 0$.
(2) Assign $x \leftarrow x \bmod y$.
(3) Exchange the values of x and y.
(4) Output x."

Algorithm R uses E as a subroutine to solve *REL-PRIME*.

$R = $ "For input $<x,y>$, x and y are natural numbers in binary representation:

(1) Run E on $<x,y>$.
(2) If the result is 1, accept it; otherwise, reject it."

Obviously, if E runs in polynomial time and is correct, then R also runs in polynomial time and is correct. So we only need to analyze the time and correctness of E. The correctness of the algorithm is well known, so we won't discuss it further here.

In order to analyze the time complexity of E, first prove that each execution of step (2) (with the possible exception of the first time) reduces the value of x by at least half. After step (2) is executed, $x < y$ is known from the properties of the function mod. After step (3), there is $x > y$, because the two values have been exchanged. So when step 2 is subsequently executed, $x > y$. If $x/2 \geqslant y$, then $x \bmod y < y \leqslant x/2$, x is reduced by at least half. If $x/2 < y$, then $x \bmod y = x - y < x/2$, x is reduced by at least half.

Each time step (3) is executed, the values of x and y are exchanged, so the original values of x and y

同时整除 x 和 y 的最大整数. 例如 $\gcd(18,24) = 6$. 显然, x 和 y 互素的充分必要条件是 $\gcd(x,y) = 1$. 在证明中把欧几里得算法描述为 E 算法, 它使用函数 mod, $x \bmod y$ 等于用 y 去整除 x 所得的余数.

证明 欧几里得算法 E 如下:

$E = $ "对输入 $<x,y>$, x 和 y 是二进制表示的自然数.

(1) 重复下面的操作, 直到 $y = 0$.
(2) 赋值 $x \leftarrow x \bmod y$.
(3) 交换 x 和 y 的值.
(4) 输出 x."

算法 R 以 E 为子程序求解 *REL-PRIME*.

$R = $ "对输入 $<x,y>$, x 和 y 是二进制表示的自然数:

(1) 在 $<x,y>$ 上运行 E.
(2) 若结果为 1, 就接受; 否则, 就拒绝."

显然, 若 E 在多项式时间内运行且正确, 则 R 也在多项式时间内运行且正确. 所以只需分析 E 的时间和正确性. 该算法的正确性是众所周知的, 这里不进一步讨论它.

为了分析 E 的时间复杂度, 首先证明步骤(2)的每一次执行(除了第一次有可能例外)都把 x 的值至少减少一半. 执行步骤(2)以后, 由函数 mod 的性质可知 $x < y$. 步骤(3)后, 有 $x > y$, 因为这两个值已经交换. 于是当步骤(2)随后执行时有 $x > y$. 若 $x/2 \geqslant y$, 则 $x \bmod y < y \leqslant x/2$, x 至少减少一半. 若 $x/2 < y$, 则 $x \bmod y = x - y < x/2$, x 至少减少一半.

每一次执行步骤(3)都使 x 和 y 的值相互交换, 所以每两次循环就使得

are reduced by at least half every two cycles. Therefore, the maximum number of executions of steps (2) and (3) is the smaller of $2\log_2 x$ and $2\log_2 y$. These two logarithms are proportional to the length of the representation, and the number of executions of the step is $O(n)$. Each step of E only consumes polynomial time, so the entire running time is polynomial.

The last example of a polynomial-time algorithm shows that every context-free language is polynomial-time decidable.

Theorem 7.2.3 Every context-free language is a member of P.

Proof idea Theorem 5.1.7 proves that every CFL is decidable, and gives a decidable algorithm for each CFL. If that algorithm runs in polynomial time, then this theorem must be established as a corollary. Recall that algorithm and see if it runs fast enough.

Let L be a CFL produced by CFG G, and G be the Chomsky paradigm. Since G is in Chomsky paradigm, any derivation to get the string w has $2n-1$ steps, and n is the length of w. When inputting a string of length n to the decider of L, it determines L by trying all possible $2n-1$ steps derivations. If one of them gets the derivation of w, the decider will accept it; otherwise, the decider will reject it.

Analyze the algorithm, we know that it cannot run in polynomial time. The number of k-step derivations may reach an exponential of k, so the algorithm may require exponential time.

In order to obtain a polynomial time algorithm, a powerful technology called **dynamic programming** is introduced here. This technology solves big problems by accumulating information about small sub-problems. Record all the solutions to the sub-problems so that you only need to solve them once. To this end, all sub-problems are compiled into a table, and when they are encountered, their solutions are systematically filled in

x 和 y 原先的值至少减少一半. 于是步骤(2)和(3)执行的最大次数是 $2\log_2 x$ 和 $2\log_2 y$ 中较小的那一个. 这两个对数与表示的长度成正比,执行步骤的次数是 $O(n)$. E 的每一步仅消耗多项式时间,所以整个运行时间是多项式的.

最后一个多项式时间算法的例子表明,每一个上下文无关语言是多项式时间可判定的.

定理7.2.3 每一个上下文无关语言都是 P 的成员.

证明思路 定理5.1.7证明了每一个CFL都是可判定的,并且为每一个CFL给出了可判定算法. 如果那个算法在多项式时间内运行,那么本定理作为推论就必然成立. 回忆一下那个算法,看它运行得是否够快.

令 L 是一个由CFG G 产生的CFL,G 是乔姆斯基范式. 因 G 是乔姆斯基范式,故任何得到字符串 w 的推导都有 $2n-1$ 步,n 是 w 的长度. 当给 L 的判定机输入长为 n 的字符串时,它通过试遍所有可能的 $2n-1$ 步推导来判定 L. 如果其中有一个得到 w 的推导,该判定机就接受;否则,就拒绝.

分析该算法可知,它不能在多项式时间内运行. k 步推导的数量可能达到 k 的指数级,所以该算法可能需要指数时间.

为了获得多项式时间算法,在此介绍一种称为**动态规划**的强有力的技术. 这种技术通过累积小的子问题的信息来解决大的问题. 把子问题的解都记录下来,这样就只需对它们求解一次. 为此,把所有子问题编成一张表,当碰到它们时,它们的解就被系统地填入这张表格.

In this example, consider the sub-problems whether each argument of G generates each substring of w. The algorithm fills the solution of the sub-problem into an $n \times n$ table. For $i \leq j$, the (i, j) item of the table contains all the arguments that produce the substring $w_i w_{i+1} \cdots w_j$. The $i > j$ table entries are not used.

The algorithm fills in the table entry for each substring of w. First fill in the table entries for the substring of length 1, then the substring of length 2, and so on. It uses the table entry content of the short substring to assist in determining the table entry content of the long substring.

For example, assume that the algorithm has determined which arguments generate all substrings of length no more than k. In order to determine whether the argument A generates a substring of length $k+1$, the algorithm splits the substring into two non-empty segments in k possible ways. For each splitting method, the algorithm examines each rule $A \to BC$, and uses the previously calculated entries to determine whether B generates the first segment and C generates the second segment. If B and C both generate their own segments, then A generates the substring and is added to the associated table entry. The algorithm starts with a string of length 1, and examines the table for rule $A \to b$.

Proof The following algorithm D realizes this proof idea. Let G be the CFG of the Chomsky paradigm that generates CFLL. Assume S is the starting argument. The comment is written in square brackets.

$D = $ "For input $w = w_1 \ldots w_n$:

(1) If $w = \varepsilon$ and $S \to \varepsilon$ is a rule, accept it. Otherwise reject it. [In the case of $w = \varepsilon$]

(2) For $i = 1 \sim n$: [Observe each substring of length 1]

(3) For each argument A:

(4) Check whether $A \to b$ is a rule, where $b = w_i$.

(5) If yes, put A into table(i,i).

(6) For $l = 2 \sim n$:

[l is the length of the substring]

(7) To $i = 1 \sim n-l+1$:

[i is the starting position of the substring]

(8) Let $j = i+l-1$.

[j is the end position of the substring]

(9) For $k = i \sim j-1$:

[k is the split position]

(10) For each rule $A \to BC$:

(11) If table(i,k) contains B and table$(k+1,j)$ contains C, put A into table(i,j).

(12) If S is in table$(1,n)$, accept; otherwise, reject."

Next we analyze D. Each step is easily run in polynomial time. Steps (4) and (5) are run at most nv times, where v is the number of variables in G, which is a fixed constant independent of n, therefore, these two steps are run $O(n)$ times. Step (6) runs at most n times. Each time step (6) is run, step (7) is run n times at most. Each time step (7) is run, steps (8) and (9) are run at most n times. Each time step (9) runs, step (10) runs r times, where r is the rule number of G, which is another fixed constant. So step (11) (that is, the inner loop of the algorithm) runs $O(n^3)$ times. A total of D executes $O(n^3)$ steps.

7.3 Class *NP*

As revealed in Section 7.2, many problems can be solved by avoiding brute force search and obtaining polynomial time solutions. However, in some other problems, efforts to avoid brute force search have not been successful, and polynomial time algorithms to solve them have not been found.

An unusual finding on this problem is that the complexity of many problems are linked together. The polynomial time algorithm that found one of the problems can be used to solve the entire class of problems. Start with an example to understand this phenomenon.

The Hamiltonian path in the directed graph G is a directed path that passes through each node exactly once. Consider such a problem: verify whether a directed graph contains a Hamiltonian path connecting two specified nodes, as shown in Fig. 7-3.

关于该问题的一个不寻常的发现是,许多问题的复杂性是联系在一起的.发现其中一个问题的多项式时间算法可以用来解决整个一类问题.从一个例子开始来理解这一现象.

有向图 G 中的哈密顿路径是恰好一次通过每个结点的有向路径.考虑这样一个问题:验证一个有向图是否包含一条连接两个指定的结点哈密顿路径,如图 7-3 所示.

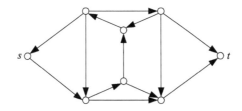

图 7-3 恰好一次通过每个结点的哈密顿路径
Fig. 7-3 The Hamiltonian path passes through each node exactly once

Let
$HAMPATH = \{<G, s, t>| G$ is a directed graph containing the Hamiltonian path from s to $t\}$

By modifying the brute force algorithm of *PATH* given in Theorem 7.2.1, it is easy to obtain the exponential time algorithm of the *HAMPATH* problem. Just add a check to verify that the possible path is the Hamiltonian path. No one knows whether *HAMPATH* can be solved in polynomial time.

The *HAMPATH* problem also has a feature called **polynomial verifiability**, which is very important for understanding its complexity. Although a fast (polynomial time) method is not known to determine whether a Hamiltonian path is included in the graph, if such a path is found in some way (possibly using an exponential time algorithm), it is easy to believe its existence, this only needs to give it. In other words, verifying the existence of the Hamiltonian path may be much easier than determining its existence.

令
$HAMPATH = \{<G,s, t>| G$ 是包含从 s 到 t 的哈密顿路径的有向图$\}$

通过修改定理 7.2.1 给出的 *PATH* 的蛮力算法,很容易获得 *HAMPATH* 问题的指数时间算法.只需增加一项检查,就可验证可能的路径是哈密顿路径.没有人知道 *HAMPATH* 是否能在多项式时间内求解.

HAMPATH 问题还具有一个特点,称为**多项式可验证性**,这对于理解其复杂性很重要.虽然还不知道一种快速(即多项式时间)的方法来确定图中是否包含了哈密顿路径,但是如果以某种方式(可能采用指数时间算法)找到这样的路径,就能很容易让人相信它的存在,这只需给出它即可.换言之,验证哈密顿路径的存在性可能比确定它的存在性容易得多.

Another problem with verifiable polynomials is compositeness. When a natural number is the product of two integers greater than 1, the natural number is called a composite number (that is, a composite number is a non-prime number). Let

$COMPOSITES = \{x \mid x = pq, \text{integer } p, q > 1\}$

Although we don't know the polynomial time algorithm to determine the problem, it is easy to verify that a number is composite—only a factor of the number is required. Recently, a polynomial time algorithm has been discovered that can verify whether a certain number is prime or composite, but it is more complicated than the aforementioned method of verifying compositeness.

Some problems may not be polynomial verifiable. For example, $\overline{HAMPATH}$, which is the complement of the $HAMPATH$ problem. Although it can be determined that there is no Hamiltonian path in the graph, if the exponential time algorithm used in the original determination is not used, there is no other way for others to verify its non-existence. The following is a formal definition.

Definition 7.3.1 The **verifier** of language A is an algorithm V, here

$A = \{w \mid \text{For a string } c, V \text{ accepts } <w, c>\}$

Since the time of the verifier is only measured according to the length of w, the polynomial time verifier runs in polynomial time of the length of w. If language A has a polynomial time verifier, it is called **polynomial verifiable**.

The verifier uses additional information (denoted by the symbol c in Definition 7.3.1) to verify that the string c is a member of A. This information is called A's membership **certificate** or **proof**. Note that for the polynomial verifier, the certificate has the length of the polynomial (the length of w), because this is the length of all information that the verifier can access within

另一个多项式可验证的问题是合数性. 当一个自然数是两个大于1的整数的乘积时,该自然数被称为合数(即合数就是非素数的数). 令

$COMPOSITES = \{x \mid x = pq, 整数\ p, q > 1\}$

虽然不知道判定该问题的多项式时间算法,但是能轻易地验证一个数是合数——只需要该数的一个因子即可. 最近,发现了一个可验证某数是素数还是合数的多项式时间算法,但它比前面提到的合数性验证方法更复杂.

有些问题可能不是多项式可验证的. 例如, $\overline{HAMPATH}$, 即 $HAMPATH$ 问题的补问题. 尽管能够判定图中没有哈密顿路径,但如果不采用原先做判定时用的那个指数时间算法,就没有其他办法让别人验证它的不存在性. 下面是形式化的定义.

定义 7.3.1 语言 A 的**验证机**是一个算法 V,这里

$A = \{w \mid 对某个字符串\ c, V\ 接受 <w, c>\}$

因为只根据 w 的长度来度量验证机的时间,所以多项式时间验证机在 w 的长度的多项式时间内运行. 若语言 A 有一个多项式时间验证机,则称它为**多项式可验证的**.

验证机利用额外的信息(在定义7.3.1中用符号 c 表示)来验证字符串 c 是 A 的成员. 该信息称为 A 的成员资格**证书**或**证明**. 注意,对于多项式验证机,证书具有多项式的长度(w 的长度),因为这是该验证机在其时间限度内所能访问的全部信息长度. 把该定义

its time limit. Apply this definition to the languages *HAMPATH* and *COMPOSITES*.

For the *HAMPATH* problem, the certificate of the string $<G, s, t> \in$ *HAMPATH* is just a Hamiltonian path from s to t. For the *COMPOSITES* problem, the certificate of composite x is only a factor of it. In both cases, when the certificate is handed over to the verifier, it can check whether the input is in the language in polynomial time.

Definition 7.3.2 *NP* is a language class with a polynomial time verifier.

The *NP* class is important because it contains many practical problems. From the previous discussion, both *HAMPATH* and *COMPOSITES* are members of *NP*. *COMPOSITES* is also a member of the subset *P* of *NP*, but it is very difficult to prove this stronger conclusion. The term *NP* stands for **non-deterministic polynomial time**, which is derived from another feature of Turing machines using non-deterministic polynomial time. Problems in *NP* are sometimes called *NP* problems.

The following is a non-deterministic Turing machine (NTM) that determines the *HAMPATH* problem in non-deterministic polynomial time. Recall that in Definition 7.1.5, the time to define a non-deterministic machine is the time taken by the longest calculation branch.

$N_1 = $ "For input $<G, s, t>$, where G is a directed graph containing nodes s and t:

(1) Write a list of m numbers, $p_1, p_2, \dots, p_m$ is the number of G nodes. Each number in the column is selected indefinitely from 1 to m.

(2) Check the repeatability in the column, and reject it if any duplication is found.

(3) Check whether $s = p_1$ and $t = p_m$ are both true. If one is not established, reject it.

(4) For each i from 1 to $m-1$, check whether (p_i, p_{i+1}) is an edge of G. If one is not, reject it.

Otherwise, all checks are passed, accept it."

In order to analyze the algorithm and verify that it runs in non-deterministic polynomial time, examine every step of it. In step (1), the non-deterministic choice obviously runs in polynomial time. In steps (2) and (3), each step is a simple check, so together they are still running in polynomial time. Finally, step (4) obviously also runs in polynomial time. Therefore, the algorithm runs in non-deterministic polynomial time.

Theorem 7.3.1 A language in *NP*, if and only if it can be determined by a certain non-deterministic polynomial time Turing machine. In proof idea, we prove how to convert a polynomial time verifier into an equivalent polynomial time NTM and how to reverse the conversion. The NTM simulates the verifier by guessing the certificate, and the verifier uses the accepted branch as the certificate to simulate the NTM.

Proof For the left-to-right direction of the theorem, assume $A \in NP$ to prove that A is determined by the polynomial time NTM N. According to the definition of *NP*, there is a polynomial time verifier V of A. Assuming that V is a Turing machine running in time n^k, construct N as follows:

N = "For the input w of length n:

(1) Non-deterministically choose the character string c with the longest n.

(2) Run V on the input $<w,c>$.

(3) If V accepts, then accept; otherwise, reject."

In order to prove another direction of the theorem, assume A is determined by the polynomial time NTM N, and the polynomial time proof machine V is constructed as follows:

V = "For input $<w,c>$, where w, c are strings:

(1) Simulate N on the input w, and treat each symbol of c as a description of the non-deterministic choice made at each step.

(2) If the current calculation branch of N accepts, then accept; otherwise, reject."

Similar to the deterministic time complexity class $TIME(t(n))$, the non-deterministic time complexity class $NTIME(t(n))$ is defined.

Definition 7.3.3 $NTIME(t(n)) = \{L | L$ is a language determined by an non-deterministic Turing machine with $O(t(n))$ time$\}$.

Corollary 7.3.1 $NP = U_k NTIME(n^k)$.

The NP class is not sensitive to the choice of a reasonable non-deterministic calculation model, because all these models are polynomial equivalent. When describing and analyzing the non-deterministic polynomial time algorithm, follow the habit of the previous deterministic polynomial time algorithm. Each step of the non-deterministic polynomial time algorithm must be based on a reasonable non-deterministic calculation model, and it should be clearly realized in the non-deterministic polynomial time. Analyze the algorithm to prove that each branch uses at most polynomial steps.

7.3.1 Examples of problems in NP

A **clique** in an undirected graph is a subgraph in which every two nodes are connected by edges. The k-clique is a clique containing k nodes. Fig. 7-4 shows a graph containing 5 cliques.

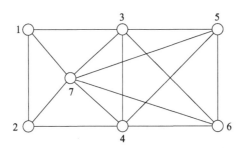

Fig. 7-4 A diagram with 5 cliques

The clique problem is to determine whether a graph contains cliques of a specified size. Let

$CLIQUE = \{<G, k> \mid G$ is an undirected graph containing k-groups$\}$

Theorem 7.3.2 $CLIQUE$ belongs to NP.

Proof idea A clique is a certificate.

Proof The following is the verifier V of $CLIQUE$.

$V =$ "For input $<<G,k>,c>$:

(1) Check whether c is a subgraph of k nodes in G.

(2) Check whether G contains all the edges connecting the nodes in c.

(3) If both checks are passed, accept; otherwise, reject."

Another kind of proof If you would like to understand the NP class from the perspective of a nondeterministic polynomial time Turing machine, then you can prove this theorem by giving a Turing machine that determines $CLIQUE$. Note the similarity between these two proofs.

$N =$ "For input $<G, k>$, where G is a graph:

(1) Indeterminately select a subset c of k nodes in G.

(2) Check whether G contains all the edges connecting the nodes in c.

(3) If yes, accept; otherwise, reject."

Next, consider the problem $SUBSET\text{-}SUM$ related to integer arithmetic. Given a number set $x_1, x_2, \ldots, x_k$ and a target number t, it is necessary to determine whether there is a subset in this set that adds up to t. That is

$SUBSET\text{-}SUM = \{<s, t> \mid s = \{x_1, \ldots, x_k\}$, and there is $\{y_1, \ldots, y_l\} \subseteq \{x_1, \ldots, x_k\}$ makes $\sum y_i = t\}$

For example, $<\{4, 11, 16, 21, 27\}, 25> \in$ SUBSET-SUM, $4 + 21 = 25$. Note that $\{x_1, \ldots, x_k\}$ and $\{y_1, \ldots, y_l\}$ are regarded as **multiple sets**, so element

团问题旨在判定一个图是否包含指定大小的团. 令

$CLIQUE = \{<G, k> \mid G$ 是包含 k-团的无向图$\}$

定理 7.3.2 $CLIQUE$ 属于 NP.

证明思路 团即证书.

证明 下面是 $CLIQUE$ 的验证机 V.

$V =$ "对输入 $<<G,k>,c>$:

(1) 检查 c 是否是 G 中 k 个结点的子图.

(2) 检查 G 是否包含连接 c 中结点的所有的边.

(3) 若两项检查都通过,则接受;否则,拒绝."

另一种证明 如果愿意从非确定型多项式时间图灵机的角度来理解 NP 类,那么可以通过给出判定 $CLIQUE$ 的图灵机来证明本定理. 注意这两种证明的相似性.

$N =$ "对输入 $<G, k>$,这里 G 是一个图:

(1) 非确定地选择 G 中 k 个结点的子集 c.

(2) 检查 G 是否包含连接 c 中结点的所有边.

(3) 如是,则接受;否则拒绝."

下面考虑与整数算术有关的问题 $SUBSET\text{-}SUM$. 给定一个数集 $x_1, x_2, \cdots, x_k$ 和一个目标数 t,要判定在这个集合中是否有一个加起来等于 t 的子集. 即

$SUBSET\text{-}SUM = \{<s,t> \mid s = \{x_1, \cdots, x_k\}$,且存在 $\{y_1, \cdots, y_l\} \subseteq \{x_1, \cdots, x_k\}$ 使得 $\sum y_i = t\}$

例如,$<\{4, 11, 16, 21, 27\}, 25> \in SUBSET\text{-}SUM, 4+21 = 25$. 注意 $\{x_1, \cdots, x_k\}$ 和 $\{y_1, \cdots, y_l\}$ 被看作**多重集**,因此

duplication is allowed.

Theorem 7.3.3 *SUBSET-SUM* belongs to *NP*.

Proof idea The subset is the certificate.

Proof The following is a verifier *V* of *SUBSET-SUM*.

$V=$ "For input $<<S,t>,c>$:

(1) Check if c is a set of numbers that add up to t.

(2) Check whether S contains all the numbers in c.

(3) If both checks are passed, accept; otherwise, reject."

Another kind of proof We can also prove this theorem by giving a non-deterministic polynomial-time Turing machine that determines *SUBSET-SUM*, as shown below:

$N=$ "For input $<S,t>$:

(1) Indefinitely choose a subset c of the numbers in S.

(2) Check whether c is a set of numbers that add up to t.

(3) If the inspection passes, then accept; otherwise, reject."

Note that the complements ($\overline{CLIQUE}$) and ($\overline{SUBSET\text{-}SUM}$) of these sets are not clearly belong to *NP*. It seems to be more difficult to verify that something does not exist than to verify that it exists. We define another complexity class, called *coNP*, which includes the complement language of the language in *NP*. It is not yet known whether *coNP* is different from *NP*.

7.3.2 *P* and *NP* problems

As mentioned earlier, *NP* is a language class that is solvable in polynomial time on a non-deterministic Turing machine, or equivalently, a language class whose membership can be verified in polynomial time. *P* is the language class whose membership can be determined in polynomial time. These contents are summarized as follows, among them, the polynomial time solvable is roughly called "quickly" decidable.

P = Language class whose membership can be quickly determined.

NP = Language class whose membership can be quickly verified.

We have already given examples of languages, such as *HAMPATH* and *CLIQUE*. They are members of *NP*, but we don't know if they belong to *P*. The ability of polynomial verifiability seems to be much greater than the ability of polynomial decidability. But it is hard to imagine that P and NP may also be equal. It is not yet possible to prove that there is a language that does not belong to P in NP.

The question of whether $P = NP$ is true is one of the biggest unanswered questions in theoretical computer science and contemporary mathematics. If these two classes are equal, then all polynomial verifiable problems will be polynomial decidable. Most researchers believe that these two classes are not equal, because people have invested a lot of energy in finding polynomial time algorithms for some problems in *NP*, but no one has succeeded. The researchers also tried to prove that the two classes are not equal, but this requires proving that there is no fast algorithm to replace brute force search. At present, scientific research has not been able to achieve this step. Fig. 7-5 shows two possibilities.

这些内容总结如下,其中,把多项式时间可解粗略地称为"快速地"可解.

P = 成员资格可以快速判定的语言类.

NP = 成员资格可以快速验证的语言类.

前面已经给出了语言的例子,如 *HAMPATH* 和 *CLIQUE*,它们是 *NP* 的成员,但不知道是否属于 *P*. 多项式可验证性的能力似乎比多项式可判定性的能力大得多. 但难以想象的是, P 和 NP 也有可能是相等的. 现在还无法证明在 *NP* 中存在一个不属于 P 的语言.

$P = NP$ 是否成立的问题是理论计算机科学和当代数学中最大的悬而未决的问题之一. 如果这两个类相等,那么所有多项式可验证的问题都将是多项式可判定的. 大多数研究者相信这两个类是不相等的,因为人们已经投入了大量的精力为 *NP* 中的某些问题寻找多项式时间算法,但没人取得成功. 研究者还试图证明这两个类是不相等的,但是这要求证明不存在快速算法来代替蛮力搜索. 目前,科学研究还无法做到这一步. 图 7-5 显示了两种可能性.

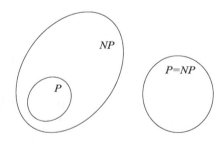

图 7-5 这两个可能性中有一个是正确的
Fig. 7-5 One of these two possibilities is correct

The best known decision language is *NP*'s deterministic method using exponential time. In other words, it can be proved

$$NP \subseteq EXPRIME = \bigcup_k TIME(2^{n^k})$$

However, it is not known whether *NP* is included in a smaller deterministic time complexity class.

7.4 *NP*-completeness

A major development on the *P* and *NP* problems was accomplished by Stephen Cook and Leonid Levin in the early 1970s. They found that the complexity of some problems in *NP* is related to the complexity of the entire class. If there is a polynomial time algorithm for any of these problems, then all *NP* problems are polynomial time solvable. These problems are called **NP-complete**. The phenomenon of *NP*-completeness is of great significance to both theory and practice.

In theory, researchers who try to prove that *P* is not equal to *NP* can focus on an *NP*-complete problem. If a problem in *NP* requires more than polynomial time, then the *NP*-complete problem must also be the same. Moreover, researchers who are trying to prove that *P* is equal to *NP* can achieve their goal only by finding a polynomial-time algorithm for an *NP*-complete problem.

In practice, the phenomenon of *NP*-completeness can prevent wasting time for a specific problem to find a polynomial time algorithm that does not exist. Although there may be insufficient mathematical evidence to prove that the problem is undecidable in polynomial time, we believe that *P* is not equal to *NP*. Therefore, proving that a problem is *NP*-complete becomes a strong evidence of its non-polynomiality.

The first *NP*-complete problem given is called the **satisfiability problem**. Recall that variables whose values are TRUE and FALSE are called **Boolean variables**.

已知最好的判定语言是使用指数时间的 *NP* 确定型方法. 换言之, 可以证明

$$NP \subseteq EXPRIME = \bigcup_k TIME(2^{n^k})$$

但是, 不知道 *NP* 是否包含在某个更小的确定型时间复杂性类中.

7.4 *NP* 完全性

在 *P* 与 *NP* 问题上的一个重大进展是由斯蒂芬·库克和列奥尼德·列文在 20 世纪 70 年代初完成的. 他们发现 *NP* 中某些问题的复杂性与整个类的复杂性相关联. 这些问题中的任何一个如果存在多项式时间算法, 那么所有 *NP* 问题都是多项式时间可解的. 这些问题称为 **NP 完全的**. *NP* 完全性现象对于理论与实践都具有重要意义.

在理论方面, 试图证明 *P* 不等于 *NP* 的研究者会把注意力集中到一个 *NP* 完全问题上. 如果 *NP* 中的某个问题需要多于多项式时间, 那么 *NP* 完全问题也一定如此. 而且, 试图证明 *P* 等于 *NP* 的研究者只需为一个 *NP* 完全问题找到多项式时间算法就可以达到目的.

在实践方面, *NP* 完全性现象可以防止为某一具体问题浪费时间去寻找本不存在的多项式时间算法. 虽然可能缺乏足够的数学依据来证明该问题在多项式时间内不可解, 但是我们相信 *P* 不等于 *NP*. 所以, 证明一个问题是 *NP* 完全的就成为它的非多项式性的一个强有力的证据.

给出的第一个 *NP* 完全问题称为**可满足性问题**. 回忆一下, 取值为 TRUE 和 FALSE 的变量称为**布尔变量**.

Usually 1 denotes TRUE and 0 denotes FALSE. The **Boolean operations** AND, OR, NOT are denoted as $\wedge$, $\vee$, and $\neg$ respectively. These operations are described in the table below. The upper horizontal line is used as the abbreviation of the symbol $\neg$, so $\bar{x}$ denotes $\neg x$.

$0 \wedge 0 = 0 \quad 0 \vee 0 = 0 \quad \bar{0} = 1$

$0 \wedge 1 = 0 \quad 0 \vee 1 = 1 \quad \bar{1} = 0$

$1 \wedge 0 = 0 \quad 1 \vee 0 = 1$

$0 \wedge 1 = 1 \quad 1 \vee 1 = 1$

Boolean formulas are expressions that include Boolean variables and operations. E. g.,

$$\varphi = (\bar{x} \wedge y) \vee (x \wedge \bar{z})$$

is a Boolean formula. If a value of 0, 1 is assigned to a variable such that the value of a formula is equal to 1, then the Boolean formula is **satisfactory**. The above formula is satisfactory, because the assignment $x = 0$, $y = 1$, and $z = 0$ makes the value of φ be 1. It is said that the assignment satisfies φ. **The satisfiability problem** is to determine whether a Boolean formula is satisfiable. Let

$SAT = \{<\varphi> | \varphi$ is a satisfiable Boolean formula$\}$

Now express a theorem that connects the complexity of the SAT problem with the complexity of all the problems in NP.

Theorem 7.4.1 $SAT \in P$, if and only if $P = NP$.

The core method of proving the theorem is described below.

7.4.1 Polynomial time reducibility

Definition 7.4.1 If there is a polynomial time Turing machine M, so that on any input w, $f(w)$ is exactly on the tape when M stops, then the function $f: \Sigma^* \to \Sigma^*$ is called **polynomial time computable function**.

Definition 7.4.2 Language A is called **a polynomial time map** that can be reduced to language B, or

polynomial time can be reduced to B, denoted as $A \leqslant_P B$. If there is a polynomial time, the function $f: \Sigma^* \to \Sigma^*$ can be calculated. For each w, there is
$$w \in A \Leftrightarrow f(w) \in B$$
function f called the **polynomial time reduction** from A to B.

Fig. 7-6 shows the polynomial time reducibility. Like the general mapping reduction, the polynomial time reduction from A to B provides a way to transform the membership determination of A into the membership determination of B, but now this conversion is effectively completed. In order to determine whether $w \in A$, use reduction f to map w to $f(w)$, and then determine whether $f(w) \in B$.

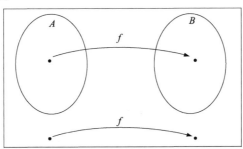

Fig. 7-6 Polynomial time reduction function f from A to B

If a language can reduce polynomial time to another language that is known to have a polynomial time algorithm, then the polynomial time algorithm of the first language can be obtained, as described in the following theorem.

Theorem 7.4.2 If $A \leqslant_P B$ and $B \in P$, then $A \in P$.

Proof Let M be the polynomial time algorithm for determining B, and f be the polynomial time reduction from A to B. The description of the polynomial time algorithm N for determining A is as follows:

$N=$ "For input w:

(1) Calculate $f(w)$.

(2) Run M on the input $f(w)$ and output the output of M."

If $w \in A$, then $f(w) \in B$, because f is a reduction from A to B. Thus, as long as $w \in A$, M accepts $f(w)$. In addition, because both steps of N run in polynomial time, N runs in polynomial time. Note that step (2) runs in polynomial time because the composition of two polynomials is still a polynomial.

Before explaining the polynomial time reduction further, let us introduce $3SAT$, which is a special case of satisfiability problem, because all formulas in it have a special form. A **literal** is a Boolean variable or the negation of a Boolean variable, such as x or $\overline{x}$. A **clause** is a number of literals connected together by $\vee$, such as $(x_1 \vee \overline{x_2} \vee \overline{x_3} \vee x_4)$. If a Boolean formula is composed of several clauses connected by $\wedge$, it is in **conjunctive normal form**, and it is called **the *cnf* formula**, such as

$$(x_1 \vee \overline{x_2} \vee \overline{x_3} \vee \overline{x_4}) \wedge (x_3 \vee \overline{x_5} \vee \overline{x_6}) \wedge (x_3 \vee \overline{x_6})$$

If all clauses have three literals, it is a **3*cnf* formula**, such as

$$(x_1 \vee \overline{x_2} \vee \overline{x_3}) \wedge (x_3 \vee \overline{x_5} \vee x_6) \wedge (x_3 \vee \overline{x_6} \vee x_4) \wedge (x_4 \vee x_5 \vee x_6)$$

Let $3SAT = \{<\varphi> | \varphi$ can satisfy the $3cnf$ formula$\}$. If an assignment satisfies a cnf formula, then each clause must contain at least one literal value of 1.

The following theorem gives the polynomial time reduction from the $3SAT$ problem to the $CLIQUE$ problem.

Theorem 7.4.3 $3SAT$ polynomial time can be reduced to $CLIQUE$.

Proof Idea Given the polynomial time reduction f from $3SAT$ to $CLIQUE$, it transforms the formula into a graph. In the constructed graph, the clique of the specified size corresponds to the satisfying assignment of the formula. The structure in the figure is designed to simulate the effects of variables and clauses.

Proof Let φ be a formula of k clauses, such as $\varphi = (a_1 \vee b_1 \vee c_1) \wedge (a_2 \vee b_2 \vee c_2) \wedge \ldots \wedge (a_k \vee b_k \vee c_k)$

Reduce f to generate a string $<G, k>$, where G is an undirected graph defined as follows.

The nodes in G are divided into k groups, each group of three nodes, called **triple** $t_1, \ldots, t_k$. Each triple corresponds to a clause in φ, and each node in the triple corresponds to a literal of the corresponding clause. Each node of G is marked with the literal in its corresponding φ.

Except for two cases, the edge of G connects all pairs of nodes. The nodes in the same triple are boundlessly connected, on the contrary the two marked nodes are boundlessly connected, such as x_2 and $\overline{x_2}$.

For example, when $\varphi = (x_1 \vee x_1 \vee x_2) \wedge (\overline{x_1} \vee \overline{x_2} \vee \overline{x_2}) \wedge (\overline{x_1} \vee x_2 \vee x_2)$, Fig. 7-7 shows this structure.

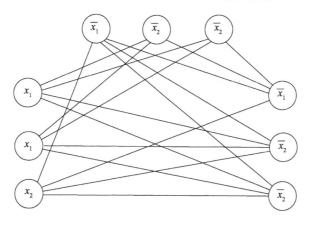

Fig. 7-7 Reduce the graph generated from $\varphi = (x_1 \vee x_1 \vee x_2) \wedge (\overline{x_1} \vee \overline{x_2} \vee \overline{x_2}) \wedge (\overline{x_1} \vee x_2 \vee x_2)$

Now explain why this structure can work, and prove that φ is satisfiable if and only if G has k-cliques.

Assume that φ has a satisfying assignment. Under satisfying assignment, at least one literal of each clause is true. In each triple of G, select the node corresponding to the literal that is true under the satisfying assignment. If more than one literal is true in a certain clause, just choose any true literal. The selected node will exactly form a k-clique. Since a node is selected from each of the k triples, the number of nodes selected is k. Each pair of selected nodes are connected by edges, and neither of them is the two exceptions described above. They cannot be from the same triple, because only one node is selected from each triple. It is also impossible for them to have the opposite label, because their associated literals are all true under the satisfactorily assigned value. So G contains k-cliques.

Assume that G has k-cliques. Because the nodes in the same triple are boundlessly connected, no two nodes in the clique are in the same triple. Therefore, each of the k triples contains exactly one node of the clique. Assign a true value to the variable of φ, so that every literal that marks the clique node is true. This can be done, because two nodes with opposite labels are boundlessly connected, so it is impossible for both to be in the clique. This assignment to the variable satisfies φ, because each triple contains a clique node, so each clause contains a literal assigned to TRUE. φ can be satisfied.

Theorem 7.4.2 and Theorem 7.4.3 state that if *CLIQUE* is solvable in polynomial time, so is 3*SAT*. At first glance, the connection between these two problems seems unusual, because on the surface they are very different. But the time reducibility of polynomials allows their complexity to be connected. Turning now to a definition that allows the complexity of a whole class of problems to be connected in a similar way.

假定 φ 有满足赋值. 在满足赋值下,每个子句至少一个文字为真. 在 G 的每个三元组中,选择在该满足赋值下为真的文字对应的结点. 如果在某一子句中不止一个文字为真,任意选择一个真文字即可. 选择出来的结点将恰好形成一个 k-团. 因为是从 k 个三元组中的每一个中挑选一个结点,所以选择的结点数为 k. 每一对选中的结点都有边相连,它们都不是前面描述的两种例外情形. 它们不可能来自同一三元组,因为从每个三元组中只选一个结点. 它们也不可能有相反标记,因为它们关联的文字在该满足赋值下都为真. 所以 G 包含 k-团.

假定 G 有 k-团. 因为在同一个三元组中的结点都无边相连,所以团中的任何两个结点都不在同一个三元组中. 因此 k 个三元组中的每一个都恰好包含团的一个结点. 给 φ 的变量赋真值,使得标记团结点的每个文字都为真. 这可以办到,因为具有相反标记的两个结点是无边相连,所以不可能两个都在团中. 给变量的这种赋值满足 φ,因为每个三元组包含一个团结点,所以每个子句包含一个赋值为 TRUE 的文字. φ 可满足.

定理 7.4.2 和定理 7.4.3 说明,如果 *CLIQUE* 在多项式时间内可解,那么 3*SAT* 也如此. 乍一看,这两个问题之间的联系显得很不寻常,因为表面上它们是非常不同的. 但是多项式时间的可归约性允许把它们的复杂性联系起来. 现在转向一个定义,它允许用类似的方式把一整类问题的复杂性联系起来.

7.4.2 Definition of NP-completeness

Definition 7.4.3 If language B satisfies the following two conditions, it is called **NP-complete**:

(1) B belongs to NP.

(2) Each A in NP can be reduced to B in polynomial time.

Theorem 7.4.4 If the above B is NP-complete, and $B \in P$, then $P = NP$.

Proof It is directly available from the definition of polynomial time reducibility.

Theorem 7.4.5 If the above B is NP-complete, and $B \leq_P C$, C belongs to NP, then C is NP-complete.

Proof Knowing that C belongs to NP, it must be proved that each A in NP can be reduced to C in polynomial time. Because B is NP-complete, each language in NP can be reduced to B in polynomial time, and B can be reduced to C in polynomial time. Polynomial time reduction can be compounded, that is, if A can be polynomial-time reduced to B, and B can be polynomial-time reduced to C, then A can be polynomial-time reduced to C. Therefore, each language in NP can be reduced to C in polynomial time.

7.4.3 Cook-Levin theorem

Once you have an NP-complete problem, you can start from it and get other NP-complete problems through polynomial time reduction. However, it is more difficult to establish the first NP-complete problem. Now, complete this step by proving that the SAT is NP-complete.

Theorem 7.4.6 SAT is NP-complete.

This theorem describes Theorem 7.4.1 in another form.

Proof idea It is simple to prove that SAT belongs to NP, and we will prove it soon. The difficulty of the proof is to prove that any language in NP can reduce polynomial time to SAT.

For this reason, a polynomial time reduction to SAT is constructed for each language A in NP. The reduction of A produces the Boolean formula φ on the string w, which is used to simulate the operation of A's NP machine on the input w. If the machine accepts, then φ has a satisfying assignment corresponding to the acceptance calculation. If the machine does not accept it, then no assignment can satisfy φ. Therefore, w belongs to A if and only if φ can be satisfied.

In fact, although many details must be dealt with, constructing a reduction that operates in this way is conceptually simple. A Boolean formula can contain the Boolean operations AND, OR, and NOT, which form the basis of circuits used in electronic computers. Therefore, the fact that a Boolean formula can be designed to simulate a Turing machine is not surprising. The detail lies in the realization of this idea.

Proof First, prove that SAT belongs to NP. The non-deterministic polynomial time machine can guess an assignment of the given formula φ, and accept it when the assignment satisfies φ.

Next, choose any language A from NP to prove that the polynomial time of A can be reduced to SAT. Let N be a non-deterministic Turing machine that determines A within time n^k, and k is a certain constant. (For convenience, it is actually assumed that N runs in time n^k-3, but only those readers who are interested in detail may worry about this minor point.) The following concepts help describe the reduction.

On w, the corresponding **screen** of N is a $n^k \times n^k$ table, where the rows represent the configuration of a calculation branch of N on the input w, as shown in Fig. 7-8.

为此,给 NP 中的每一个语言 A 构造一个到 SAT 的多项式时间归约. A 的归约在字符串 w 上产生布尔公式 φ,用它模拟在输入 w 上 A 的 NP 机器的运行. 如果机器接受,那么 φ 有一个对应于接受计算的满足赋值. 如果机器不接受,那么没有赋值能满足 φ. 所以,当且仅当可满足 φ 时,w 属于 A.

实际上,虽然必须处理很多细节,但是构造一种以这种方式运算的归约在概念上是简单的. 一个布尔公式可以包含布尔操作 AND,OR 和 NOT,这些操作形成了电子计算机中使用的电路的基础. 因此,可以设计布尔公式来模拟图灵机这一事实毫不令人奇怪. 细节在于这一思想的实现.

证明 首先证明 SAT 属于 NP. 非确定型多项式时间机器可以猜测给定的公式 φ 的一个赋值,当赋值满足 φ 时接受.

下面从 NP 中取任一个语言 A,证明 A 的多项式时间可归约到 SAT. 设 N 是在时间 n^k 内判定 A 的非确定型图灵机,k 是某个常数. (为了方便,实际上假定 N 在时间 n^k-3 内运行,但只有那些对细节感兴趣的读者可能会担心这个次要的地方.) 下面的概念有助于描述该归约.

在 w 上,N 的对应**画面**是一张 $n^k \times n^k$ 的表格,其中行代表 N 在输入 w 上的一个计算分支的格局,如图 7-8 所示.

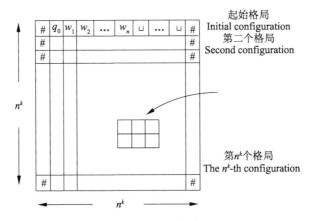

Fig. 7-8 The corresponding screen is the configuration table of $n^k \times n^k$

For convenience, we assume that each configuration starts and ends with the symbol #, so that the first and last columns of the screen are #. The first line of the screen is the initial configuration of N on w, and each line is obtained from the previous line according to the transfer function of N. If a certain line of the screen accepts the configuration, the screen is said to be **accepted**.

Each accepted screen of N on w corresponds to a calculation branch of N on w. Therefore, the problem of determining whether N accepts w is equivalent to the problem of determining whether there is an acceptance screen where N is on w.

Now start to describe the polynomial time reduction f from A to SAT. On input w, this reduction produces a formula φ. Start with describing the variables of φ. Let Q and Γ be the state set and tape alphabet of N, respectively. Let $C = Q \cup \Gamma \cup \varphi$. For every i and j between 1 and n^k and every s in C, there is a variable $x_{i,j,s}$.

Each grid in $(n^k)^2$ screen grids is called a **cell**. The cell in the i-th row and j-th column is called cell $[i, j]$ and contains a symbol in C. Use φ variables to denote the contents of the cell. If the value of $x_{i,j,s}$ is 1, it means that cell $[i, j]$ contains s.

Now design φ, so that a satisfying assignment of a variable does correspond to an acceptance screen where N is on w. The formula φ is a four-part AND operation: $\varphi_{\text{cell}} \land \varphi_{\text{start}} \land \varphi_{\text{move}} \land \varphi_{\text{accept}}$, which describes each part in turn.

As mentioned earlier, turning on the variable $x_{i,j,s}$ corresponds to putting the symbol s into cell $[i,j]$. In order to obtain the correspondence between the assignment and the screen, the first thing that must be ensured is that the assignment exactly opens a variable for each cell. The formula φ_{cell} ensures this requirement. It expresses this in the language of Boolean operations.

$$\varphi_{\text{cell}} = \bigwedge_{1 \leq i,j \leq n^k} \left[\left(\bigvee_{s \in C} x_{i,j,s} \right) \land \left(\bigwedge_{\substack{s,t \in C \\ s \neq t}} (\overline{x_{i,j,s}} \lor \overline{x_{i,j,t}}) \right) \right]$$

The symbols $\land$ and $\lor$ denote repeated AND and OR respectively. For example, part of the above formula

$$\bigvee_{s \in C} x_{i,j,s}$$

is an abbreviation of the following formula:

$$x_{i,j,s_1} \lor x_{i,j,s_2} \lor \cdots \lor x_{i,j,s_l}$$

Where $C = \{s_1, s_2, \cdots, s_l\}$. Therefore, φ_{cell} is actually a long expression, which contains a segment for each cell in the screen, because i and j change from 1 to n^k. The first part of each segment says that at least one variable is turned on in the corresponding cell. The second part of each segment says that at most one variable is turned on in the corresponding cell. These parts are connected by the operation $\land$.

Part 1 of φ_{cell} in square brackets stipulates that at least one variable associated with each cell is turned on; and part 2 stipulates that only one variable is turned on for each cell. Thus, any satisfying assignment assigns a symbol to each cell in the table. The φ_{start}, φ_{move} and φ_{accept} guarantee that the form is indeed an acceptance screen, as shown below.

现在设计 φ,使得变量的一个满足赋值确实对应 N 在 w 上的一个接受画面. 公式 φ 是四部分的 AND 运算: $\varphi_{\text{cell}} \land \varphi_{\text{start}} \land \varphi_{\text{move}} \land \varphi_{\text{accept}}$,依次描述每一部分.

如前所述,开启变量 $x_{i,j,s}$ 相当于把符号 s 放进单元 $[i,j]$. 为了获得赋值与画面之间的对应关系,必须保证的第一件事是赋值恰好为每个单元开启一个变量. 公式 φ_{cell} 确保了这一要求,它用布尔运算的语言来表达这一点.

$$\varphi_{\text{cell}} = \bigwedge_{1 \leq i,j \leq n^k} \left[\left(\bigvee_{s \in C} x_{i,j,s} \right) \land \left(\bigwedge_{\substack{s,t \in C \\ s \neq t}} (\overline{x_{i,j,s}} \lor \overline{x_{i,j,t}}) \right) \right]$$

符号 $\land$ 和 $\lor$ 分别代表反复出现的 AND 和 OR. 例如,上面公式中的部分

$$\bigvee_{s \in C} x_{i,j,s}$$

是下式的缩写:

$$x_{i,j,s_1} \lor x_{i,j,s_2} \lor \cdots \lor x_{i,j,s_l}$$

其中 $C = \{s_1, s_2, \cdots, s_l\}$. 因此,$\varphi_{\text{cell}}$ 实际上是一个长的表达式,它包含了画面中每个单元的一个片段,因为 i 和 j 从 1 变到 n^k. 每一片段的第一部分称为在相应单元中至少有一个变量被开启. 每一片段的第二部分称为在相应单元中最多有一个变量被开启. 这些片段通过运算 $\land$ 连接起来.

φ_{cell} 在方括号中的第 1 部分规定,至少开启一个与每个单元相关联的变量;而第 2 部分规定对每个单元只开启一个变量. 于是,任何满足的赋值都给表中的每个单元指定了一个符号. φ_{start},φ_{move} 和 φ_{accept} 等部分保证该表格确实是一个接受画面,如下面所示.

The formula φ_{start} guarantees that the first row of the table is the starting configuration of N on w, which clearly stipulates that the corresponding variable is turned on:

$$\varphi_{\text{start}} = x_{1,1,\#} \wedge x_{1,2,q_0} \wedge x_{1,3,w_1} \wedge x_{1,4,w_2} \wedge \cdots \wedge x_{1,n+2,w_n} \wedge x_{1,n+3,\sqcup} \wedge \cdots \wedge x_{1,n^k-1,\sqcup} \wedge x_{1,n^k,\#}$$

The formula φ_{accept} guarantees that the acceptance configuration appears in the screen. It ensures that q_{accept} (that is, the symbol that represents the acceptance state) appears in a certain cell of the screen by stipulating that one of the corresponding variables is turned on:

$$\varphi_{\text{accept}} = \bigvee_{1 \leq i,j \leq n^k} x_{i,j,q_{\text{accept}}}$$

Finally, the formula φ_{move} guarantees that each line of the screen corresponds to the configuration obtained by legally transferring from the configuration of the previous line according to the rule of N. It guarantees this by ensuring that every 2×3 window cell is legal. If a 2×3 window does not violate the action specified by the N transfer function, the window is said to be **legal**. In other words, if it can appear in the process of correctly transferring from one configuration to another, the window is called legal.

For example, let a, b, and c be the members of the tape alphabet, and q_1, and q_2 are the states of N. Assume that in state q_1, when the read-write head reads a, N writes one b, which is still in state q_1, and moves to the right. In state q_1, when the read-write head reads b, N is non-deterministic:

(1) Write one c, enter state q_2 and move left.

(2) Write one a, enter state q_2 and move right.

Formally denoted as $\delta(q_1, a) = \{(q_1, b, R)\}$, $\delta(q_1, b) = \{(q_2, c, L), (q_2, a, R)\}$. An example of the legal window of the machine is shown in Fig. 7-9.

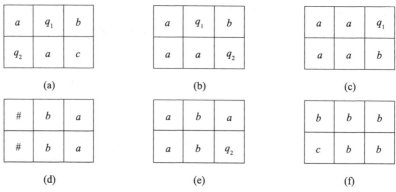

图 7-9 合法窗口示例
Fig. 7-9 Example of a legal window

In Fig. 7-9, windows (a) and (b) are legal because the transfer function allows N to move in the specified way. Window (c) is legal, because q_1 appears on the right side of the top row, and we don't know what symbol the read-write head is on. That symbol may be a, and q_1 can change it to b and move it to the right. It is possible to generate the window, so it does not violate the rule of N. Window (d) is obviously legal, because the top row and bottom row are the same, which occurs when the position of the read-write head and the window are not adjacent. Note that in the legal window, # can appear on the left or right of the top and bottom rows. Window (e) is legal, because it may be the state q_1 immediately to the right of the top row, read the symbol b, and then move left so that the state q_2 appears on the right side of the bottom row. Finally, window (f) is legal, because the state q_1 may be immediately to the left of the top row, it changes b to c, and then moves left.

The window shown in Fig. 7-10 is not legal for machine N.

在图 7-9 中,窗口(a)、(b)是合法的,因为转移函数允许 N 以指定的方式移动。窗口(c)是合法的,因为 q_1 出现在顶行的右边,我们不知道读写头在什么符号上边.那个符号可能是 a,q_1 可以把它变为 b,并且向右移.这就有可能产生该窗口,所以它不违反 N 的规则.窗口(d)显然是合法的,因为顶行与底行是相同的,当读写头与窗口的位置不相邻时就会出现这种情况.注意在合法窗口中,#可以出现在顶行和底行的左边或右边.窗口(e)是合法的,因为紧靠顶行的右边可能就是状态 q_1,读取符号 b,然后向左移,使状态 q_2 出现在底行的右边.最后,窗口(f)是合法的,因为状态 q_1 可能紧挨着顶行的左边,它把 b 变为 c,然后向左移.

图 7-10 所示的窗口对于机器 N 不是合法的.

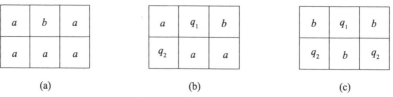

图 7-10 非法窗口示例
Fig. 7-10 Example of an illegal window

In window (a), the symbol in the middle of the top row will not change because there is no state adjacent to it. Window (b) is not legal, because the transfer function specifies that b should be changed to c instead of a. Window (c) is not legal, because two states appear on the bottom row.

Assertion 7.4.1 If the top line of the screen is the starting configuration, and every window in the screen is legal, then each row of the screen is the configuration legally transferred from the previous row.

To prove this assertion, consider any two adjacent configurations in the screen, called the upper configuration and the lower configuration. In the upper configuration, every cell that contains a tape subsymbol and is not adjacent to a status symbol is a middle cell in the top row of a certain window and the top row of the window does not contain status. Therefore, the symbol must remain unchanged and appear in the middle of the bottom row of the window, that is, it appears in the same position on the bottom row configuration.

The middle cell of the top row of the window contains the status symbol, which makes the corresponding three positions consistent update according to the requirements of the transfer function. Therefore, if the upper configuration is a legal configuration, so is the lower configuration, and the lower configuration is transferred from the upper configuration according to the rules of N. Note that this proof is obviously easy to understand, but its key depends on choosing a 2×3 window.

Now turn to the structure of φ_{move}, which stipulates that all windows in the screen are legal. Each window contains 6 cells, which can be set as legal windows in a fixed number of ways. The formula φ_{move} points out that the setting of these 6 cells must be one of these ways, namely

在窗口(a)中,顶行中间的符号不会改变,因为没有状态与它相邻.窗口(b)不是合法的,因为转移函数指明 b 应变为 c 而不是 a. 窗口(c)不是合法的,因为在底行出现了两个状态.

断言 7.4.1 如果画面的顶行是起始格局,画面中的每一个窗口都是合法的,那么画面的每一行都是从上一行合法转移得到的格局.

为证明该断言,考虑画面中任意两个相邻格局,称为上格局和下格局.在上格局中,每一个包含带子符号且不与状态符号相邻的单元都是某个窗口顶行的中间单元,且窗口顶行不含状态.所以,该符号必定保持不变,出现在窗口的底行中间,即出现在底行格局的同一位置.

窗口顶行的中间单元包含状态符号,这就使相应的三个位置按照转移函数的要求一致更新.因此,如果上格局是合法格局,那么下格局也是合法格局,并且下格局是根据 N 的规则从上格局转移得到的.注意,这个证明显然且易懂,但它的关键依赖于选择了大小为 2×3 的窗口.

现在来看 φ_{move} 的构造,它规定画面中的所有窗口都是合法的.每个窗口包含 6 个单元,它们可以用固定数目的方式设置为合法窗口.公式 φ_{move} 指出,这 6 个单元的设置必须是这几种方式之一,即

$$\varphi_{\text{move}} = \bigwedge_{1 \leq i < n^k, 1 \leq j < n^k} ((i,j)\text{-window is legal})$$

The cell $[i,j]$ is located at the top center of the (i,j)-window. Replace the text "(i,j)-window is legal" in this formula with the following formula, and write the contents of the 6 cells of the window as $a_1, \ldots, a_6$:

$$\bigvee_{\substack{a_1,\ldots,a_6 \\ \text{is a legal window}}} (x_{i,j-1,a_1} \wedge x_{i,j,a_2} \wedge x_{i,j+1,a_3} \wedge x_{i+1,j-1,a_4} \wedge x_{i+1,j,a_5} \wedge x_{i+1,j+1,a_6})$$

The following analyzes the complexity of reduction and proves that it is completed in polynomial time. For this reason, the size of φ is examined. First, estimate the number of its variables. Recall that the screen is an $n^k \times n^k$ table, so it contains n^{2k} cells. Each cell has l variables associated with it, and l is the number of symbols in C. Because l only depends on the Turing machine N and not on the input length n, the total number of variables is $O(n^{2k})$.

Estimate the size of each part of φ. For each cell of the screen, the formula φ_{cell} contains a fixed-length formula segment, so the length is $O(n^{2k})$. The formula φ_{start} contains one segment for each cell in the top row, so the length is $O(n^k)$. Formula φ_{move}, φ_{accept} contains fixed-length formula segments for each cell of the screen, so their length is $O(n^{2k})$. So the total length of φ is $O(n^{2k})$. This result is in full compliance with the goal, because it shows that the length of φ is a polynomial of n. If the length exceeds the polynomial relationship, then the reduction will not be able to generate it in polynomial time.

To see that the formula can be generated in polynomial time, pay attention to its high degree of repeatability. Each part of the formula consists of many almost the same segments, but there are simple changes in the subscripts. Therefore, a reduction can be easily constructed to generate φ from the input w in polynomial time.

This completes the proof of the Cook-Levin theorem and proves that the *SAT* is *NP*-complete. Proving the *NP*-completeness of other languages usually does not require such a long proof. *NP*-completeness can also be proved by polynomial time reduction starting from a language known to be *NP*-complete. *SAT* can be used for this, but 3*SAT* is usually easier. Recall that the formula of 3*SAT* is in conjunctive normal form, with three literals in each clause. First, it is necessary to prove that 3*SAT* itself is *NP*-complete. Prove it as a corollary of Theorem 7.4.6.

Corollary 7.4.1 3*SAT* is *NP*-complete.

Proof Obviously 3*SAT* belongs to *NP*, so it is only necessary to prove that all languages in *NP* are reduced to 3*SAT* in polynomial time. To prove this, one method is to prove that the *SAT* polynomial time is reduced to 3*SAT*. Here we use another method to modify the proof of Theorem 7.4.6 so that it directly generates a conjunctive normal form formula with three literals in each clause.

The formula produced by Theorem 7.4.6 is almost in conjunctive normal form. The formula φ_{cell} is a general conjunction of sub-formulas, and each sub-formula contains a general disjunction and a general conjunction of disjunctions. Therefore, φ_{cell} is a conjunction of clauses and is already in the form of *cnf*. The formula φ_{accept} is a general conjunctive of variables. Think of each variable as a clause of length 1, and you can see that φ_{start} is *cnf*. The formula φ_{accept} is a general disjunction of variables, so it is a single clause. The formula φ_{start} is the only formula that is not yet *cnf*, but it can be easily converted into a *cnf* formula, as described below.

Recall that φ_{move} is a general conjunctive of sub-formulas, and each sub-formula is the disjunction of conjunction, describing all possible legal windows. You can replace the disjunction of conjunction with the equivalent conjunction of disjunction. Doing so may

这样就完成了库克-列文定理的证明,证明 *SAT* 是 *NP* 完全的. 证明其他语言的 *NP* 完全性通常不需要这样长的证明. *NP* 完全性还可以通过从一个已知为 *NP* 完全的语言出发的多项式时间归约来证明. 为此可以用 *SAT*, 但是用 3*SAT* 通常更加容易. 回忆一下, 3*SAT* 的公式是合取范式形式的, 每个子句有三个文字. 首先, 必须证明 3*SAT* 本身是 *NP* 完全的. 把它作为定理 7.4.6 的推论来证明.

推论 7.4.1 3*SAT* 是 *NP* 完全的.

证明 显然 3*SAT* 属于 *NP*, 所以只需证明 *NP* 中的所有语言都在多项式时间内归约到 3*SAT*. 为证明这一点, 一种方法是证明 *SAT* 多项式时间归约到 3*SAT*. 这里改用另一种方法, 修改定理 7.4.6 的证明, 使得它直接产生每个子句中有三个文字的合取范式形式的公式.

定理 7.4.6 产生的公式已经几乎是合取范式形式的了. 公式 φ_{cell} 是子公式的大合取, 每个子公式包含一个大析取以及析取的大合取. 因此, φ_{cell} 是子句的合取, 它已经是 *cnf* 形式了. 公式 φ_{accept} 是变量的大合取. 把每个变量看作长为 1 的子句, 就能看出 φ_{start} 是 *cnf*. 公式 φ_{accept} 是变量的大析取, 因此是单个子句. 公式 φ_{move} 是唯一一个还不是 *cnf* 的公式, 但是可以容易地把它转化为 *cnf* 形式的公式, 如下所述.

回忆一下, φ_{move} 是子公式的大合取, 每个子公式是合取的析取, 描述了所有可能的合法窗口. 可以用替换等价的析取的合取的析取. 这么做可能会极大地增加每个子公式的长度, 但是

greatly increase the length of each sub-formula, but the total length of φ_{move} can only be increased by a constant multiple, because the length of each sub-formula only depends on N. The result is a formula in conjunctive normal form.

Now that the formula is written in *cnf* form, it is transformed into a form in which each clause has three literals. In the current clause with one or two literals, copy one of the literals so that the total number of literals reaches 3. In a clause with more than 3 literals, split it into several clauses and add some additional variables to maintain the satisfiability or unsatisfiability of the original formula.

For example, replace the clause $(a_1 \vee a_2 \vee a_3 \vee a_4)$ (where each a_i is a literal) with the expression $(a_1 \vee a_2 \vee z) \wedge (\bar{z} \vee a_3 \vee a_4)$ of two clauses, where z is the new argument. If a certain assignment of a_i satisfies the original clause, then a certain assignment of z can be found so that these two new clauses are satisfied, and vice versa. Generally speaking, if the clause contains l literals, such as
$$(a_1 \vee a_2 \vee \ldots \vee a_l)$$
then we can replace it with $l-2$ clauses, such as
$(a_1 \vee a_2 \vee z_1) \wedge (\bar{z_1} \vee a_3 \vee z_2) \wedge (\bar{z_2} \vee a_4 \vee z_3) \wedge \ldots \wedge (\bar{z_{l-3}} \vee a_{l-1} \vee a_l)$

It is easy to verify that the new formula is satisfiable if and only if the original formula is satisfiable, the proof is complete.

7.5 Typical *NP*-complete problems

The phenomenon of *NP*-completeness is very extensive, and there are *NP*-completeness problems in many fields. Due to some reasons that have not been deeply understood, most naturally occurring *NP* problems are either to be in P or to be *NP*-complete. If you are looking for a polynomial-time algorithm for a new *NP* problem, it is wise to spend some energy trying to prove that it is *NP*-complete, because it can prevent you from looking for a polynomial-time algorithm that does not exist.

In this section, several theorems are given to prove that several different languages are *NP*-complete. These theorems provide examples of proof techniques for similar problems. The general strategy is to give a polynomial time reduction from 3*SAT* to the language, if it is more convenient, sometimes other *NP*-complete languages can be reduced.

When constructing a polynomial time reduction from 3*SAT* to a language, we look for the structure of variables and clauses in the language that can simulate Boolean formulas. This structure is sometimes called a **component**. For example, in the reduction from 3*SAT* to *CLIQUE* given in Theorem 7.4.3, the node simulates the variable and the node's triple simulates the clause. A specific node may or may not be a member of a clique, which corresponds to a variable that can be true or not true in satisfying assignment. Each clause must contain literals that assigns a value of true. Correspondingly, each triple must contain a node of the clique. The following is a corollary of Theorem 7.4.3, which shows that *CLIQUE* is *NP*-complete.

Corollary 7.5.1 *CLIQUE* is *NP*-complete.

7.5.1 Vertex cover problem

If G is an undirected graph, then the **vertex cover** of G is a subset of nodes, so that each edge of G is associated with one of the nodes in the subset. The vertex cover problem is to determine whether there is a vertex cover of a specified scale in the graph:

$VERTEX\text{-}COVER = \{<G,k> | G$ is an undirected graph with a vertex cover of k nodes$\}$

Theorem 7.5.1 *VERTEX-COVER* is *NP*-complete.

这样可以防止去寻找一个并不存在的多项式时间算法.

本节再给出几个定理,证明几个不同的语言是 *NP* 完全的. 这些定理为同类问题的证明技巧提供了示例. 一般策略是给出从 3*SAT* 到该语言的多项式时间归约,如果更加方便的话,有时也从其他 *NP* 完全语言归约.

在构造从 3*SAT* 到一个语言的多项式时间归约时,我们寻找这个语言中能模拟布尔公式的变量和子句的结构,这种结构有时称为**构件**. 例如,在定理 7.4.3 中给出的从 3*SAT* 到 *CLIQUE* 的归约中,结点模拟变量,结点的三元组模拟子句. 一个具体的结点可以是也可以不是团的成员,这对应于在满足赋值中可以是真也可以不是真的一个变量. 每个子句必须包含赋值为真的文字. 相应地,每个三元组必须包含团的一个结点. 下面是定理 7.4.3 的推论,表明 *CLIQUE* 是 *NP* 完全的.

推论 7.5.1 *CLIQUE* 是 *NP* 完全的.

7.5.1 顶点覆盖问题

若 G 是无向图,则 G 的**顶点覆盖**是结点的一个子集,使得 G 的每条边都与子集中的结点之一相关联. 顶点覆盖问题旨在确定图中是否存在指定规模的顶点覆盖:

$VERTEX\text{-}COVER = \{<G,k> | G$ 是一个无向图,其顶点覆盖有 k 个结点$\}$

定理 7.5.1 *VERTEX-COVER* 是 *NP* 完全的.

Proof idea To prove that *VERTEX-COVER* is *NP*-complete, it must be proved that it belongs to *NP* and all problems in *NP* can be reduced to it in polynomial time. The first part is easier: the certificate is a vertex cover of scale k. Prove the second part by proving that $3SAT$ polynomial time can be reduced to *VERTEX-COVER*. This reduction converts a $3cnf$ formula φ into a graph G and a value k, and as long as there are k nodes in G covering the vertices, φ can be satisfied. The conversion is done without knowing whether the φ can be satisfied. In fact, G simulates φ. The figure contains components, variables and clauses in the simulation formula. Designing these components requires a bit of original ingenuity.

For variable components, looking for a structure in G that participates in the vertex cover in one of two possible ways, which corresponds to the two possible real assignments of variables. The variable component contains two possible real assignments of variables. The variable component contains two nodes connected by an edge. This structure is effective because one of the two nodes must appear in the vertex cover. Arbitrarily associate two nodes with TRUE and FALSE respectively.

For clause components, we need to look for such a structure: it makes at least one of the variable component nodes included in the vertex cover corresponding to at least one literal in the clause that has a value of true. This component contains three nodes and their connected edges, so that any vertex cover must contain at least two of its nodes, or all three nodes. If one of the component nodes only helps to cover only one edge, then only the other two nodes need to be included in the vertex cover, that is, the corresponding literal satisfies the condition of the clause. Otherwise, all three nodes must be covered. Finally, select the value of k so that the found vertices cover one node corresponds to one variable component, and two nodes correspond to one clause component.

证明思路 要证明 *VERTEX-COVER* 是 *NP* 完全的,必须证明它属于 *NP* 且 *NP* 中的所有问题都能在多项式时间里归约到它. 第一部分较容易:证书就是一个规模为 k 的顶点覆盖. 通过证明 $3SAT$ 多项式时间可归约到 *VERTEX-COVER* 来证明第二部分. 该归约将一个 $3cnf$ 公式 φ 转换为一个图 G 和数值 k,且只要 G 中有 k 个结点的顶点覆盖,φ 就能够被满足. 转换是在不知道 φ 能否被满足的情况下完成的. 实际上,G 模拟了 φ. 该图包含着构件、模拟公式中的变量和子句. 设计这些构件需要一点独出心裁的巧思.

对于变量构件,在 G 中寻找一种结构,它以两种可能的方式之一参与到顶点覆盖中,正好对应变量的两种可能的真实赋值. 变量构件包含两个可能的变量实际赋值. 变量构件包含被一条边连接的两个结点. 这种结构之所以有效是因为两个结点之一一定会出现在顶点覆盖中. 任意地将两个结点分别与 TRUE 和 FALSE 关联起来.

对于子句构件,要寻找这样的结构:它使得顶点覆盖所包含的变量构件结点中,至少有一个结点对应着该子句中至少一个取值为真的文字. 这个构件包含三个结点及它们相连的边,这样任何一个顶点覆盖都一定会包含它的至少两个结点,或者是全部三个结点. 如果构件结点中的一个只是有助于覆盖仅一条边,那么顶点覆盖中只需包括另外两个结点,也就是对应的文字满足了该子句的情况. 否则,所有三个结点都必须覆盖. 最后,选择 k 值使得找到的顶点覆盖一个结点对应一个变量构件,两个结点对应一个子句构件.

Proof Here we give the details of a reduction from 3*SAT* to *VERTEX-COVER* that operates in polynomial time. The reduction maps the Boolean formula φ to a graph G and the value k. For each variable x in φ, an edge connecting two nodes is generated. Mark the two nodes in this component as x and $\overline{x}$. Assigning x to TRUE corresponds to the vertex covering selecting the left node of the edge, and assigning it to FALSE corresponds to the node marked as $\overline{x}$.

The component corresponding to the clause is a bit more complicated. The component of each clause is a triple of nodes marked with the three literals of the clause. These three nodes are connected to each other and connected to the node with the same label in the variable component. Therefore, the total number of nodes appearing in G is $2m+3l$, where φ has m variables and l clauses. Let k be $m+2l$.

For example, if $\varphi = (x_1 \vee x_1 \vee x_2) \wedge (\overline{x_1} \vee \overline{x_2} \vee \overline{x_2}) \wedge (\overline{x_1} \vee x_2 \vee x_2)$, the reduction generates $<G,k>$ from φ, where $k=8$, and the shape of G is shown in Fig. 7-11.

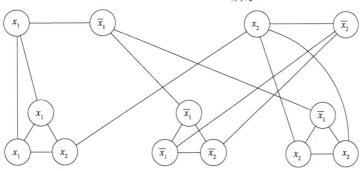

Fig. 7-11 Reduce the graph generated from $\varphi = (x_1 \vee x_1 \vee x_2) \wedge (\overline{x_1} \vee \overline{x_2} \vee \overline{x_2}) \wedge (\overline{x_1} \vee x_2 \vee x_2)$

In order to prove that the reduction satisfies the requirements, it is necessary to prove that it can be satisfied if and only if G has a vertex cover of k nodes. Starting from a satisfying assignment, first put the node in the variable component corresponding to the true literal in the assignment into the vertex cover. Then, select a true literal in each clause, and put the remaining two nodes in each clause component into the vertex cover. There are now k nodes in total. They cover all edges, because obviously the edges of each variable component are covered, all three edges in each clause component are also covered, and all edges between the variable component and the clause component are also covered. So G has a vertex covering of k nodes.

Secondly, if G has the vertex covering of k nodes, it is proved that φ is satisfiable by constructing satisfying assignment. In order to cover the edges of the variable component and the three edges of the clause component, the vertex cover must include one node of each variable component and two nodes of each clause component. This occupies all the nodes covered by the vertices, and there is no remaining share. Select the node in the vertex cover in the variable component, and assign the corresponding literals as true. This assignment satisfies φ, because the three edges connecting the variable component and each clause component are covered, and only two nodes in the clause component are in the vertex cover, so one of the edges must be covered by a node in the variable component. So the assignment satisfies the corresponding clause.

7.5.2 Hamiltonian path problem

The Hamiltonian path problem is whether the input graph contains a path that passes through each node exactly once from s to t.

Theorem 7.5.2 *HAMPATH* is *NP*-complete.

Proof idea In section 7.3, it has been proved that *HAMPATH* belongs to *NP*. To prove that the polynomial time of each *NP* problem can be reduced to *HAMPATH*, we prove that the 3*SAT* polynomial time can be reduced to *HAMPATH*.

Give a method to transform the 3*cnf* formula into a graph, so that the Hamiltonian path in the graph corresponds to the satisfying assignment of the formula. The figure contains components that simulate variables and clauses. The variable component is a diamond structure that can be passed through in one of two ways, corresponding to two true value assignments. The clause component is a node, that guarantees the path through each clause component corresponding to the guarantee that each clause is satisfied in the satisfaction assignment.

Proof It has been proved that *HAMPATH* belongs to *NP*. What needs to be done is to prove $3SAT \leq_p HAMPATH$. For each 3*cnf* formula φ, we show how to construct a directed graph G containing two nodes s and t, so that there is a Hamiltonian path between s and t if and only if φ can be satisfied.

Constructed from the 3*cnf* formula φ containing k clauses:

$$\varphi = (a_1 \lor b_1 \lor c_1) \land (a_2 \lor b_2 \lor c_2) \land \ldots \land (a_k \lor b_k \lor c_k)$$

Each of a, b, c is the literal x_i or $\overline{x_i}$. Let $x_1,\ldots,x_l$ be l variables of φ.

Now explain how to convert φ into graph G. The constructed graph G uses different parts to denote variables and clauses that appear in φ.

Denote each variable x_i as a diamond-shaped structure containing a row of horizontal nodes, as shown in Fig. 7-12. The number of nodes included in the horizontal row will be explained later.

Denote each clause of x_i as a single node, as shown in Fig. 7-13.

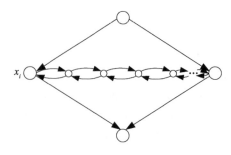

图 7-12 变量 x_i 表示为一个钻石形结构
Fig. 7-12 The variable x_i is denoted as a diamond-shaped structure

图 7-13 把子句 c_j 表示为结点
Fig. 7-13 Denoting the clause c_j as a node

Fig. 7-14 depicts the global structure of G. It shows all the elements of G and their interrelationships, except that the edges denoting the relationship between variables and clauses (which contain these variables) are not drawn.

图 7-14 描绘了 G 的全局结构. 除了表示变量与子句(包含着这些变量)关系的边没有画出以外,它展示了 G 的所有元素及其相互关系.

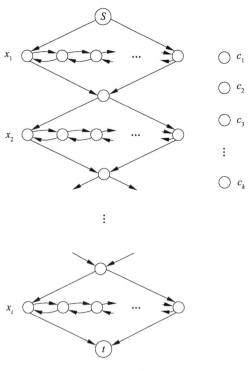

图 7-14 G 的高层结构
Fig. 7-14 High-rise structure of G

The following explains how to connect the diamond representing the variable with the node representing the clause. Each diamond structure contains a row of horizontal nodes, which are connected by edges in two directions. In addition to the two nodes on the two ends of the diamond, the horizontal row also contains $3k+1$ nodes. These nodes are divided into adjacent pairs, one pair for each clause, and these node pairs are separated by other nodes, as shown in Fig. 7-15.

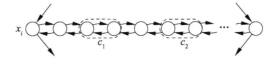

图 7-15 钻石结构中的水平结点
Fig. 7-15 Horizontal nodes in the diamond structure

If the variable x_i appears in the clause c_j, add the two edges from the j-th pair of nodes of the i-th diamond to the j-th clause node as shown in Fig. 7-16.

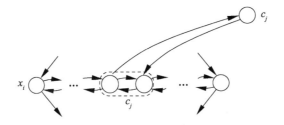

图 7-16 当子句 c_j 包含 x_i 时添加的边
Fig. 7-16 Edge added when clause c_j contains x_i

If $\overline{x_i}$ appears in the clause c_j, add the two edges from the j-th pair of nodes of the i-th diamond to the j-th clause node as shown in Fig. 7-17.

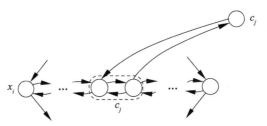

图 7-17 当子句 c_j 包含 $\overline{x_i}$ 时添加的边
Fig. 7-17 Edge added when clause c_j contains $\overline{x_i}$

After adding the edges corresponding to each x_i or $\overline{x}_i$ appearing in each clause, the construction of G is completed. In order to show that this structure satisfies the requirements, we assert that if φ is satisfiable, there is a Hamiltonian path from s to t; conversely, if there is such a path, φ is satisfiable.

Assume that φ is satisfiable. In order to show the Hamiltonian path from s to t, first ignore the clause nodes. The path starts at s, passes through each diamond in turn, and ends at t. In order to pass through the horizontal nodes in the diamond, the path zigzags from left to right (left-right style), or from right to left (right-left style), which method is determined by the satisfaction assignment of φ. If x_i is assigned a value of TRUE, the corresponding diamond will be passed in a left-right style. If x_i is assigned FALSE, it takes the right-left form. Fig. 7-18 shows these two possibilities.

把每一子句中出现的每个 x_i 或 $\overline{x}_i$ 所对应的边都添加进去以后，G 的构造就完成了. 为了说明这种构造满足要求，我们断言，若 φ 是可满足的，则从 s 到 t 存在一条哈密顿路径；反之，若存在这样的路径，则 φ 是可满足的.

假设 φ 是可满足的. 为了展示从 s 到 t 的哈密顿路径，首先忽略子句结点. 路径从 s 开始，依次经过每个钻石，到 t 终止. 为了经过钻石中的水平结点，该路径从左到右（左-右式），或者从右到左（右-左式）曲折前进，由 φ 的满足赋值决定采用哪一种方式. 如果 x_i 赋值为 TRUE，就以左-右式通过相应的钻石. 如果 x_i 赋值为 FALSE，就以右-左式通过相应的铅石. 图 7-18 展示了这两种可能.

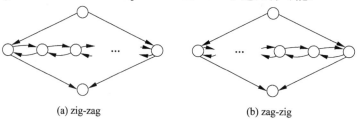

(a) zig-zag　　　　　　(b) zag-zig

图 7-18　由满足赋值决定采用左-右式，还是右-左式通过钻石

Fig. 7-18　Take the left-right form and the right-left form to pass through the diamond, determined by satisfying assignments

So far, the path covers all nodes in G except for clause nodes. By adding loops to the horizontal nodes, clause nodes can be easily included in the path. In each clause, select a literal that satisfies the assignment TRUE.

If you select $\overline{x}_i$ in the clause c_j, you can detour on the j-th pair of nodes of the i-th diamond. This can be done because x_i must be TRUE, and the path passes through the corresponding diamond from left to right. Therefore, the order of the edges connected to the node c_j is just allowed to detour and return. Simi-

迄今为止，该路径覆盖了 G 中除子句结点以外的所有结点. 通过在水平结点上增加回路，可以轻易地把子句结点纳入路径. 在每个子句中，选择一个满足赋值 TRUE 的文字.

如果在子句 c_j 中选择 $\overline{x}_i$，就能在第 i 个钻石的第 j 对结点上绕行. 可以这样做是因为 x_i 必定是 TRUE，该路径从左到右通过相应的钻石. 所以，连到结点 c_j 的边的次序正好允许绕行并返回. 类似地，如果在子句 c_j 中选择了

larly, if you select $\overline{x_i}$ in the clause c_j, you can detour on the j-th pair of nodes of the i-th diamond, because x_i must be FALSE, and the path passes through the corresponding diamond from right to left. Therefore, the order of the edges connected to the node c_j is also allowed to detour and return (Note that each true literal in the clause provides a choice for the detour through the node of the clause. As a result, if several literals in the clause are true, only one circuitous route is selected). The required Hamiltonian path is constructed.

For the proof in the opposite direction, if G has a Hamiltonian path from s to t, a satisfactory assignment of φ is given. If the Hamiltonian path is normal, that is, except for the detour to the clause node, it passes through each diamond in turn from top to bottom, and it is easy to obtain a satisfactory assignment. If it passes through the diamond in a left-right form, assign the corresponding variable to TRUE; if it is a right-left form, assign it to FALSE. Because each clause node appears on the path, by observing the circuitous route passing through it, you can determine which literal in the corresponding clause is TRUE.

What needs to be proved now is that the Hamiltonian path must be normal. The only way to violate normality is to enter the clause node from one diamond, but return to another diamond, as shown in Fig. 7-19.

$\overline{x_i}$,就能在第 i 个钻石的第 j 对结点上绕行,因为 x_i 必定是 FALSE,该路径从右到左通过相应的钻石.所以,连到结点 c_j 的边的次序正好也允许绕行并返回(注意,子句中的每个真文字都给经过子句结点的绕行提供了一种选择.结果是,如果子句中有几个文字为真,那么只选取一条迂回路线).这样就构造好了所需的哈密顿路径.

对于相反方向的证明,若 G 有一条从 s 到 t 的哈密顿路径,给出一个 φ 的满足赋值.若该哈密顿路径是正规的,即除了到子句结点的绕行以外,它从上到下依次通过每个钻石,则容易获得满足赋值.若它以左-右式通过钻石,则把相应变量指定为 TRUE;若是右-左式,则指定为 FALSE.因为每个子句结点都出现在路径上,所以通过观察经过它的迂回路线的情况,可以确定相应的子句中哪个文字为 TRUE.

现在还需证明的就是哈密顿路径必须是正规的.违反正规性的唯一途径是路径从一个钻石进入子句结点,却返回另一个钻石,如图 7-19 所示.

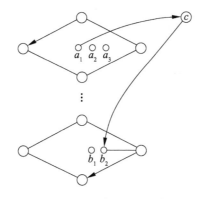

图 7-19 这种情况不可能发生
Fig. 7-19 This situation cannot happen

The path goes from node a_1 to c, but instead of returning to a_2 of the same diamond, it returns to b_2 of a different diamond. If so, then a_2 or a_3 must be the separation node. If a_2 is a separation node, the edges entering a_2 can only come from a_1 and a_3. If a_3 is a separation node, then a_1 and a_2 are in the same clause pair, so the edges that enter a_2 can only come from a_1, a_3, and c. In both cases, the path cannot contain node a_2. It cannot enter a_2 from c or a_1, because it starts from these two nodes and leads to other places. It is also impossible to enter a_2 from a_3, because a_3 is the only remaining node that a_2 is aimed at, and it must pass a_2 after exiting a_3. So the Hamiltonian path must be normal. The reduction obviously operates in polynomial time, and the proof is complete.

Next, consider an undirected Hamiltonian path problem, called *UHAMPATH*. In order to prove that *UHAMPATH* is *NP*-complete, a polynomial time reduction starting from the directed Hamiltonian path problem is given.

Theorem 7.5.3 *UHAMPATH* is *NP*-complete.

Proof For a directed graph G containing nodes s and t, reduce to construct an undirected graph G' containing nodes s' and t'. Graph G' has a Hamiltonian path from s to t if and only if G' has a Hamiltonian path from s' to t'. Describe G' as follows.

Except for s and t, each node u of G is replaced with three nodes of G': u^{in}, u^{mid}, and u^{out}. The nodes s and t of G are replaced with the nodes $s^{out} = s'$ and $t^{in} = t'$ of G'. G' has two types of edges. First, there are edges connecting u^{mid} and u^{in} and u^{mid} and u^{out}. Second, if there is an edge from u to v in G, then u^{out} and v^{in} are connected by an edge. This completes the construction of G'.

By proving that G has a Hamiltonian path from s to t if and only if G' has a Hamiltonian path from s^{out} to t^{in}, it can be shown that this construction satisfies the requirements. To prove a direction, notice the

Hamiltonian path P in G:

$$s, u_1, u_2, \ldots, u_k, t$$

There is a corresponding Hamiltonian path P' in G':

$$s^{out}, u_1^{in}, u_1^{mid}, u_1^{out}, u_2^{in}, u_2^{mid}, u_2^{out}, \ldots, t^{in}$$

In order to prove the other direction, we assert that any Hamiltonian path from s^{out} to t^{in} in G' is like the path P' just described, and must be from one triple of the node to another triple, except where it starts and ends. This will complete the proof, because such a path has a corresponding Hamiltonian path in G. Prove this assertion by starting from the node s^{out} and tracing the path. Note that the next node on the path must be u_i^{in} (for some i), because only those nodes are connected to s^{out}. The next node must be u_t^{mid}, because in the Hamiltonian path, no other methods can include u_t^{mid}. After u_t^{mid} is u_t^{out}, because this is the only node that u_t^{mid} connects to. The next node must be u_j^{in} (for a certain j), because no other node can be connected to u_t^{out}. Repeat this reasoning until t^{in} is reached.

7.5.3 The subset-sum problems

Recall the *SUBSET-SUM* problem defined in Theorem 7.3.3. In this problem, there is a number set $x_1, \ldots, x_k$ and a target number t. It is necessary to determine whether the number set contains a subset that adds up to t. Now prove that the problem is *NP*-complete.

Theorem 7.5.4 *SUBSET-SUM* is *NP*-complete.

Proof idea Theorem 7.3.3 has proved that *SUBSET-SUM* belongs to *NP*. By reducing the *NP*-complete language 3*SAT* to it, it is proved that all languages in *NP* can be reduced to *SUBSET-SUM* in polynomial time. Given a 3*cnf* formula, construct an instance of the problem *SUBSET-SUM*, so that it contains a subset that add up to the target t if and only if φ can be satisfied. This subset is called T.

Use two numbers y_i and z_i to denote the variable x_i. Prove that for each i, y_i or z_i must be in T, thus establishing a encoding for the true value of x_i in satisfying the assignment.

Each clause position contains a certain value in the target t, which places certain requirements on the subset T. Prove that this requirement is consistent with the requirement in the corresponding clause, that is, one of the clause's literal is required to be assigned a value of TRUE.

Proof *SUBSET-SUM* $\in$ *NP* is known, prove $3SAT \leq_p SUBSET\text{-}SUM$.

Assume φ is a Boolean formula whose variables are $x_1, \ldots, x_l$, and the clauses are $c_1, \ldots, c_k$. Reduction transforms φ into an instance $<S, t>$ of the problem of *SUBSET-SUM*, where the elements of S and the number t are the lines denoted in the usual decimal notation in Fig. 7-20. The line above the double line is marked as:

$y_1, z_1, y_2, z_2, \ldots, y_l, z_l$ and $g_1, h_1, g_2, h_2, \ldots, g_k, h_k$

They make up the elements of S. The line below the double line is t.

Thus, corresponding to each variable x_i of φ, S contains a pair of numbers y_i and z_i. The decimal representation of these numbers is divided into two parts, as shown in Fig. 7-20.

用两个数 y_i 和 z_i 来表示变量 x_i. 证明对于每个 i, y_i 或 z_i 必定在 T 中, 以此建立起在满足赋值中 x_i 的真值的编码.

每个子句位置都包含目标 t 中的某一值, 这就对子集 T 提出一定的要求. 证明这种要求与相应子句中的要求是一致的, 即要求该子句的文字之一赋值为 TRUE.

证明 已知 *SUBSET-SUM* $\in$ *NP*, 证明 $3SAT \leq_p SUBSET\text{-}SUM$.

设 φ 是一个布尔公式, 其变量是 $x_1, \ldots, x_l$, 子句是 $c_1, \ldots, c_k$. 归约把 φ 转化为 *SUBSET-SUM* 问题的一个实例 $<S, t>$, 其中 S 的元素和数 t 是图 7-20 中以通常的十进制记法表示的行. 双线上面的行标记为:

$y_1, z_1, y_2, z_2, \ldots, y_l, z_l$ 和 g_1, $h_1, g_2, h_2, \ldots, g_k, h_k$

它们组成 S 的元素. 双线下面的行是 t.

于是, 对应于 φ 的每个变量 x_i, S 包含一对数 y_i 和 z_i. 这些数的十进制表示分为两部分, 如图 7-20 所示.

	1	2	3	4	$\cdots$	l	c_1	c_1	$\cdots$	c_k
y_1	1	0	0	0	$\cdots$	0	1	0	$\cdots$	0
z_1	1	0	0	0	$\cdots$	0	0	0	$\cdots$	0
y_2		1	0	0	$\cdots$	0	0	1	$\cdots$	0
z_2		1	0	0	$\cdots$	0	1	0	$\cdots$	0
y_3			1	0	$\cdots$	0	1	1	$\cdots$	0
z_3			1	0	$\cdots$	0	0	0	$\cdots$	1
$\vdots$					$\ddots$	$\vdots$	$\vdots$	$\vdots$		$\vdots$
y_l						1	0	0	$\cdots$	0
z_l						1	0	0	$\cdots$	0
g_1							1	0	$\cdots$	0
h_1							1	0	$\cdots$	0
g_2								1	$\cdots$	0
h_2								1	$\cdots$	0
$\vdots$									$\ddots$	$\vdots$
g_k										1
h_k										1
t	1	1	1	1	$\cdots$	1	3	3	$\cdots$	3

图 7-20 从 *3SAT* 到 *SUBSET-SUM* 的归约
Fig. 7-20 Reduction from *3SAT* to *SUBSET-SU*

The left part consists of 1 and the following $l-i$ 0s. The right part corresponds to each clause with a number. When the clause c_j contains the literal x_i, y_i is in c_j and the column is 1; when the clause c_j contains the literal $\overline{x_i}$, z_i is in c_j and the column is 1. Bits that are not designated as 1 are all 0s.

Fig. 7-20 fills in part of the sample sentences c_1, c_2, $\cdots$, c_k:

$$(x_1 \vee \overline{x_2} \vee x_3) \wedge (x_2 \vee x_3 \vee \ldots) \wedge \ldots \wedge (\overline{x_3} \vee \ldots \vee \ldots)$$

In addition, S contains a pair of numbers g_j, h_j for each clause c_j. These two numbers are equal and consist of 1 and the following $k-j$ 0s.

Finally, the target number t (that is, the bottom row of the table) consists of l 1s followed by k 3s.

Next, explain why this structure can meet the requirements, and prove that φ can be satisfied if and only if a certain subset of S adds up to t.

Assuming that φ can be satisfied, a subset of S is constructed as follows. In the satisfaction assignment, if x_i is assigned TRUE, then y_i is selected; if x_i is assigned FALSE, then z_i is selected. If you add up the selected numbers, each bit of the first l is 1, because y_i or z_i is selected for each i. Moreover, each of the last k bits is between 1 and 3, because each clause is satisfied, so it contains 1 to 3 true literals. Further select enough g and h so that each of the last k bits is added to 3, so as to reach the target value.

Let a subset of S add up to t. After paying attention, construct an φ satisfying assignment. First, all bits of the members in S are 0 or 1. Second, each column describing S in the table contains at most five 1s. Therefore, when a certain subset of S is added, there will be no carry to the next column. In order to get 1 in each of the first l columns, the subset must contain either y_i or z_i for each i, but not both.

Now the construction satisfies the assignment. If the subset contains y_i, assign z_i to TRUE, otherwise assign it to FALSE. This assignment must satisfy φ, because the sum of each of the last k columns is always 3. In the c_j column, g_j and h_j provide at most 2, so y_i or z_i in the subset must provide at least 1 in this column. If it is y_i, then x_i appears in c_j and is assigned the value TRUE, so c_j is satisfied. If it is z_i, then $\overline{x_i}$ appears in c_j and x_i is assigned FALSE, so c_j is also satisfied. Therefore φ is satisfied.

Finally, it must be ensured that the reduction can be completed in polynomial time. The size of the table is approximately $(k+l)^2$, and the content of each grid can be easily calculated from any φ. So the whole time is $O(n^2)$ simple steps.

现在构造满足赋值. 如果子集包含 y_i, 就赋 z_i 为 TRUE, 否则赋它为 FALSE. 该赋值一定满足 φ, 因为后 k 列的每一列之和总是 3. 在 c_j 列, g_j 和 h_j 最多提供 2, 所以子集中的 y_i 或者 z_i 在该列必须至少提供 1. 如果是 y_i, 那么 x_i 出现在 c_j 中, 而且赋值为 TRUE, 所以 c_j 被满足. 如果是 z_i, 那么 $\overline{x_i}$ 出现在 c_j 中, 而且 x_i 赋为 FALSE, 所以 c_j 也被满足. 因此 φ 被满足.

最后, 必须保证该归约可以在多项式时间内完成. 表的尺寸大约是 $(k+l)^2$, 每一格的内容都可以从任何 φ 中轻易地计算出来. 所以全部时间是 $O(n^2)$ 个简单步骤.

Exercise 7

1. Are the following items true or false?
   (a) $2n = O(n)$;
   (b) $n^2 = O(n)$;
   (c) $n^2 = O(n\log^2 n)$;
   (d) $n\log n = O(n^2)$;
   (e) $3^n = 2^{O(n)}$;
   (f) $2^{2^n} = O(2^{2^n})$.

2. Are the following items true or false?
   (a) $n = o(2n)$;   (b) $2n = o(n^2)$;
   (c) $2n = o(3^n)$;   (d) $1 = o(n)$;
   (e) $n = o(\log n)$;   (f) $1 = o(1/n)$.

3. Which of the following pairs of numbers are relatively prime? Write down the calculation process to reach the conclusion.
   (a) 1274 and 10505;
   (b) 7289 and 8029.

4. For the string $w = \text{baba}$ and the following CFG G, try to fill in the table described in the polynomial time algorithm for identifying context-free languages in Theorem 7.2.3:
   $S \to RT$
   $R \to TR \mid a$
   $T \to TR \mid b$

5. Is the following formula satisfiable?
   $(x \vee y) \wedge (x \vee \bar{y}) \wedge (\bar{x} \vee y) \wedge (\bar{x} \vee \bar{y})$

6. Prove that P is closed under union, join, and complement operations.

7. Prove that NP is closed under union and join operations.

8. Let $CONNECTED = \{<G> \mid G \text{ is a connected undirected graph}\}$. Prove that this language belongs to P.

9. The triangle in the undirected graph is a 3-clique. Prove $TRIANGLE \in P$, where $TRIANGLE = \{<G> \mid G \text{ contains a triangle}\}$.

10. Prove that ALL_{DFA} belongs to P.

11. Analyze the time complexity of your algorithm.

(a) Prove $EQ_{DFA} \in P$.

(b) For language A, if $A = A'$, then A is called star-closed. Give a polynomial time algorithm to test whether a DFA recognizes a star-closed language (Note: EQ_{NFA} belongs to P is not known).

12. If the nodes of graph G are reordered, G can become exactly the same as H, then G and H are said to be isomorphic. Let
$ISO = \{<G, H> \mid G$ and H are isomorphic graphs$\}$. Prove $ISO \in NP$.

13. If and only if $P = NP$, prove that P is closed under homomorphis.

14. In a directed graph, the in-degree of a node is the total number of incoming edges, and the out-degree is the total number of outgoing edges. Prove that the following problem is NP-complete. Given an undirected graph G and a subset C of G nodes, can G be converted into a directed graph by assigning directions to each edge of G and satisfy the in-degree or out-degree of the nodes belonging to C is 0, the in-degree of the node that does not belong to C is at least 1?

15. Let $f: \mathbf{N} \to \mathbf{N}$ be any function of $f(n) = o(n \log n)$. Prove that $TIME(f(n))$ contains only regular language.

16. If two Boolean formulas have the same set of variables and the same set of satisfying assignments (that is, they describe the same Boolean function), they are said to be equivalent. The smallest Boolean formula means that there is no equivalent formula shorter than it. Let MIN-FORMULA be the latest set of Boolean formulas. Prove: If $P = NP$, then MIN-FORMULA $\in P$.

17. For a cnf formula with m variables and c clauses, it is proved that an NFA with $O(cm)$ states can be constructed in polynomial time, and it accepts

10. 证明 ALL_{DFA} 属于 P.

11. 分析你的算法的时间复杂度.

(a) 证明 $EQ_{DFA} \in P$.

(b) 对语言 A, 如果 $A = A'$, 则称 A 是星闭的. 给出测试一个 DFA 是否识别一个星闭的语言的多项式时间算法.(注意:并未知晓 EQ_{NFA} 属于 P).

12. 若图 G 的结点重新排序后, G 可以变得与 H 完全相同, 则称 G 与 H 是同构的. 令
$ISO = \{<G, H> \mid G$ 和 H 是同构的图$\}$. 证明 $ISO \in NP$.

13. 证明 P 在同态下封闭当且仅当 $P = NP$.

14. 在有向图中, 一个结点的入度为所有射入边的总数, 出度为所有射出边的总数. 证明如下问题是 NP 完全的. 给定一个无向图 G 和一个 G 结点的子集 C, 是否可以通过给 G 的每条边赋予方向, 将 G 转换为一个有向图并且满足属于 C 的结点的入度或出度为 0, 不属于 C 的结点的入度至少为 1?

15. 令 $f: \mathbf{N} \to \mathbf{N}$ 为任意 $f(n) = o(n \log n)$ 的函数. 证明 $TIME(f(n))$ 中只含有正则语言.

16. 如果两个布尔公式有相同的变量集且有同样的满足赋值集(就是说它们描述了同样的布尔功能), 那么称它们等价. 最小布尔公式是指没有比它更短的等价公式. 令 MIN-FORMULA 是最新布尔公式集. 证明: 如果 $P = NP$, 则 MIN-FORMULA $\in P$.

17. 对于一个含有 m 个变量及 c 个子句的 cnf 公式, 证明可在多项式时间内构造一个有 $O(cm)$ 个状态的 NFA,

all unsatisfied assignments denoted by a Boolean string of length m. It is concluded that the problem of minimizing NFA cannot be solved in polynomial time, unless $P = NP$.

18. The $2cnf$ formula is the AND of clauses, where each clause is the OR of at most two literals. Let $2SAT = \{<\varphi> | \varphi$ is a satisfiable $2cnf$ formula$\}$. Prove $2SAT \in P$.

19. Let $SET\text{-}SPLITTING = \{<S, C> | S$ is a finite set, $C = \{C_1, \ldots, C_k\}$ is a set composed of certain subsets of S, $k > 0$, so that the elements of S can be dyed red or blue, and for all C_i, the elements in C_i will not be dyed the same color. Prove that $SET\text{-}SPLITTING$ is NP-complete.

20. Let G denote an undirected graph, let $SPATH = \{<G, a, b, k> | G$ contains a simple path from a to b with a length of at most $k\}$, as well as $LPATH = \{<G, a, b, k> | G$ contains a simple path from a to b with a length of at least $k\}$

（a）Prove $SPATH \in P$.
（b）Prove that $LPATH$ is NP-complete.

习题7答案
Key to Exercise 7

Chapter 8 Space Complexity

8.1 Savitch's theorem

Theorem 8.1 Any non-deterministic Turing machine that consumes $f(n)$ space can be transformed into a truly deterministic Turing machine that consumes only $f^2(n)$ space. For any function f, where $f(n) \geq n$ $NSPACE(f(n)) \subseteq SPACE(f(n))$

The theorem shows that a deterministic machine can simulate a non-deterministic machine with very little space. For time complexity, this simulation seems to require an exponential increase in time.

The space complexity of an algorithm is $S(n)$ when the size of the current problem increases from 1 to $S(n)$ in a certain unit, and the storage space used by the algorithm to solve the problem also increases from 1 to $S(n)$ in a certain unit.

Measures of space complexity:

Example 8.1.1 A function that evaluates an expression

```
float abc(float a, float b, float c){
    return a+b+b*c+(a+b-c)/(a+b);
}
```

In this function, the problem size of the function is determined by a, b and c, and a, b and c each occupies a unit of storage space, so the storage space required by this function during operation is a constant.

Example 8.1.2 An iteration that adds up the first n terms.

```
float sum(float a[], const int n){
    float s = 0.0;
    for(int i=0; i<n; i++)
        s+=a[i];
    return s;
}
```

In Example 8.1.2, the algorithm has a problem size of n. The program uses an integer n to store the number of accumulated items, uses a floating point number s as the storage space to store the accumulated value, and for array $a[\ ]$, only one space unit is used to store the address of its first element $a[0]$. Therefore, the required storage space of the algorithm is also a constant.

Example 8.1.3 A recursion that adds up the first n terms.

```
float rsum(float a[], const int n){
    if(n<=0)
        return 0
    else
        return rsum(a, n-1)+a[n-1];
}
```

The algorithm is recursive, and the problem size is also n. A recursive work stack is used to implement the recursion, adding one work record to the recursion for each recursion level.

In the work stack, the work records are the formal parameters ($a[0]$ and n for the first address of $a[\ ]$), the return value of the function, and the return address, and four storage units are reserved. Since the recursion depth of the algorithm is $n+1$, the required stack space is $4(n+1)$.

8.2 Class *PSPACE*

8.2.1 What is *P*, *NP* and *PSPACE*?

P: For polynomial time, all problems can be easily solved with classic computers. The algorithm in class *P* must stop and give the correct answer in n^c time, where n is the size of the input and c is a constant.

NP: Non-deterministic polynomial time can use the classic computer to quickly verify the answer to all the questions. A problem is an *NP* problem if an answer to a certain question is given and there is a short proof of the correctness of the answer. If you enter a string X and you need to verify that whether the answer is "YES", then the short proof above refers to another string Y that can be used to verify that whether the answer is "YES" in polynomial time.

PSPACE: Polynomial space, *PSPACE* contains all the problems that can be solved with reasonable memory. In the class *PSPACE* problem, you don't care about the time, you only care about the memory space required for an algorithm.

Computer scientists have shown that *PSPACE* contains *NP*, and *NP* also contains the *P* class. That is, all problems in *P* and *NP* belong to the *PSPACE* class.

8.2.2 *NP*-complete problem

NP-complete problem (*NP-C* problem), that is, the non-deterministic problem of polynomial complexity, is one of the Millennium Prize Problems. The easy way to write it is "*NP*=*P*?". The question lies in the "?", whether *NP* equals *P* or *NP* does not equal *P*.

Specifically, the complexity of certain problems in *NP* is associated with the complexity of the class as a whole. All *NP* problems are solvable in polynomial time if polynomial-time algorithms exist for any of these problems. These problems are called *NP*-complete problems (Fig. 8-1).

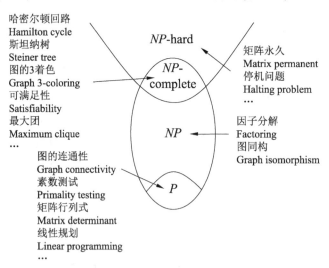

Fig. 8-1 Relationship between *P*, *NP* and *NP-C*

Example 8.2.1 It's a saturday night and you attend a big party. Because feeling embarrassed, you wonder if there is anyone in the hall you have already known. The host suggests that you must know Rose, the lady in the corner near the dessert plate. In less than a second you can scan the place and find that the host is right. However, if there is no such hint, you have to look around the hall, one by one, to see if there is anyone you know.

Generating a solution to a problem usually takes much more time than verifying a given solution. This is an example of this general phenomenon. Similarly, if someone tells you that the number 13717421 can be written as the product of two smaller numbers, you might not know whether to believe them, but if he tells you 13717421 can be factored into 3607 times 3803, then you can easily verify this with a pocket calculator. It is found that all completely polynomial non-deterministic problems can be transformed into a class of logical operation problems called satisfiability problems. And since all possible answers to these kinds of problems can be computed in polynomial time, people wonder is there a deterministic algorithm for these kinds of problems that can directly compute or search for the right answer in polynomial time? That's the famous "$NP=P?$" conjecture. Determining an answer can be quickly verified by using internal knowledge, regardless of our programming dexterity. But without such hints, it takes a lot of time to solve, which is seen as one of the most prominent problems in logic and computer science.

8.3 *PSPACE*-completeness

Definition 8.3.1 A polynomial space is said to be complete, if a language A recognized by the polynomial space also satisfies the condition that any language X recognized by the polynomial space is logarithmic space many-one reducible to A. The deterministic problem *QBF* of quantifier Boolean formula is a *PSPACE* complete problem under logarithmic space many-one reduction.

The *PSPACE* complete problem is the most difficult type of *PSPACE* problem. That is, a problem A can be called a *PSPACE* complete problem if A satisfies that, for any other *PSPACE* problem B, there exists a polynomial time method of converting an instance of B into an instance of A, such that the two instances have the same answer (both "yes" or both "no") in both questions. From this definition, this algorithm can be used to solve all *PSPACE* problems as long as it can give a fast algorithm for a *PSPACE* complete problem.

Example 8.3.1 True quantified boolean formula (TQBF) problem: An example of this problem is a Boolean logic formula with quantifiers (existent, arbitrary) for each variable. Each variable in the formula can only be "true" or "false". For example: (any x) (exists y)(exists z)((x or y) and z), we need to determine whether such a formula is always "true".

Example 8.3.2 LBA problem, namely linear bounded automata problem. An example of this problem is: given a linear bounded automaton and an input, whether the automaton accepts the input or not. The *PSPACE*-completeness proof of this problem is very simple.

Therefore, there is little difference between using this problem to prove that the pushing box problem is a *PSPACE* complete problem and directly proving that the pushing box problem is a *PSPACE* complete problem by definition.

8.4 Class L and NL

So far, we have only considered the case where the time and space complexity are at least linear, that is, the bound $f(n)$ is at least n. Now consider the smaller sub-linear space limit. In sublinear space complexity,

空间完全的. 量词布尔公式的判定性问题 QBF 在对数空间多一归约下就是一个 PSPACE 完全问题.

PSPACE 完全问题是 PSPACE 问题中最困难的一类. 即一个问题 A 能够被称为 PSPACE 完全问题, 如果问题 A 满足: 对任何其他的 PSPACE 问题 B, 都存在一个把 B 的实例转化成 A 的实例的多项式时间的方法, 使得这两个实例在这两个问题中有相同的答案(都为"是"或者都为"否"). 由此定义, 只要能给出某个 PSPACE 完全问题的快速算法, 那么这个算法就可以用来解决所有的 PSPACE 问题.

例 8.3.1 TQBF 问题. 这个问题的一个实例就是一个每个变量都带有量词(存在、任意)的布尔逻辑公式. 公式中的每个变量只能取值为"真"或者"假". 如: (任意 x)(存在 y)(存在 z)((x 或者 y)并且 z), 我们需要判断这样的一个公式是否总是"真"的.

例 8.3.2 LBA 问题, 即线性有界自动机问题. 这个问题的一个实例是: 给出一个线性有界自动机和一个输入, 该自动机是否接受该输入. 这个问题的 PSPACE 完全性的证明非常简单.

因此, 利用此问题证明推箱子问题是 PSPACE 完全问题, 和直接用定义证明推箱子问题是 PSPACE 完全问题, 几乎没有多大区别.

8.4 L 类和 NL 类

到目前为止, 我们只考虑了时间和空间复杂性至少是线性的情况, 即界限 $f(n)$ 至少是 n. 现在考虑更小的亚线性空间界限. 在亚线性空间复杂度

the machine can read the entire input, but it does not have enough space to store the input. In order to make the consideration of this situation meaningful, the calculation model must be modified.

Introduce a Turing machine with two tapes: a read-only input tape and a read-write working tape. The input head on the read-only tape can read symbols, but cannot change them. This head must rest on the part of the tape that contains the input. Provide a path to the machine so that it can detect when the read-write head is at the left and right ends of the input. The working tape can be read and written in the usual way. Only the scanned units on the working tape constitute the space complexity of this form of Turing machine.

Think of the read-only input tape as a CD-ROM, a device used for input on many personal computers. Usually, a CD-ROM contains more data than the computer can store in the main memory. The sub-linear space algorithm allows the computer to process data that is not all stored in the main memory.

For at least linear space boundaries, the dual-tape TM model is equivalent to the standard single-tape model. For the sub-linear space boundary, only the dual-tape model is used.

Definition 8.4.1 L is the decidable language class of the deterministic Turing machine in logarithmic space. In other words,

$$L = SPACE(\log n)$$

NL is the decidable language class of the non-deterministic Turing machine in logarithmic space. In other words,

$$NL = NSPACE(\log n)$$

We pay attention to $\log n$ space instead of $\sqrt{n}$ or $\log^2 n$ space. The reason is similar to the reason for choosing polynomial space-time boundary. The logarithmic space is sufficient to solve many interesting computational problems, and its mathematical properties are also attractive, such as maintaining robustness when the machine model and the coding method of input are changed. The pointer to the input can be denoted in logarithmic space, so one way to consider the computing power of the logarithmic space algorithm is to consider the computing power of a fixed number of input pointers.

Example 8.4.1 Language $A = \{0^k 1^k \mid k \geq 0\}$ is a member of L. In Section 8.1, a Turing machine that determines A is described. It scans the input back and forth and deletes the matched 0 and 1. The algorithm uses linear space to record which positions have been deleted, but it can be modified to only use logarithmic space.

It is determined that the logarithmic space TM of A cannot delete the matched 0 and 1 on the input tape, because the tape is read-only. The machine instead counts the number of 0s and 1s in binary system on the working tape. The only space required is to record these two counters. In the form of binary system, each counter consumes only logarithmic space, so the algorithm runs in $O(\log n)$ space. So $A \in L$.

Definition 8.4.1 If M is a Turing machine with a separated read-only input tape, and w is an input, then the configuration of M on w is a setting of state, working tape, and two read-write head positions. The input of w is not part of the configuration of M on w.

8.5 *NL*-completeness

Similar to the question of whether $P = NP$ is established, there is also the question of whether $L = NL$ is established. As a step to solve the problem of L and NL, prove that certain languages are NL-complete. Just like other complete languages of complexity, NL-complete languages are, in a certain sense, examples of the most difficult language in NL. If L and NL are not equal, then all NL-complete languages do not belong to L.

入的编码方法改变时保持稳健性. 指向输入的指针可以在对数空间内表示, 所以考虑对数空间算法的计算能力的一种方式是考虑固定数目的输入指针的计算能力.

例 8.4.1 语言 $A = \{0^k 1^k \mid k \geq 0\}$ 是 L 的成员. 在 8.1 节中描述了一个判定 A 的图灵机, 它来回扫描输入, 删掉匹配的 0 和 1. 该算法用线性空间记录哪些位置已经被删掉了, 但是它可以修改为只使用对数空间.

判定 A 的对数空间 TM 不能删除输入带上已经匹配的 0 和 1, 因为该带是只读的. 机器转而在工作带上用二进制分别计算 0 和 1 的数量. 唯一需要的空间是用来记录这两个计数器的. 以二进制形式, 每个计数器只消耗对数空间, 因此算法在 $O(\log n)$ 空间内运行. 所以 $A \in L$.

定义 8.4.1 若 M 是一个有单独的只读输入带的图灵机, w 是输入, 则 M 在 w 上的格局是状态、工作带和两个读写头位置的一种设置. 输入 w 不作为 M 在 w 上的格局的一部分.

8.5 *NL* 完全性

类似于 $P = NP$ 是否成立的问题, 也有 $L = NL$ 是否成立的问题. 作为解决 L 与 NL 问题的一个步骤, 证明某些语言是 NL 完全的. 正如其他复杂性类的完全语言一样, NL 完全语言在一定意义上是 NL 中最困难的语言的案例. 如果 L 与 NL 不相等, 那么所有 NL 完全语言就不属于 L.

Like the previous definition of completeness, *NL*-complete language is defined as belonging to *NL*, and all other languages in *NL* can be reduced to it. But polynomial time reducibility is not used here, because, as we will see, all the problems in *NL* are solvable in polynomial time. Therefore, except for φ and Σ^*, any two problems in *NL* are mutually polynomial time reducible. Therefore, the polynomial time reducibility is too strong to distinguish the problems in *NL* from each other. Use a new type of reducibility, called logarithmic space reducibility.

Definition 8.5.1 **Logarithmic space converter** is a Turing machine with a read-only input tape, a write-only output tape, and a read-write working tape. The working tape can contain $O(\log n)$ symbols. The logarithmic space converter M calculates a function $f: \Sigma^* \to \Sigma^*$, where $f(w)$ is the string stored on the output tape when w is placed on the input tape of M from start to stop of M. We call f a **calculable function in logarithmic space**. If language A can be reduced to language B through the logarithmic space computable function f mapping, then we call A is **logarithmic space reducible** to B, denoted as $A \leq_L B$.

Definition 8.5.2 Language B is **NL-complete**, if

(1) $B \in NL$.

(2) Each A in *NL* is logarithmic space reducible to B.

If a language is logarithmic space reducible to another language known to belong to L, then this language also belongs to L, as explained by the following theorem.

Theorem 8.5.1 If $A \leq_L B$ and $B \in L$, then $A \in L$.

Corollary 8.5.1 If there is an *NL*-complete language belonging to L, then $L = NL$.

Theorem 8.5.2 *PATH* is *NL*-complete.

Corollary 8.5.2 $NL \subseteq P$.

8.6 NL equals coNL

This section contains one of the most surprising results of the known results regarding the interrelationships between complexity classes. It is generally believed that *NP* and *coNP* are not equal. At first glance, the same result seems to hold for *NL* and *coNL*. In fact, as will be proved, *NL* is equal to *coNL*. This shows that there are still many gaps in our intuition about computing.

Theorem 8.6.1 *NL = coNL*.

Proof idea In order to prove that every problem in *coNL* is also in *NL*, first prove that *PATH* belongs to *NL*, because *PATH* is *NL*-complete. The given *NL* algorithm *M* for determining the *PATH* must have an acceptance calculation when the graph *G* does not contain the path from *s* to *t*.

Solve an easier problem first. Let *c* be the number of nodes in *G* reachable from *s*. Assuming that *c* is provided as input to *M*, first explain how to use *c* to solve *PATH*, and then explain how to calculate *c*.

Given *G*, *s*, *t*, and *c*, the machine *M* operates as follows. *M* walks through all the nodes of *G* one by one, guessing indefinitely whether each node is reachable from *s*. Once node *u* is guessed to be reachable, *M* verifies this by guessing a path from *s* to *u*. If a calculation branch fails to verify this guess within *m* steps, and *m* is the number of nodes in *G*, it is rejected. In addition, if a branch guesses that *t* is reachable, it is rejected. Machine *M* counts the number of nodes that have been verified as reachable. When a branch traverses all the nodes of *G*, it checks whether the number of nodes reachable from *s* is equal to *c*, that is, the number of nodes that are actually reachable, and rejects it if they are not equal. Otherwise, the branch accepts it.

In other words, if *M* uncertain chooses exactly *c* nodes reachable from *s*, excluding *t*, and verifies that each of them is reachable from s by guessing the path, then M knows that the remaining nodes, including t, are all unreachable, so it is acceptable.

Next, explain how to calculate c, that is, the number of nodes reachable from s. Describe a nondeterministic logarithmic space process, it has at least one calculation branch with the correct c value, and all other branches reject it.

For each value i from 0 to m, define A_i as the set of nodes in G whose distance from s does not exceed i (that is, there is a path from s whose length does not exceed i). So $A_0 = \{s\}$, each $A_i \subseteq A_{i+1}$, A_m contains all nodes reachable from s. Let c_i be the number of nodes in A_i. The following describes a process to calculate c_{i+1} from c_i. Repeated application of this process will obtain the desired value $c = c_m$.

Calculating c_{i+1} from c_i, the idea used is similar to that given in the previous proof idea. The algorithm goes through all the nodes of G, determines whether each node is a member of A_{i+1}, and then counts the number of members.

In order to determine whether the node v is in A_{i+1}, an inner loop is used to traverse all the nodes of G and guess whether each node is in A_i. Every successful guess is confirmed by guessing a path that starts from s, whose length is at most i. For each node u verified to be in A_i, the algorithm checks whether (u, v) is an edge of G. If it is an edge, then v is in A_{i+1}. In addition, it is confirmed that the number of nodes in A_i is also calculated. At the end of the inner loop, if it is confirmed that the total number of nodes belonging to A_i is not equal to c_i, then all the nodes of A_i have not been found, so the calculation branch is rejected. If the total is indeed equal to c_i, and v has not been confirmed to belong to A_{i+1}, then it can be concluded that it is not in A_{i+1}. Then go to the next v and start the outer loop.

Proof The algorithm for determining *PATH* is as follows.

$M=$ "For input $<G, s, t>$"

(1) Let $c_0 = 1$

(2) For $i = 0$ to $m-1$:

(3) Let $c_{i+1} = 0$

(4) Let $d = 0$

(5) For each node v in G:

(6) For each node u in G:

(7) Perform or skip the following steps non-deterministically:

(8) Non-deterministically follow the path starting from s, whose length is i, if it does not encounter node u, just reject.

(9) d plus 1.

(10) If (u, v) is an edge of G, add 1 to c_{i+1}. v becomes the next node, go to step (6).

(11) If $d \neq c_j$, reject.

(12) Let $d = 0$

(13) For each node u in G:

(14) Perform or skip the following steps non-deterministically:

(15) Non-deterministically follow the path starting from s whose length is m. If the node u is not encountered, just reject.

(16) If $u = t$, reject it.

(17) d plus 1.

(18) If $d \neq c_m$, reject; otherwise, accept.

At any time, this algorithm only needs to store $m, u, v, c_i, c_{i+1}, d, i$ and a pointer to the end of the path, so it runs in logarithmic space.

Summarize the current known knowledge about the interrelationship between several complexity classes as follows:

$$L \subseteq NL = coNL \subseteq P \subseteq PSPACE$$

Exercise 8

1. Prove that for any function $f: \mathbf{N} \to \mathbf{N}_+$, where $f(n) \geq n$, no matter if you use a single-tape TM model or a two-tape read-only input TM model, the defined space complexity class $SPACE(f(n))$ is always the same.

2. Consider the following standard children's game chessboard (Fig. 8-2). Assume it is the × player's turn to go in the next step. Please describe the player's winning strategy. (Recall that the winning strategy is not only the best move in the current game, it also includes all the moves that the player must take in order to win, no matter how the opponent moves)

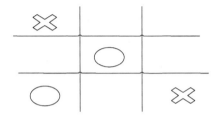

Fig. 8-2 Chess game

3. In the generalized geography game (Fig. 8-3) below, the starting one is the one pointed to by a directed passive arrow without an origin. There are two players, and each player can only move one step at a time, and cannot violate the direction of the arrows. Is there a winning strategy for player Ⅰ? What about player Ⅱ? Give reasons.

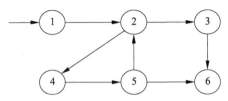

Fig. 8-3 Generalized geography game diagram

4. Prove that PSPACE is closed under union,

complement, and asterisk operations.

5. Prove that *NL* is closed under union, intersection, and asterisk operations.

6. Prove that *PSPACE*-hard languages are also *NP*-hard.

7. Prove $A_{\text{DFA}} \in L$.

8. Prove that *ANA* is *NL*-complete.

9. Prove that if every *NP*-hard language is also *PSPACE*-hard, then *PSPACE* = *NP*.

5. 证明 *NL* 在并、交和星号运算下封闭.

6. 证明 *PSPACE* 难的语言也是 *NP* 难的.

7. 证明 $A_{\text{DFA}} \in L$.

8. 证明 *ANA* 是 *NL* 完全的.

9. 证明如果每一个 *NP* 难的语言也是 *PSPACE* 难的,那么 *PSPACE* = *NP*.

习题 8 答案
Key to Exercise 8

Part V Algorithm Design and Analysis

Chapter 9 Divide and Conquer Strategy

9.1 Basic idea of divide and conquer

The computational time required for any problem that can be solved by a computer depends on its size. The smaller the size of the problem, the easier it is to solve directly and the less computation time it takes to solve the problem. For example, for the sorting problem of n elements, when $n=1$, no calculation is required. When $n=2$, it becomes a little more difficult to deal with. **Divide and conquer strategies** are needed when size n becomes too large to solve directly. The idea behind divide and conquer is to divide a big problem that is difficult to solve directly into some smaller problems so that they can be broken down individually. As shown in Fig. 9-1, the original problem of size n is decomposed into k ($1<k\leqslant n$) smaller problems. These subproblems are independent of each other and have the same form as the original problem. These problems are solved recursively, and then the solution of the original problem is obtained by combining the solutions of each subproblem.

第 V 部分 算法设计与分析

第 9 章 分治策略

9.1 分治法的基本思想

任何一个可以用计算机求解的问题所需的计算时间都与其规模有关. 问题的规模越小越容易直接求解, 解题所需的计算时间也越少. 例如, 对于 n 个元素的排序问题, 当 $n=1$ 时, 无须任何计算. 当 $n=2$ 时, 问题就变得不容易处理了. 当规模 n 大到使问题变得难以直接解决时, 需要采用**分治策略**. 分治法的设计思想就在于将一个难以直接解决的大问题, 分割成一些规模较小的问题, 以便各个击破, 分而治之. 如图 9-1, 将规模为 n 的原问题分解成 $k(1<k\leqslant n)$ 个规模较小的问题, 这些子问题互相独立且与原问题形式相同. 通过递归地解这些问题, 将各子问题的解合并得到原问题的解.

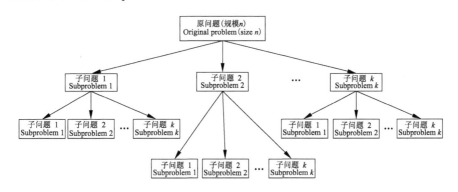

图 9-1 原问题分解

Fig. 9-1 Decomposition of original problem

Among them, if the decomposed k subproblem is still not small enough, the k subproblem is divided again, and the recursion continues until the size of the problem is small enough to find its solution easily (Fig. 9-1). After that, the solution of the solved small-scale problem is merged into the solution of a larger scale problem, and the solution of the original problem is gradually solved from the bottom up.

In short, the divide and conquer method consists of three steps at each level of recursion:

(1) **Divide**: Decompose the original problem into several smaller, independent and structurally identical subproblems.

(2) **Conquer**: If the subproblem is small, it can be solved directly; otherwise, recursively solve the subproblems.

(3) **Merge**: Merge the solutions of each subproblem into the solution of the original problem.

Of course, if you use divide and conquer to solve a problem, the problem needs to have the following four characteristics:

(1) The problem can be easily solved when the scale is reduced to a certain extent.

(2) The problem can be decomposed into several smaller identical problems, that is, the problem has the property of optimal substructure.

(3) The solutions of subproblems decomposed by this problem can be merged into the solutions of this problem.

(4) Each subproblem decomposed by this problem is independent of each other, that is, there is no common subproblem among the subproblems.

The first characteristic above is that most problems can be satisfied, because the computational complexity of problems generally increases with the increase of problem size. The second characteristic is the prerequisite for the application of divide and conquer, which can be satisfied by most problems. The

其中,如果分解后的 k 个子问题的规模仍然不够小,则再划分 k 个子问题,如此递归地进行下去,直到问题的规模足够小,很容易求出其解为止(图9-1).之后,将求出的小规模的问题的解合并为一个更大规模的问题的解,自下向上逐步求出原问题的解.

简而言之,分治法在每一层递归上由三个步骤组成:

(1) **划分**:将原问题分解为若干个规模较小、相互独立、与原问题形式相对的子问题.

(2) **解决**:若子问题较小,则直接求解;否则递归求解各个子问题.

(3) **合并**:将各个子问题的解合并为原问题的解.

当然,如果使用分治法解决问题,则该问题需要具备以下四个特征:

(1) 该问题的规模缩小到一定的程度就可以轻易地解决.

(2) 该问题可以分解为若干个规模较小的相同问题,即该问题具有最优子结构性质.

(3) 由该问题分解出的子问题的解可以合并为该问题的解.

(4) 该问题所分解出的各个子问题是相互独立的,即子问题之间不包含公共的子问题.

上述第一个特征是绝大多数问题都可以被满足,因为问题的计算复杂性一般是随着问题规模的增加而增加.第二个特征是应用分治法的前提,它可以被大多数问题所满足,此特征反映了递归思想的应用.第三个特征是

second characteristic reflects the application of recursion. The third characteristic is the key, and whether or not you can use divide and conquer depends entirely on whether or not the problem has the third characteristic, and if you have the first and the second character, but not the third, then you can consider the **greedy algorithm** or the **dynamic programming algorithm**. The fourth characteristic relates to the efficiency of the divide and conquer. If each subproblem is not independent, the divide-and-conquer method will do a lot of unnecessary work to repeatedly solve the common subproblem. At this time, although the divide and conquer can be used, the dynamic programming method is generally preferred.

The general algorithm design paradigm of divide and conquer is as follows:

Algorithm 9-1:
```
divide_and_conquer(P){
    if(|P|<=c)
        return dsolve(P);
    else
        divide P into k into P_1, P_2, ... , P_k subproblems;

    for(i=1; i<k; i++)
        s_i = divide_and_conquer(P_i);
    S = combine(s_1, s_2, ... , s_k);
    return S;
}
```

Among them, $|P|$ denotes the scale of problem P, c is a threshold value, when the scale of problem P is less than c, the problem is easy to work out and we don't have to continue to break down. $dsolve(P)$ is the basic sub-algorithm of divide and conquer, and is used to directly solve the small scale problem P. When the scale of P does not exceed c, use $dsolve(P)$ to solve it directly. The algorithm $combine(s_1, s_2, ... , s_k)$ is the combined subalgorithm of the divide and conquer, which is used to merge the solutions $s_1, s_2, ... , s_k$ of the subproblems $P_1, P_2, ... , P_k$ of P into the solutions of P.

关键,能否利用分治法完全取决于问题是否具有第三个特征,如果具备了第一个和第二个特征,而不具备第三个特征,则可以考虑**贪心算法**或**动态规划算法**.第四个特征涉及分治法的效率,如果各子问题是不独立的,则分治法要做许多不必要的工作,重复地解公共的子问题,此时虽然可用分治法,但一般选择使用动态规划法较好.

分治法的一般算法设计范型如下:

算法 9-1:
```
divide_and_conquer(P){
    if(|P|<=c)
        return dsolve(P);
    else
        divide P into k into P_1, P_2, ···, P_k subproblems;

    for(i=1; i<k; i++)
        s_i = divide_and_conquer(P_i);
    S = combine(s_1, s_2, ···, s_k);
    return S;
}
```

其中,$|P|$表示问题P的规模,c为一个阈值,表示当问题P的规模小于c时,问题容易解出,不必再继续分解. $dsolve(P)$是该分治法的基本子算法,用于直接求解小规模的问题P.当P的规模不超过c时,直接使用$dsolve(P)$求解.算法$combine(s_1, s_2, ···, s_k)$是该分治法的合并子算法,用于将$P$的子问题$P_1, P_2, ···, P_k$的解$s_1, s_2, ···, s_k$合并为$P$的解.

It can be seen from the general design pattern of divide and conquer that the algorithm directly designed by it is a **recursive algorithm**. We can use a recursive equation to describe the running time of a recursive algorithm. Let $T(n)$ denote the computational time required to solve a problem of size n using divide and conquer. If the problem size is small enough, such as $n \leq c$, then the problem can be solved directly and $T(n) = O(1)$. Assuming that the original problem is divided into k subproblems, and the size of each subproblem is $1/m$ of the original problem, if the time to divide and merge the problem is $D(n)$ and $C(n)$ respectively, then the calculation time $T(n)$ of the algorithm can be denoted as the following recursive equation:

$$T(n) = \begin{cases} O(1), & n \leq c \\ kT(n/m) + D(n) + C(n), & n > c \end{cases}$$

If n is a power of m and the time to divide and merge the problem is $f(n)$, then the solution of the recursive equation is

$$T(n) = n^{\log_m k} + \sum_{i=0}^{\log_m n - 1} k^i f(n/m^i)$$

9.2 Binary search

Binary search is a typical application of divide and conquer to solve problems.

Given $a[0:n-1]$ of n elements that have been sorted in ascending order, we now want to find a particular element x out of these n elements.

First, if $n = 1$, that is, there is only one element, then simply comparing this element with x can determine whether x is in the table. Therefore, this problem satisfies the first condition of divide and conquer, that is, the problem can be easily solved when the size of the problem is reduced to a certain degree. If n is greater than 1, compare x to the middle element $a[\text{mid}]$ of array a:

(1) If $x = a[\text{mid}]$, the position of x in array a is mid;

(2) If $x < a[\text{mid}]$, since a is in ascending order, then x must rank before $a[\text{mid}]$ if x is in a, so we only need to find x before $a[\text{mid}]$.

Finding x before or after $a[\text{mid}]$ is done in the same way as finding x in a, but on a smaller scale. Finally, it is clear that the subproblems decomposed by this problem are independent of each other, that is, finding x before or after $a[\text{mid}]$ is an independent subproblem, thus satisfying the fourth applicable condition of the divide and conquer.

We want to find elements with the same keyword value as a particular element x in an ordered table of length $n(a_0, a_1, \ldots, a_{n-1})$ sorted by keyword value in non-decreasing order. The search results can return the entire data element or indicate the position of the element in the table.

If the number of elements in the table is $n = 0$, obviously the search fails; if $n > 0$, the ordered table can be decomposed into several subtables (the simplest decomposition into two subtables, that is, binary search of the ordered table). Assuming the element a_m is the split point, there are three possible results of a_m and x comparison:

(1) If $x < a_m$, the element whose keyword value is the same as x must be in the subtable $(a_0, a_1, \ldots, a_{m-1})$;

(2) If $x = a_m$, search is successful;

(3) If $x > a_m$, the element whose keyword value is the same as x must be in the subtable $(a_{m+1}, a_{m+2}, \ldots, a_{n-1})$.

The binary search algorithm framework using divide and conquer to search ordered tables is as follows:

Algorithm 9-2:
template<class T>
int SortableList < T >: BSearch (const T& x, int left, int right) const
{

（2）如果$x<a[\text{mid}]$，由于a是递增排序的，因此假如x在a中，x必然排在$a[\text{mid}]$的前面，所以只要在$a[\text{mid}]$的前面查找x即可。

无论是在$a[\text{mid}]$前面还是后面查找x，其方法都和在a中查找x一样，只不过查找的规模缩小了。最后，很显然此问题分解出的子问题相互独立，即在$a[\text{mid}]$的前面或者后面查找x是独立的子问题，因此满足分治法的第四个适用条件。

要在按关键字值非减排序的长度为n的有序表$(a_0, a_1, \cdots, a_{n-1})$中找出与特定元素$x$有相同关键字值的元素，搜索结果可以返回整个数据元素，也可以指示该元素在表中的位置。

若表中元素的个数为$n = 0$，显然搜索失败；若$n > 0$，则可以将有序表分解为若干个子表（最简单的分解为两个子表，即为有序表的二分搜索）。假设元素a_m为分割点，a_m与x比较的结果有三种可能：

（1）若$x < a_m$，则关键字值与x相同的元素必在子表$(a_0, a_1, \cdots, a_{m-1})$中；

（2）若$x = a_m$，则搜索成功；

（3）若$x > a_m$，则关键字值与x相同的元素必在子表$(a_{m+1}, a_{m+2}, \cdots, a_{n-1})$中。

采用分治法搜索有序表的二分搜索算法框架如下：

算法9-2：
template<class T>
int SortableList<T>: BSearch(const T& x, int left, int right) const
{

```
    if (left <= right) {
        int m = Divide(left, right);//The Divide function finds the partition point m according to some rule
        if (x<a[m])
            return BSearch(x, left, m-1);//Search the left subtable
        else if (x>a[m])
            return BSearch(x, m+1, right);//Search the right subtable
        else
            return m;//Search success
    }
    return -1;//Search failed
}
```

Using different rules to find the partition point m, we can get different binary search methods, such as **half-and-half search**, **Fibonacci search**, etc.

Half-and-half search recursive algorithm:

Algorithm 9-3:

```
template<class T>
int SortableList<T>::BSearch(const T& x, int left, int right) const {
    if (left <= right) {
        int m = (left+right)/2; //Split in half: m = (left+right)/2
        if (x<a[m]) return BSearch(x, left, m-1); //Search the left subtable
        else if (x>a[m]) return BSearch(x, m+1, right); //Search the right subtable
        else return m;
    }
    return -1;
}
```

The half-and-half search algorithm divides the table into two subtables of nearly equal size (easy to prove its correctness by mathematical induction). The recursive call statement of a program is the last executable statement, which is context-free. This is tail recursion, so it can be converted to an iterative function.

Recursive functions tend to be less efficient, so the above programs can be converted to iterative algorithms to implement.

Half-and-half search iterative algorithm:

Algorithm 9-4:
```
template<class T>
int SortableList<T>::BSearch(const T& x) const
{
    int m, left=0, right=n-1;
    while (left <= right) {
        m=(left+right)/2;
        //Split in half
        if (x<a[m])           //Search the left subtable
            right=m-1;
        else if (x>a[m])      //Search the right subtable
            left=m+1;
        else                  //Search success
            return m;
    }
    return 1; //Search failed
}
```

It's easy to see that each time the algorithm's while loop is executed, the size of the array to be searched is halved. Thus, in the worst case, the while loop is executed $O(\log n)$ times. Each loop, the inner loop takes a constant time of $O(1)$. Therefore, the computational time complexity of the whole algorithm in the worst case is $O(\log n)$.

9.3 Matrix multiplication

Matrix multiplication is one of the most common problems in linear algebra, and it is widely used in numerical calculation. Let A and B be two n by n matrices, and their product AB is also an n by n matrix. Element $C[i][j]$ in the product matrix C of A and B is defined as $C[i][j] = \sum_{k=1}^{n} A[i][k]B[k][j]$

If the product matrix C of A and B is calculated according to this definition, each element $C[i][j]$ of

C needs to be calculated n times of multiplication and $n-1$ times of addition. Therefore, the computation time required to calculate n^2 elements of matrix C is $O(n^2)$. In the late 1960s, Strassen adopted a divide and conquer strategy similar to that used in large integer multiplication, and improved the calculation time required to calculate the product of two matrices of order n to $O(n^{\log 7}) = O(n^{2.81})$. The basic idea was to use the divide and conquer. First of all, let's assume that n is a power of two. Each of the matrices A, B, and C is divided into four equally sized submatrices, each of which is a square matrix of $(n/2)$ by $(n/2)$. The equation $C = AB$ can thus be rewritten as

$$\begin{bmatrix} C_{11} & C_{12} \\ C_{21} & C_{22} \end{bmatrix} = \begin{bmatrix} A_{11} & A_{12} \\ A_{21} & A_{22} \end{bmatrix} \begin{bmatrix} B_{11} & B_{12} \\ B_{21} & B_{22} \end{bmatrix}$$

and follows that

$$C_{11} = A_{11}B_{11} + A_{12}B_{21}$$
$$C_{12} = A_{11}B_{12} + A_{12}B_{22}$$
$$C_{21} = A_{21}B_{11} + A_{22}B_{21}$$
$$C_{22} = A_{21}B_{12} + A_{22}B_{22}$$

If $n = 2$, then the product of two square matrices of order 2 can be calculated directly, requiring 8 multiplications and 4 additions. When the order of the submatrix is greater than 2, in order to take the product of the 2 submatrices, you can continue to block the submatrix until the order of the submatrix is 2. The result is a **divide and conquer order reduction** recursive algorithm. According to this algorithm, computing the product of two square matrices of order n is transformed into computing the product of eight square matrices of order $n/2$ and the addition of four square matrices of order $n/2$. The addition of two $(n/2)$ by $(n/2)$ matrices can obviously be done in $O(n^2)$ time. Therefore, the calculation time $T(n)$ of the above divide and conquer, should satisfy:

$$T(n) = \begin{cases} O(1), & n = 2 \\ 8T(n/2) + O(n^2), & n > 2 \end{cases}$$

需要做 n 次乘法运算和 $n-1$ 次加法运算. 因此, 矩阵 C 的 n^2 个元素所需的计算时间为 $O(n^2)$. 20 世纪 60 年代末, Strassen 采用了类似于在大整数乘法中用过的分治策略, 将 2 个 n 阶矩阵乘积所需的计算时间改进到 $O(n^{\log 7}) = O(n^{2.81})$, 其基本思想还是分治法的思想. 首先, 仍假设 n 是 2 的幂. 将矩阵 A, B 和 C 中每一矩阵都分块成 4 个大小相等的子矩阵, 每个子矩阵都是 $(n/2) \times (n/2)$ 的方阵. 由此可将方程 $C = AB$ 重写为

$$\begin{bmatrix} C_{11} & C_{12} \\ C_{21} & C_{22} \end{bmatrix} = \begin{bmatrix} A_{11} & A_{12} \\ A_{21} & A_{22} \end{bmatrix} \begin{bmatrix} B_{11} & B_{12} \\ B_{21} & B_{22} \end{bmatrix}$$

由此可得

$$C_{11} = A_{11}B_{11} + A_{12}B_{21}$$
$$C_{12} = A_{11}B_{12} + A_{12}B_{22}$$
$$C_{21} = A_{21}B_{11} + A_{22}B_{21}$$
$$C_{22} = A_{21}B_{12} + A_{22}B_{22}$$

若 $n = 2$, 则 2 个二阶方阵的乘积可以直接计算出来, 共需 8 次乘法和 4 次加法. 当子矩阵的阶大于 2 时, 为求两个子矩阵的积, 可以继续将子矩阵分块, 直到子矩阵的阶降为 2. 由此产生**分治降阶**的递归算法. 依此算法, 计算 2 个 n 阶方阵的乘积转化为计算 8 个 $n/2$ 阶方阵的乘积和 4 个 $n/2$ 阶方阵的加法. 2 个 $(n/2) \times (n/2)$ 矩阵的加法显然可以在 $O(n^2)$ 时间内完成. 因此, 计算上述分治法的时间耗费 $T(n)$ 应满足:

$$T(n) = \begin{cases} O(1), & n = 2 \\ 8T(n/2) + O(n^2), & n > 2 \end{cases}$$

The solution to this recursive equation is still $T(n) = O(n^3)$. Therefore, this method is no more efficient than direct calculation with the original definition. The reason is that the method does not reduce the number of matrix multiplications. And matrix multiplication takes a lot more time than matrix plus or minus. In order to improve the computational time complexity of matrix multiplication, multiplication must be reduced.

According to the idea of divide and conquer above, it can be seen that to reduce the time of multiplication operations, the key lies in whether the multiplication operations less than 8 can be used when calculating the product of two square matrices of order 2. Strassen proposed a new algorithm to compute the product of two square matrices of order 2. His algorithm used only seven multiplications, but increases the number of operations of addition and subtraction. So the 7 multiplications are

$$M_1 = A_{11}(B_{12} - B_{22})$$
$$M_2 = (A_{11} + A_{12})B_{22}$$
$$M_3 = (A_{21} + A_{22})B_{11}$$
$$M_4 = A_{22}(B_{21} - B_{11})$$
$$M_5 = (A_{11} + A_{22})(B_{11} + B_{22})$$
$$M_6 = (A_{12} - A_{22})(B_{21} + B_{22})$$
$$M_7 = (A_{11} - A_{21})(B_{11} + B_{12})$$

After doing these 7 times of multiplication, plus a few more times of addition and subtraction, the following equations are obtained

$$C_{11} = M_5 + M_4 - M_2 + M_6$$
$$C_{12} = M_1 + M_2$$
$$C_{21} = M_3 + M_4$$
$$C_{22} = M_5 + M_1 - M_3 - M_7$$

The correctness of the above calculation is easy to verify.

In Strassen matrix multiplication, there are 7 recursive calls for $n/2$ order matrix multiplication and 18

这个递归方程的解仍然是 $T(n) = O(n^3)$. 因此,该方法并不比用原始定义直接计算更有效. 究其原因,乃是该方法并没有减少矩阵的乘法计算次数. 而矩阵乘法耗费的时间要比矩阵加(减)法耗费的时间多得多. 要想改进矩阵乘法的计算时间的复杂性,必须减少乘法运算.

按照上述分治法的思想可以看出,要想减少乘法运算次数,关键在于计算两个二阶方阵的乘积时,能否使用少于 8 次的乘法运算. Strassen 提出了一种新的算法来计算两个二阶方阵的乘积. 他的算法只用了 7 次乘法运算,但增加了加、减法的运算次数. 这 7 次乘法运算分别是

$$M_1 = A_{11}(B_{12} - B_{22})$$
$$M_2 = (A_{11} + A_{12})B_{22}$$
$$M_3 = (A_{21} + A_{22})B_{11}$$
$$M_4 = A_{22}(B_{21} - B_{11})$$
$$M_5 = (A_{11} + A_{22})(B_{11} + B_{22})$$
$$M_6 = (A_{12} - A_{22})(B_{21} + B_{22})$$
$$M_7 = (A_{11} - A_{21})(B_{11} + B_{12})$$

做了这 7 次乘法运算后,再做若干次加、减法运算就可以得到

$$C_{11} = M_5 + M_4 - M_2 + M_6$$
$$C_{12} = M_1 + M_2$$
$$C_{21} = M_3 + M_4$$
$$C_{22} = M_5 + M_1 - M_3 - M_7$$

以上计算的正确性很容易验证.

Strassen 矩阵乘法中,用了 7 次对于 $n/2$ 阶矩阵乘法的递归调用和 18 次

addition and subtraction operations for $n/2$ order matrix multiplication. It can be seen that the calculation time $T(n)$ required by the algorithm satisfies the following recursive equation:

$$T(n) = \begin{cases} O(1), & n=2 \\ 7T(n/2) + O(n^2), & n>2 \end{cases}$$

Solving this recursive equation yields $T(n) = O(n^{\log 7}) \approx O(n^{2.81})$. It can be seen that the computational time complexity of Strassen matrix multiplication is better than that of ordinary matrix multiplication.

Someone has listed 36 different ways to calculate the multiplication of two matrices of order 2×2, but all of them do at least 7 multiplications. The lower bound of the computation time for the matrix product is lower than $O(n^{2.81})$ unless an algorithm is found to compute the product of the second order square matrix in such a way that the number of multiplications is less than 7. However, Hopcroft and Kerr had shown in 1971 that 7 multiplications are necessary to calculate the product of two 2×2 matrices. Therefore, to further improve the time complexity of matrix multiplication, instead of relying on methods like calculating 7 multiplications of a 2×2 matrix, perhaps better algorithms for 3×3 or 5×5 matrices should be investigated.

Since Strassen, many algorithms have improved the computational time complexity of matrix multiplication. The current best upper bound for computing time is $O(n^{2.376})$. So far, the best lower bound of matrix multiplication is still its trivial lower bound $\Omega(n^2)$. Therefore, it is not possible to know exactly the time complexity of matrix multiplication until now.

9.4 Merge sort

Merge sort is a typical divide and conquer sorting algorithm, which is used to merge two or more ordered sequences into one ordered sequence. The most basic merge sort algorithm is a two-way merge sort algorithm, which can combine two ordered sequences into

$n/2$ 阶矩阵的加减运算. 由此可知,该算法所需的计算时间 $T(n)$ 满足如下递归方程：

$$T(n) = \begin{cases} O(1), & n=2 \\ 7T(n/2) + O(n^2), & n>2 \end{cases}$$

解此递归方程得 $T(n) = O(n^{\log 7}) \approx O(n^{2.81})$. 由此可见,Strassen 矩阵乘法的计算时间复杂性比普通矩阵乘法有较大改进.

有人曾列举了计算 2 个 2×2 阶矩阵乘法的 36 种不同方法,但所有的方法都至少做 7 次乘法. 除非能找到一种计算 2 阶方阵乘积的算法,使乘法的计算次数少于 7 次,计算矩阵乘积的时间下界才有可能低于 $O(n^{2.81})$. 但是,Hopcroft 和 Kerr 在 1971 年已经证明,计算 2 个 2×2 矩阵的乘积,7 次乘法是必需的. 因此,要想进一步改进矩阵乘法的时间复杂性,就不能再基于计算 2×2 矩阵的 7 次乘法这样的方法了,或许应当研究 3×3 或 5×5 矩阵的更好算法.

在 Strassen 之后,又有许多算法改进了矩阵乘法的计算时间复杂性. 目前最好的计算时间上界是 $O(n^{2.376})$. 而目前所知道的矩阵乘法的最好下界仍是它的平凡下界 $\Omega(n^2)$. 因此,到目前为止还无法确切知道矩阵乘法的时间复杂性.

9.4 合并排序

合并排序是一种典型的符合分治策略的排序算法,用于实现把两个或多个有序序列合并成一个有序序列. 其中最基本的合并排序算法为两路合并排序算法,实现将两个有序序列合并

an ordered sequence. The following will introduce the implementation steps of the two-way merge sort algorithm:

(1) Divide the sequence of elements to be sorted into two parts to get two subsequences of basically equal length and sort them respectively.

(2) If the subsequence is long, the subsequence can be further subdivided until the length of the subsequence does not exceed 1.

(3) When the decomposed subsequence has been arranged in order, the two ordered subsequences are merged into one ordered sequence to obtain the solution of the original problem.

The basic implementation is to compare the minimum values of the two sequences and output the smaller one, then repeat the process until one of the queues is empty, and then output the remaining elements in turn if there are elements left in another queue that have not been output. Two-way merge sort can be described as follows:

Algorithm 9-5:
void MergeSort(int left, int right){
  if(left<right){ //When the sequence length is greater than 1, further partition
  int mid=(left+right)/2; //Split in two
  MergeSort(left, right); //Sort the left subsequence element
  //Sort the right subsequence element
  MergeSort(mid+1, right);
  //Combine sorted left and right subsequences into an ordered sequence
  Merge(left, mid, right);
  }
}

The algorithm MergeSort splits a sequence into two subsequences of nearly equal length, sorts them separately, and then invokes the Merge function to complete merge sort by combining two ordered subsequences into one ordered sequence.

成一个有序序列.下面介绍两路合并排序算法的实现步骤:

(1)将待排序的元素序列一分为二,得到长度基本相等的两个子序列,分别排序.

(2)如果子序列较长,还可以继续细分,直到子序列的长度不超过1为止.

(3)当分解所得的子序列已排列有序时,将两个有序子序列合并成一个有序序列,从而得到原问题的解.

基本实现方法是通过比较两序列的最小值,输出其中较小者,然后重复此过程,直到其中一个队列为空,如果另一个队列还有元素没有输出,那么将剩余元素依次输出.两路合并排序可以描述如下:

算法9-5:
void MergeSort(int left, int right){
  if(left<right){ //序列长度大于1时,进一步划分
  int mid=(left+right)/2; //一分为二
  MergeSort(left, right); //对左子序列元素排序
  //对右子序列元素排序
  MergeSort(mid+1, right);
  //将已排好序的左、右子序列合并成一个有序序列
  Merge(left, mid, right);
  }
}

算法MergeSort将一个序列分解成两个长度几乎相等的子序列,对它们分别排序,然后调用Merge函数将两个有序子序列合并成一个有序序列,从而完成合并排序.

The Merge function in this algorithm is responsible for two ordered subsequences (a[left], ..., a[mid]) and (a[mid+1], ..., a[right]) to form an ordered sequence. A temporary array temp[ ] is used to temporarily store the merged results during the merging process. After the merging is completed, the merged ordered sequence will be re-copied to (a[left], ..., a[right]).

Algorithm 9-6:

```
void Merge(int left, int mid, int right)
{
    T * temp = new T[right-left+1];
    //k is the current position in the array temp
    int k = 0;
    //i is the current position of the left subsequence; j is the current position of the right subsequence
    int i = left, j = mid+1;
    while ((i <= mid) && (j <= right))
        if (a[i]<=a[j])
            temp[k++]=a[i++];
        else
            temp[k++]=a[j++];
//If there are still elements in a subsequence that are not output, the remaining elements are output in turn
    while (i <= mid) temp[k++] = a[i++];
    while (j <= right) temp[k++] = a[j++];
    for(i=0, k = left; k <= right;)
        a[k++] = temp[i++];//Write back from temp to sequence a
}
```

Example 9.4.1 Use the two-way merge algorithm to sort the following sequence and merge it into an ordered sequence.

[35　40　72　49　39　80　49]

First, the MergeSort algorithm was used to decompose the 7-element sequence into two subsequences [35,40,72,49] and [39, 80, 49]. Then, the two subsequences were decomposed to obtain four subsequences [35 40], [72 49], [39 80] and [49]. The decomposition continues until the length of each subsequence is 1, and the final result is [35],

[40], [72], [49], [39] and [80].

Next, use the Merge function to complete the merge sort of all subsequences in turn:

First, the sorting of [35] and [40] is merged into [35 40], and the sorting of [72] and [49] is merged into [49 72]. Then, the two merged subsequences are merged into the sorting result of the left subsequence [35 40 49 72].

The right subsequence is merged and sorted in the same way until the two subsequences are merged into a complete ordered sequence.

The recursion process of MergeSort algorithm can be eliminated. The specific sorting process is:

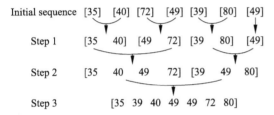

In conclusion, two-way merge sort is to treat n data as n tables of length 1, and then merge adjacent tables in pairs to get $n/2$ ordered tables of length 2. Further, the adjacent tables are combined in pairs to obtain $n/4$ ordered tables of length 4, …; this is repeated until all data is merged into an ordered table of length n, that is, the sorting is completed.

Each of these merge processes is called a **Pass**, and the entire sorting process is a two-way merge sort.

In terms of time complexity, the Merge function merges two ordered subsequences whose sum length is n into an ordered sequence. During the execution of the Merge function, keyword values are compared at most $n-1$ times. The time complexity is $O(n)$.

Thus, the time complexity recursive function of MergeSort recursion algorithm can be obtained:

$$T(n) = \begin{cases} d, & n \leq 1 \\ 2T(n/2) + cn, & n > 1 \end{cases}$$

[49], [39], [80].

接着,使用 Merge 函数依次完成所有子序列的合并排序:

先将[35]和[40]排序合并成[35 40],[72]和[49]排序合并成[49 72],再对合并后的这两个子序列进行合并排序,然后得到左子序列的排序结果[35 40 49 72].

按同样的方式对右子序列进行合并、排序,直到两个子序列合并成一个完整有序的序列.

算法 MergeSort 的递归过程可以消去,具体的排序过程为

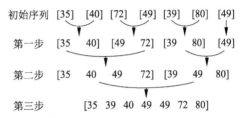

综上,两路合并排序实际就是先将 n 个数据看成 n 个长度为 l 的表,将相邻的表成对合并,得到长度为 2 的 $n/2$ 个有序表;进一步,再将相邻表成对合并,得到长度为 4 的 $n/4$ 个有序表;……;如此重复做下去,直至所有数据均合并到一个长度为 n 的有序表为止,即完成排序.

上述每一次的合并过程称为一**趟**,整个排序过程就是两路合并排序.

从时间复杂度来看,Merge 函数将长度之和为 n 的两个有序子序列合并成一个有序序列,执行过程中最多需进行 $n-1$ 次关键字值间的比较,其时间复杂度为 $O(n)$.

由此,可得到合并排序递归算法 MergeSort 的时间复杂度递归函数:

$$T(n) = \begin{cases} d, & n \leq 1 \\ 2T(n/2) + cn, & n > 1 \end{cases}$$

Therefore, the time complexity of two-way merge sort is $T(n) = O(n\log n)$. Two-way merge sort generally requires an auxiliary array temp of the same length as the original sequence, so the extra space required is $O(n)$.

9.5 Quick sort

9.5.1 Quick sort algorithm

Quick sort, like merge sort, is a design paradigm based on divide and conquer algorithm, which consists of the following three steps:

(1) **Divide**: Divide array $A[p:r]$ into two subarrays $A[p:q-1]$ and $A[q+1:r]$ (one of which may be empty), such that each element in array $A[p:q-1]$ does not exceed each element in array $A[q+1:r]$. Calculate the subscript q as part of the division process.

(2) **Conquer**: Recursively call quick sort to sort two subarrays $A[p:q-1]$ and $A[q+1:r]$.

(3) **Merge**: The entire array $A[p:r]$ is ordered because the elements in the subarray have been sorted and no combination operation is required.

Here is the QuickSort procedure to implement quick sort.

Algorithm 9-7:
```
void QuickSort(int A[ ], int p, int r){
    if (p<r){
        int q=Partition(A, p, r);
        QuickSort(A, p, q-1);
        QuickSort(A, q+1, r);
    }
}
```

Perform quick sort on an array $A[0:n-1]$ with n elements, simply call QuickSort$(A, 1, n)$ at the beginning. The key to the algorithm is the division process Partition, if you don't count the space of the stack, the space required by quick sort is $O(1)$.

Algorithm 9-8:

int Partition(int A[ ], int p, int r){
    int x = A[r]; //The right-most element is the pivot element
        int i = p−1;
        for(int j = p; j<r; j++){
            if (A[j]≤x){
                i = i+1;
                exchange(A, i, j);
            }
        }
        exchange(A, i+1, r);
    return i+1;
}

算法 9-8:

int Partition(int A[ ], int p, int r){
    int x = A[r];//最右端元素作为枢轴元素
        int i = p−1;
        for(int j = p; j<r; j++){
            if (A[j]≤x){
                i = i+1;
                exchange(A, i, j);
            }
        }
        exchange(A, i+1, r);
    return i+1;
}

Fig. 9-2 shows the Partition running process of eight elements. The Partition always selects $x = A[r]$ as the pivot element to divide the array $A[p:r]$. When the procedure runs, the array is divided into four (possibly empty) regions (Fig. 9-3).

图 9-2 显示了 8 个元素的 Partition 运行过程. Partition 总是选择元素 $x = A[r]$ 作为枢轴元素,对数组 $A[p:r]$ 进行划分. 当过程运行时,数组被划分成四个区域(可能为空)(图 9-3).

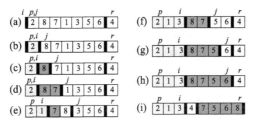

图 9-2 Partition 运行过程

Fig. 9-2 Partition running process

图 9-3 划分的四个区域

Fig. 9-3 Divided four regions

At the beginning of each iteration of the for loop, for any index k,

(1) If $p \leq k \leq i$, then $A[k] \leq x$;
(2) If $i+1 \leq k \leq j-1$, then $A[k] > x$;
(3) If $k = r$, then $A[k] = x$.

在 for 循环的每次迭代的开始,对于任一下标 k,

(1) 如果 $p \leq k \leq i$,那么 $A[k] \leq x$;
(2) 如果 $i+1 \leq k \leq j-1$,那么 $A[k] > x$;
(3) 如果 $k = r$,那么 $A[k] = x$.

For the subarray $A[p:r]$, the value in $A[p:i]$ is less than or equal to x, and the value in $A[i+1:j-1]$ is greater than x, $A[r]=x$. $A[j:r-1]$ can be any value.

Next, we will show that the loop invariant holds up until the first iteration, remains in each iteration of the loop, and we will also use the loop invariant to prove the correctness of the algorithm when it terminates.

Initial: Before the first iteration of the loop, $i=p-1$, $j=p$. There is no value between p and i, and no value between $i+1$ and $j-1$, so the first two conditions of the loop are trivial true. The assignment in line 1 of Partition algorithm satisfies property (3).

Maintain: Refer to Fig. 9-4 to consider the two cases. Related to the judgment conditions in line 4 of Partition algorithm. Fig. 9-4a shows the case where $A[j]>x$. The only thing to do in the loop is to increase the variable j. After the value of variable j increases, condition (2) holds for $A[j-1]$. All other elements stay the same. Fig. 9-4b shows that when $A[j]\leq x$, the variable is increased, and $A[i]$ and $A[j]$ are swapped, and then j is increased. As a result of the exchange, $A[i]\leq x$ is true, and condition (1) is satisfied. Similarly, by the loop invariant, the terms swapped into $A[j-1]$ are greater than x, that is, $A[j-1]>x$.

对于子数组 $A[p:r]$，在 $A[p:i]$ 中的值小于等于 x，在 $A[i+1:j-1]$ 中的值大于 x, $A[r]=x$. $A[j:r-1]$ 可取任意值.

接下来，我们将证明循环不变式在第一次迭代之前成立，并在循环的每次迭代中保持；此外，我们也会利用循环不变式证明算法终止时的正确性.

初始：在循环的第一次迭代之前，$i=p-1$, $j=p$. p 和 i 之间没有值，$i+1$ 和 $j-1$ 之间也没有值，因此，循环前两个条件平凡成立. Partition 的第 1 行的赋值满足性质(3).

维持：参照图 9-4，考虑两种情况. 与 Partition 算法的第 4 行的判断条件有关，图 9-4a 表示 $A[j]>x$ 的情况，循环中唯一要做的是使变量 j 增加. 变量 j 值增加后，条件(2)对 $A[j-1]$ 成立. 所有其他元素保持不变. 图 9-4b 表示 $A[j]\leq x$ 的情况，变量增加，交换 $A[i]$ 和 $A[j]$，然后 j 增加. 交换使得 $A[i]\leq x$ 成立，条件(1)得到满足. 类似地，由循环不变式，被交换进入 $A[j-1]$ 的项大于 x，即 $A[j-1]>x$.

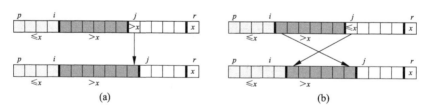

图 9-4 Partition 迭代过程的两种情况
Fig. 9-4 Two cases of the Partition iteration process

Termination: When the algorithm terminates, $j=r$. Thus, each element in the array is located in one of the three sets described by the invariant, and the element values in the array are divided into three sets,

终止：该算法终止时，$j=r$. 于是，数组中的每个元素位于不变式描述的三个集合之一，并把数组中的元素值划分成三个集合，即小于等于 x 的集合、

namely the set less than or equal to x, the set greater than x, and the lone set containing only x.

The last two lines of the procedure Partition swap the pivot element with the leftmost element greater than x, placing it in the middle of the array. The output of the Partition meets the specification given in the division step. Its running time on array $A[p:r]$ is $O(n)$, where $n=r-p+1$. The proof is left as an exercise.

The running time of quick sort depends on whether the division is balanced, which in turn depends on the division element used. if the division is balanced, the asymptotic running time of the algorithm is the same as merge sort; if the division is unbalanced, its running time is the same as that of bubble sort, which is $O(n^2)$. In the next section, we will examine the performance of the quick sort algorithm with three different divisions.

9.5.2 Performance analysis of quick sort algorithm

1. Quick sort worst-case analysis

Quick sort has the worst case when the division process produces two subproblems of size $n-1$ and 0, respectively. Assume this imbalance occurs every time a recursive call is made. The division time complexity is $O(n)$, because a recursive call to an array of size 0 will only return $T(0) = O(1)$, and the running time of the recursive equation is

$$T(n) = T(n-1) + T(0) + O(n)$$
$$= T(n-1) + O(n)$$

Directly, if we sum the cost of each level of recursion, we will get an arithmetic series whose sum is $O(n^2)$. It is also possible to prove by substitution that the solution to the recursive equation $T(n) = T(n-1) + O(n)$ is $O(n^2)$. Thus, the running time is $O(n^2)$, if the division produces the maximum imbalance at each level of recursion of the algorithm. As a result, the worst-case running time of quick sort is not better

大于 x 的集合和只含 x 的孤集.

过程 Partition 的最后两行将枢轴元素与最左边大于 x 的元素交换,将其置于数组的中间. Partition 的输出满足划分步骤给出的规范. 它在数组 $A[p:r]$ 上的运行时间为 $O(n)$,其中 $n=r-p+1$. 证明留作习题.

快速排序的运行时间与划分是否平衡有关,而是否平衡又取决于所用的划分元素. 如果划分平衡,算法的渐近运行时间与归并排序一样;如果划分不平衡,则它的运行时间与冒泡排序一样,都为 $O(n^2)$. 下一小节我们将研究在三种不同划分的情况下快速排序算法的性能.

9.5.2 快速排序算法的性能分析

1. 快速排序最坏情况分析

当划分过程产生的规模分别为 $n-1$ 和 0 的两个子问题时,快速排序出现最坏的情况. 假设每次递归调用时产生这种不平衡的情况. 划分的时间复杂度为 $O(n)$,因为对规模为 0 的数组的递归调用只会返回, $T(0)=O(1)$,递归方程的运行时间为

$$T(n) = T(n-1) + T(0) + O(n)$$
$$= T(n-1) + O(n)$$

直接地,如果对递归每一层的开销求和,会得到一个算术级数,其和为 $O(n^2)$. 也可以利用替换方法证明递归方程 $T(n)=T(n-1)+O(n)$ 的解为 $O(n^2)$. 因而,如果划分在算法的每一层递归上产生最大不平衡,则运行时间为 $O(n^2)$. 因此,快速排序最坏情况下的运行时间不比冒泡排序的运行时

than that of bubble sort, which occurs when the input is already fully ordered (ascending).

2. Quicksort best case analysis

In most uniform division cases, the Partition produces two subproblems of size not more than $n/2$, one of size $n/2$ and the other of size $n/2-1$. In this case, quick sort runs faster. In this case, the recursion equation is

$$T(n) \leq 2T(n/2) + O(n)$$

The solution of the recursive equation is $T(n) = O(n\log n)$. Therefore, if division produces two problems of the same size at each level of recursion of the algorithm, then we have the best case for quick sort.

3. Quick sort average case analysis

The average case analysis of quick sort is more like the best case analysis, and the key is to understand how the balanced division reflects the recursive equation that describes the running time. Assume that division algorithms always produce a division ratio of $9:1$, this seems rather unbalanced, and the recursive equation for quick sort is denoted as follows:

$$T(n) \leq T(9n/10) + T(n/10) + cn$$

Fig. 9-5 shows the recursion tree for this recursive equation.

间好,而它的最坏情况是在输入已经完全有序(升序)时出现的.

2. 快速排序最好情况分析

在大多数均匀划分的情况下,Partition产生两个规模不超过$n/2$的子问题,其中一个规模为$n/2$,另一个规模为$n/2-1$. 在这种情况下,快速排序运行得更快. 此时,递归方程为

$$T(n) \leq 2T(n/2) + O(n)$$

这个递归方程的解为 $T(n) = O(n\log n)$. 因此,如果划分在算法的每一层递归上产生两个相同规模的问题,那么可得快速排序算法的最佳情况.

3. 快速排序平均情况分析

快速排序算法的平均情况分析更类似于最佳情况分析,关键在于要理解平衡划分是如何反映描述运行时间的递归方程的. 假定划分算法总是产生 $9:1$ 的划分比例,这似乎是相当不平衡的,快速排序的递归方程表示如下:

$$T(n) \leq T(9n/10) + T(n/10) + cn$$

图9-5 表示这个递归方程的递归树.

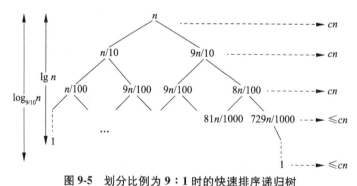

图9-5 划分比例为 $9:1$ 时的快速排序递归树

Fig. 9-5 Quick sort recursion tree in the division ratio of $9:1$

When we run quick sort on random input arrays, the division is not always the same each time. We expect some divisions to be reasonably balanced, while others are quite unbalanced. For example, about 80%

在随机输入数组上运行快速排序时,每一次的划分并不总是一样的. 我们期望某些划分合理地平衡,而其他一些划分却相当不平衡. 例如,大约80%

of Partition processes produce a balance greater than 9∶1, while about 20% of Partition processes produce a balance less than 9∶1. On average, the Partition process produces both "good" and "bad" divisions.

In the recursion tree of average execution of the Partition process, the good and bad cases are randomly distributed in the recursion tree. Assume that good and bad cases alternate in the tree, and that the good case is the best case and the bad case is the worst case. Fig. 9-6 shows the division of two successive layers in the recursion tree. At the root node, the division cost is n, resulting in two divisions of size $n-1$ and 0, which is the worst case. At the next level, the best division of a subarray of size $n-1$ produces two subarrays of size $(n-1)/2$ and $(n-1)/2-1$. Assume a subarray of size 0 has a boundary condition cost of 1.

的 Partition 过程产生平衡大于 9∶1，而大约 20% 的 Partition 过程产生平衡小于 9∶1. 在平均情况下，Partition 过程产生的划分既有"好"也有"坏".

在 Partition 过程平均情况执行的递归树中，好坏的情况随机地分布在递归树中. 假定好坏的情况在树中交替出现，并且好的情况就是最佳情况，坏的情况就是最坏情况. 图 9-6 表示递归树中两个连续层的划分. 在树根结点，划分开销为 n，产生两个规模分别为 $n-1$ 和 0 的划分，这是一种最坏情况. 在下一层，对规模为 $n-1$ 的子数组进行最佳划分，产生两个规模分别为 $(n-1)/2$ 和 $(n-1)/2-1$ 的子数组. 假设规模为 0 的子数组，其边界条件开销为 1.

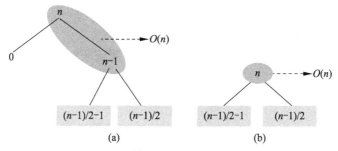

图 9-6 快速排序的双层递归树

Fig. 9-6 Two-layer recursion tree for quick sort

The elliptic region denotes the division cost of the subproblem, which is $O(n)$. The remaining subproblems to be solved in Fig. 9-6a (square shaded region) will not be greater than the remaining subproblems to be solved in Fig. 9-6b. In Fig. 9-6a, three subarrays of sizes 0, $(n-1)/2-1$ and $(n-1)/2$ generated by division are combined, and the total cost is

$$O(n)+O(n-1)=O(n)$$

To be sure, this situation is not worse than the balanced division in Fig. 9-6b, which produces two layers of division of size $(n-1)/2$ with a cost of $O(n)$. The latter situation is balanced! Intuitively, the bad

椭圆形区域表示子问题的划分开销，开销都为 $O(n)$. 在图 9-6a 中所剩下要解的子问题（方形阴影区域）不会大于图 9-6b 中所剩下要解的子问题. 在图 9-6a 中，将划分产生的三个规模分别为 0, $(n-1)/2-1$, $(n-1)/2$ 的子数组组合，总开销为

$$O(n)+O(n-1)=O(n)$$

肯定地说，这种情况不会比图 9-6b 中的平衡划分坏，即产生两层规模为 $(n-1)/2$ 的某划分，其开销为 $O(n)$. 而后者的情形是平衡的！从直觉上来看，

division $O(n-1)$ can be absorbed into the good partition $O(n)$, resulting in a good division. Therefore, when the good division and the bad division alternate between the tree levels, the running time of quick sort, as if both were good divisions, is still $O(n\log n)$, but the constant factor implied in the O is larger. Next, we will briefly present several methods to improve the performance of quick sort algorithm:

1. Improve the selection method of pivot elements

The performance of the quick sort algorithm depends on the symmetry of the division. By modifying the algorithm Partition, we can design a quick sort algorithm Randomized Partition with random selection strategy.

At each step of the quick sort algorithm, when the array is not yet divided, it can randomly select an element in $(A[p], ..., A[r])$ as the division basis, which can make the selection of division basis random, so that the division can be expected to be symmetric.

Algorithm 9-9:

```
int Randomized Partition (int p, int r){
        int i = Random(p, r);
        exchange(A, i, p);
        return Partition (A, p, r);//Call the Partition function again}
```

2. Use the direct insertion method for sorting

When the length of the subsequence to be sorted is less than a certain value, quick sort is not as fast as some simple sorting algorithms (such as direct insertion method). Therefore, subsequences of very small length, can be sorted by straight insertion without further division.

3. Use non-recursive quick sort

Recursive algorithms are often less efficient than corresponding non-recursive algorithms. For speed, non-recursive quick sort (such as using the stack instead of the system stack) can be used instead of the original recursive algorithm.

坏的划分 $O(n-1)$ 可以被吸收进好的划分 $O(n)$ 中,导致最终划分结果是好的. 因此,当好的划分与坏的划分在树的层次间交替进行时,快速排序的运行时间就像都是好的划分结果一样,仍然为 $O(n\log n)$,但是隐含在 O 中的常数因子较大. 下面,将简要给出改善快速排序算法性能的几种方法:

1. 改进主元的选择方法

快速排序算法的性能取决于划分的对称性. 通过修改算法 Partition,可以设计出采用随机选择策略的快速排序算法 Randomized Partition.

在快速排序算法的每一步中,当数组还没有被划分时,可以在$(A[p],\cdots, A[r])$中随机选出一个元素作为划分基准,这样可以使划分基准的选择是随机的,从而可以期望划分是对称的.

算法 9-9:

```
int Randomized Partition (int p, int r){
        int i = Random(p, r);
        exchange(A, i, p);
        return Partition (A, p, r);//再调用 Partition 函数}
```

2. 使用直接插入法进行排序

当待排序的子序列长度小于一定值时,快速排序的速度反而不如一些简单的排序算法(如直接插入法). 因此,对长度很短的子序列,可以不再继续划分,而使用直接插入法进行排序.

3. 使用非递归的快速排序

递归算法的效率常常不如相应的非递归算法. 为提高速度,可使用非递归的快速排序(如使用堆栈代替系统栈)来取代原来的递归算法.

Exercise 9

1. Describe the design idea of divide and conquer method.

2. Redesign the binary search algorithm with divide and conquer method.

3. Let $a[0:n-1]$ be a sorted array. Rewrite the binary search algorithm so that when searching for an element x that is not in the array, the maximum element position i that is less than x and the minimum element position j that is greater than x are returned. When searching for elements in an array, i and j are the same, both being the position of x in the array.

4. Given an array of ascending integers $nums$, and a target value, $target$. Find the start and end positions in the array for the given target value. The time complexity of the algorithm must be $O(\log n)$. Return $[-1, -1]$ if no target value exists in the array.

Example 1:
Input: nums = $[5, 7, 7, 8, 8, 10]$, target = 8

Output: $[3, 4]$
Example 2:
Input: nums = $[5, 7, 7, 8, 8, 10]$, target = 6

Output: $[-1, -1]$

5. For any non-zero even number n, we can always find the odd m and the positive integer k such that $n = m2^k$. To find the product of two n-order matrices, you can divide an n-order matrix into $m \times m$ submatrices, each of which has $2^k \times 2^k$ elements. When the product of $2^k \times 2^k$ submatrices is required, the Strassen algorithm is used. Design a matrix multiplication algorithm combining the traditional method with Strassen's algorithm, which can solve the product of two matrices of order n for any even number n. And analyze the computational time complexity of the algorithm.

6. The worst-case running time of $O(n\log n)$ is required to determine the number of reverse pairs in any permutation of n elements. (Tip: modify merge sort)

7. Use merge sort to change the following sequence to be sorted into increasing order.

[69, 81, 30, 38, 9, 2, 47, 61, 32, 79]

8. Explains how QuickSort algorithms sort the following array elements.

(1) 24,23,24,45,12,12,24,12

(2) 3,4,5,6,7

(3) 23,22,27,18,45,11,63,12,69,25,32,14

9. Show the working performance of the QuickSort when all the elements in the input array are identical.

10. Try to compare merge sort and quick sort, and give analysis.

11. Prove: The running time of the Partition on array $A[p:r]$ is $O(n)$, where $n = r-p+1$.

6. 要求用 $O(n\log n)$ 的最坏情况运行时间,确定 n 个元素的任何排列中逆序对的数目. (提示:修改合并排序)

7. 请使用合并排序算法将下面待排序的序列改为递增排列.

[69,81,30,38,9,2,47,61,32,79]

8. 说明 QuickSort 算法是如何对下面数组元素进行排序的.

(1) 24,23,24,45,12,12,24,12

(2) 3,4,5,6,7

(3) 23,22,27,18,45,11,63,12,69,25,32,14

9. 说明在输入数组的元素全部相同的情况下,QuickSort 算法的工作性能.

10. 试比较合并排序和快速排序,并给出分析.

11. 证明:Partition 在数组 $A[p:r]$ 上的运行时间为 $O(n)$,其中 $n = r-p+1$.

习题 9 答案
Key to Exercise 9

Chapter 10 Dynamic Programming

Dynamic programming is a branch of operations research, usually used to solve the optimization problem of multi-stage decision-making process. It is efficient and fast for solving the shortest route problem, machine load problem, etc. At present, dynamic programming often appears in various computer algorithm competitions or programmer written test and interviews, and rarely in mathematical modeling competitions. However, the idea of this algorithm is very practical in life. we hope that after studying the content of this chapter, you can get some inspiration in the way of thinking to solve practical problems.

10.1 Basic idea of dynamic programming

We first introduce several important concepts related to dynamic programming, before introducing the basic concepts of dynamic programming.

Stage: Appropriately divide the process of the given problem into several interrelated stages so that the problem can be solved in a certain order. The division of stages is generally carried out according to the characteristics of time and space, but it must be able to transform the process of the problem into a multi-stage decision-making problem.

State: State denotes the natural or objective condition at the beginning of each stage.

Decision-making: Decision-making refers to the decision that can be made when the process is in a certain state and in a certain stage to determine the state of the next stage. This decision is called decision-making. The variables that describe the decision-making are called decision variables. It can be a number or numbers or a vector. Commonly used $u_k(s_k)$ denotes the decision-making variable when the k-th stage is at s_k. It can be seen that the decision-making variable is a function of the state, that is, the decisions that can be made in different states are related to the current state.

By introducing the above three concepts, we derive the basic concept of dynamic programming: each decision depends on the current state, and then causes the transition of the state. A decision-making sequence is generated in a changing state, so this multi-stage optimization decision-making process to solve the problem is called **dynamic programming**.

The basic idea of dynamic programming:

(1) The key to the dynamic programming problem is to correctly write the basic recurrence relations and appropriate boundary conditions (that is, the basic equations). To achieve this, the process of the problem must be divided into several interconnected stages, selecting appropriate state variables, decision-making variables and defining the optimal value function, so as to turn a big problem into a family of sub-problems of the same type, and then solve them one by one. That is to say, starting from the boundary conditions, step by step recursive optimization. In the solution of each sub-problem, the optimization results of the previous sub-problems are used in sequence, and the optimal solution obtained from the last sub-problem is the optimal solution to the whole problem.

(2) In the process of multi-stage decision-making, the dynamic programming method is an optimization method that not only separates the current stages from the future stages, but also combines the current benefits with future benefits. Therefore, the selection of each stage of decision-making is considered from the overall situation, which is generally different from the optimal choice of that stage.

(3) When seeking the optimal strategy for the entire problem, since the initial state is known, and the strategy of each stage is a function of the state of the stage, the states of each stage passed by the optimal strategy can be obtained by successive transformations, thus, the optimal route is determined.

通过介绍上述三个概念,我们引出动态规划的基本概念:每次决策依赖于当前状态,又随即引起状态的转移.一个决策序列就是在变化的状态中产生出来的,所以,这种多阶段最优化决策解决问题的过程就称为**动态规划**.

动态规划的基本思想:

(1)动态规划问题的关键在于正确地写出基本的递推关系式和恰当的边界条件(也就是基本方程).要做到这一点,必须将问题的过程划分成几个相互联系的阶段,选取恰当的状态变量、决策变量以及定义最优值函数,从而把一个大问题化成一族同类型的子问题,然后逐个求解.即从边界条件开始,逐段递推寻优,在每一个子问题的求解中,依次利用它前面的子问题的最优化结果,从最后一个子问题得到的最优解,就是整个问题的最优解.

(2)在多阶段决策的过程中,动态规划方法是既把当前一段和未来各段分开,又把当前效益和未来效益结合起来的一种最优化方法.因此,每段决策的选取是从全局来考虑的,通常与该段的最优选择不同.

(3)在求整个问题的最优策略时,由于初始状态是已知的,而每段的策略都是该段状态的函数,因此最优策略所经过的各段状态便可以逐次变换得到,从而确定了最优路线.

After clarifying the basic concepts and basic ideas of dynamic programming, we see that when building a dynamic programming model for a practical problem, the following five points must be done:

(1) Divide the problem process into appropriate stages;

(2) Correctly choose the state variable s_k, so that it can describe the evolution of the process, but also has no aftereffect;

(3) Determine the decision variable u_k and the allowable decision set $D_k(s_k)$ at each stage;

(4) Write the state transition equation correctly;

(5) Write the index function V_k and n correctly. It should satisfy the following properties: ① It is a quantitative function defined on the whole process and all subsequent sub-processes; ② It must be separable and satisfy the recurrence relationship.

10.2 The shortest path problem of multi-segment graphs

The shortest path problem in a multi-segment graph is one of the classic problems of dynamic programming. Many optimization problems can be transformed into the shortest path problem in a multi-segment graph and then solved.

10.2.1 Problem analysis and solution

The problem of the shortest path of a multi-segment graph is described as follows:

If the graph $G = (V, E)$ is a weighted directed connected graph, if the vertex set V is divided into K ($2 \leq K \leq n$) mutually disjoint subsets $V_i (1 \leq i \leq k)$, where V_1 and V_k have only one vertex S (called source) and one vertex t (called sink) respectively. The starting and ending point of all edges (u, v) are in the two adjacent subsets V_i and V_i+1: $u \in V_i$, $v \in V_i+1$, and the edge (u,v) has a positive weight, recorded as $c(u, v)$. The shortest path algorithm for multi-segment graphs is to find the path with the smallest sum of weights from the source s to the sink t.

Since the multi-segment graph divides the vertices into k disjoint subsets, the multi-segment graph is divided into k segments, and each segment contains a subset of the vertices. The vertices of the multi-segment graph are numbered according to the order of the segments, and the order of the vertices in the same segment does not matter. Assuming that the number of vertices in the graph is n, the serial number of the source point s is 0, the serial number of the end point t is $n-1$, and for any edge (u,v) in the graph, the serial number of the vertex u is less than the serial number of vertex v. Fig. 10-1 below is a multi-segment graph with 10 vertices.

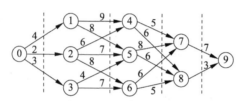

图 10-1 多段线路图
Fig. 10-1 Multi-segment circuit graph

According to the nature of the multi-segment graph, we can divide this special graph structure into multiple subsets. For example, the multi-segment graph shown in Fig. 10-1 above can be divided into 5 subsets. It can be clearly seen that if you want to reach the vertices of a certain subset, you must start from the vertices of the previous adjacent vertex set, and there are no reachable edges between non-adjacent subsets.

For this feature, we can deduce a solution to the problem. For example, if I want to reach vertex 10, then vertex 8 or vertex 9 must be reached first. In other words, the shortest distance to vertex 10 is the shortest distance to vertex 8 $d(1,8)$ plus the weight of edge $(8,10)$, and the shortest distance to vertex 9 $d(1,9)$ plus the weight of edge $(9,10)$. Because there are no edges between sets of non-adjacent vertices, there are only two ways to reach vertex 10.

Assume C is the weight of an edge, $d(m,n)$ is the shortest distance from point m to point n, then the description using mathematical language is as follows:

$$d(1,10) = \min(d(1,8)+C_{8,10}, d(1,9)+C_{9,10})$$

Let's look at another example, assume we want to analyze the shortest distance to vertex 8, there are only 3 cases. That is, the shortest distance to vertex 5 $d(1,5)$ plus the weight of edge $(5,8)$, and the shortest distance to vertex 6 $d(1,6)$ plus the weight of edge $(6,8)$, and the shortest distance to vertex 7 $d(1,7)$ plus the weight of the edge $(7,8)$, take the smallest value among the three. The description using mathematical language is as follows:

$$d(1,8) = \min(d(1,5)+C_{5,8}, d(1,6)+C_{6,8}, d(1,7)+C_{7,8})$$

According to the discussion of the above two examples, we can generalize the situation from special to general. Let C_{uv} be the weight of the directed edge $<u,v>$ of the multi-segment graph, and the shortest path length from the source point s to the end point v is $d(s,v)$, the end point is t, then the state transition equation of the problem can be obtained as

$$\begin{cases} d(s,v) = C_{s,v} & (<s,v> \in E) \\ d(s,v) = \min(d(s,u)+C_{u,v}) & (<u,v> \in E) \end{cases}$$

Proof of the optimal substructure: assume a multi-segment graph has one and only one starting point S, and one and only one end point T, $S \to S_1 \to S_2 \to \ldots \to S_n \to T$ is the shortest path from the starting point S to the end point T. Assuming that the cost of $S \to S_1$ has been calculated, the minimum cost from the starting point S to the end point T will be converted to the minimum cost from the point S_1 to the end point T.

Assuming that $S_1 \to S_2 \to S_3 \to \ldots \to S_n \to T$ is not the shortest path from point S_1 to the end point T, there must be another path $S_1 \to R_1 \to R_2 \to \ldots \to R_n \to T$ with less cost than $S_1 \to S_2 \to S_3 \to \ldots \to S_n \to T$, and then it is deduced that the shortest path from the start

且仅有上述两种. 设 C 为某条边的权重, $d(m,n)$ 为从点 m 到点 n 的最短距离, 则使用数学语言描述如下:

$$d(1,10) = \min(d(1,8)+C_{8,10}, d(1,9)+C_{9,10})$$

再看一个例子, 假设要分析到达顶点 8 的最短距离, 则只有 3 种情况. 即在到达顶点 5 的最短距离 $d(1,5)$ 加上边 $(5,8)$ 的权重、到达顶点 6 的最短距离 $d(1,6)$ 加上边 $(6,8)$ 的权重, 以及到达顶点 7 的最短距离 $d(1,7)$ 加上边 $(7,8)$ 的权重三者之间取最小值. 使用数学语言描述如下:

$$d(1,8) = \min(d(1,5)+C_{5,8}, d(1,6)+C_{6,8}, d(1,7)+C_{7,8})$$

根据上面两个例子的论述, 我们可以把该情况从特殊推广到一般, 设 C_{uv} 为多段图有向边 $<u,v>$ 的权值, 从源点 s 到终点 v 的最短路径长为 $d(s,v)$, 终点为 t, 则可以得到该问题的状态转移方程为

$$\begin{cases} d(s,v) = C_{s,v} & (<s,v> \in E) \\ d(s,v) = \min(d(s,u)+C_{u,v}) & (<u,v> \in E) \end{cases}$$

最优子结构证明: 设一个多段图有且仅有一个起点 S, 有且仅有一个终点 T, $S \to S_1 \to S_2 \to \cdots \to S_n \to T$ 为从起点 S 到终点 T 的最短路径. 设 $S \to S_1$ 的开销已经求出, 则从起点 S 到终点 T 的最小开销的求解将转换为对点 S_1 到终点 T 的最小开销进行求解.

假设 $S_1 \to S_2 \to S_3 \to \cdots \to S_n \to T$ 不是点 S_1 到终点 T 的最短路径, 则必然存在另一条路径 $S_1 \to R_1 \to R_2 \to \cdots \to R_n \to T$ 的开销小于 $S_1 \to S_2 \to S_3 \to \cdots \to S_n \to T$ 的路径, 进而推出起点 S 到终点 T 的最

point S to the end point T is $S \to S_1 \to R_1 \to R_2 \to \ldots \to R_n \to T$. How-ever, it is known that the path $S \to S_1 \to S_2 \to \ldots \to S_n \to T$ is the shortest one from the start point S to the end point T. It is impossible for the total cost of other paths to be less than the cost of this path, which creates a contradiction. The shortest path problem of the multi-segment graph satisfies the optimal substructure.

短路径为径为 $S \to S_1 \to R_1 \to R_2 \to \cdots \to R_n \to T$. 然而已知路径 $S \to S_1 \to S_2 \to \cdots \to S_n \to T$ 为起点 S 到终点 T 的最短路径,不可能存在其他路径的总开销比该路径的开销还要小,于是产生了矛盾,因此多段图的最短路径问题满足最优子结构.

Now to solve the shortest path problem in Fig. 10-1, we can build the adjacency matrix shown in Fig. 10-2.

现在解决图 10-1 中的最短路径问题,可以建立如图 10-2 所示的邻接矩阵.

0	4	2	3	0	0	0	0	0	0
0	0	0	0	9	8	0	0	0	0
0	0	0	0	6	7	8	0	0	0
0	0	0	0	0	4	7	0	0	0
0	0	0	0	0	0	0	5	6	0
0	0	0	0	0	0	0	8	6	0
0	0	0	0	0	0	0	6	5	0
0	0	0	0	0	0	0	0	0	7
0	0	0	0	0	0	0	0	0	3
0	0	0	0	0	0	0	0	0	0

图 10-2　邻接矩阵

Fig. 10-2　Adjacency matrix

Create a one-dimensional array cost[MAXV] to store the minimum cost of each vertex. The minimum cost path to vertices 1 ~ 9 is calculated and stored in the array shown in Fig. 10-3.

创建一个一维数组 cost[MAXV] 存储到每个顶点的最小开销. 通过计算求出到达顶点 1 ~ 9 的最小开销路径存储到图 10-3 所示数组.

顶点	1	2	3	4	5	6	7	8	9	10
最小开销	0	4	2	3	8	7	10	13	13	

图 10-3　记录开销的数组

Fig. 10-3　Array for recording cost

Then according to the state transition equation, it can be concluded that the minimum cost from vertex 1 to vertex 10 is 16. The calculation process is min(cost[8]+topography. edges[8][10], cost[9]+topography. edges[9][10])= 16)

再根据状态转移方程,可以得出从顶点 1 到达顶点 10 的最小开销为 16,计算过程为 min(cost[8]+topography. edges[8][10], cost[9]+topography. edges[9][10])= 16)

Another one-dimensional array path[MAXV] is used to store the predecessor vertices of each vertex to find the shortest path. The way to find the shortest path is to visit the corresponding vertices in sequence

再使用一个一维数组 path[MAXV],用于保存每个顶点的前驱顶点,用于找到最短路径. 找到最短路径的方式为从 path[10] 开始依次访问

starting from path[10], and the order of visit is the shortest path sought until the starting point is visited, as shown in Fig. 10-4.

对应的顶点，访问的次序即为所求的最短路径，直到访问了起点为止，具体如图 10-4 所示.

path[10]→path[9]→path[6]→path[4]→path[1]

图 10-4 访问过程

Fig. 10-4 Visit process

10.2.2 Algorithm realization of the shortest path problem in multi-segment graphs

First define the structure of the graph structure and use the adjacency matrix to store:

Algorithm 10-1:

```
typedef struct        //The definition of the graph
{
    int edges[MAXV][MAXV];        //Adjacency matrix
    int n;  //Number of vertices
} MGraph;
```

Define the auxiliary structure needed to solve the problem:

```
MGraph topography;        //Save the adjacency matrix of the vertex relationship
int path[MAXV] = {};    //The predecessor corresponding to the shortest path saved to the vertex
int min_cost[MAXV] = {}; //The shortest path length saved to each vertex
```

Solve the problem according to the state transition equation, and finally output the shortest path and its cost:

```
int main()
{
    int point_num=0;
    int a_cost;
    int pre;
    cout << "The number of vertices is:";
    cin >> point_num;
    topography=CreateMGraph(point_num);
    min_cost[1]=0;
    for(int i=2; i <= topography.n; i++)
    {
        min_cost[i]=99999;
    }
    for(int i=2; i <= point_num; i++)
    {
```

10.2.2 多段图的最短路径问题算法实现

首先定义图结构的结构体，使用邻接矩阵来存储：

算法 10-1：

```
typedef struct      //图的定义
{
    int edges[MAXV][MAXV]; //邻接矩阵
    int n;      //顶点数
} MGraph;
```

定义求解问题需要的辅助结构：

```
MGraph topography; //保存顶点关系的邻接矩阵
int path[MAXV] = {}; //保存到该顶点的最短路径对应的前驱
int min_cost[MAXV] = {};   //保存到每个顶点的最短路径长度
```

根据状态转移方程求解问题，最后输出最短路径及其开销：

```
int main()
{
    int point_num=0;
    int a_cost;
    int pre;
    cout << "顶点数量为:";
    cin >> point_num;
    topography=CreateMGraph(point_num);
    min_cost[1]=0;
    for(int i=2; i <= topography.n; i++)
    {
        min_cost[i] = 99999;
    }
    for(int i = 2; i <= point_num; i++)
    {
```

```
            for( int j = 1; j < i; j++)
            {
                if( topography. edges[j][i] != 0)
                {
                    a_cost = min_cost[j]+topography. edges[j][i];
                    if( a_cost < min_cost[i] )
                    {
                        min_cost[i] = a_cost;
                        path[i] = j;
                    }
                }
            }
        }

        for( int i = 1; i <= point_num; i++)
        {
            cout << "To the vertex" << i << "the minimum cost is:" << min_cost[i] << ",path:" << i;
            pre = i;
            while( path[ pre ] )
            {
                cout << "<-" << path[ pre ];
                pre = path[ pre ];
            }
            cout << endl;
        }
        return 0;
    }
```

10.3 Multiplication problem of matrix chain

Given n matrices $\{A_1, A_2, \ldots, A_n\}$, where A_i and A_{i+1} are multiplicative ($i = 1, 2, 3, \ldots, n-1$), determine the calculation order of the product of chain-multiplication of the matrix to make the number of multiplications required to calculate the product of chain-multiplication the matrix in this order the least. The input data is the number of matrices and the size of each matrix, and the output result is the calculation

10.3 矩阵连乘问题

给定n个矩阵$\{A_1,A_2,\cdots,A_n\}$,其中A_i和A_{i+1}是可乘的($i=1,2,3,\cdots,n-1$),确定计算矩阵连乘积的计算次序,使得依此次序计算矩阵连乘积需要数乘的次数最少.输入数据为矩阵的个数和每个矩阵的规模,输出结果为计算矩阵连乘积的计算次序和最小数乘次数.

order of the chain-multiplication product of the matrices and the minimum number of multiplications.

Example 10.1 There are four matrices A, B, C, and D. Their dimensions are 50×10, 10×40, 40×30, 30×5. Then chain-multiplying A, B, C, D, there are five multiplication orders, as shown below:

(1) $(A((BC)D))$, the number of multiplications is 16000;

(2) $((AB)(CD))$, the number of multiplications is 36000;

(3) $((A(BC))D)$, the number of multiplications is 34500;

(4) $(A(B(CD)))$, the number of multiplications is 10500;

(5) $(((AB)C)D)$, the number of multiplications is 87500.

From the above example, it can be seen that the matrices of different dimensions are multiplied together, and the number of multiplication obtained by the different order of multiplication is also very different. In this example, when the order of matrix multiplication is $(A(B(CD)))$, the number of multiplications is the least, 10500 times; when the order of multiplication is $(((AB)C)D)$, the number of multiplications is the most, 87500 times, and the difference between the two is as much as 8 times. Therefore, it is particularly important to find the optimal separation method for matrix chain multiplication.

There are several steps to solve the matrix chain multiplication problem by dynamic programming algorithm:

(1) Problem analysis: For convenience, the product of chain multiplication $A_i A_{i+1} \ldots A_j$ is abbreviated as $A[i:j]$, and the dimension of A_i is recorded as $p_{i-1} \times p_i$. Then the purpose of calculation is to solve the optimal solution of $A[1:n]$, and the sub-strategies of an optimal strategy should also be optimal, so the problem can be decomposed into the optimal calculation

例 10.1 现有四个矩阵 A, B, C, D, 他们的维数分别是 $50\times10, 10\times40, 40\times30, 30\times5$. 则连乘 A, B, C, D 共有五种相乘的次序, 如下所示:

(1) $(A((BC)D))$, 数乘次数为 16000;

(2) $((AB)(CD))$, 数乘次数为 36000;

(3) $((A(BC))D)$, 数乘次数为 34500;

(4) $(A(B(CD)))$, 数乘次数为 10500;

(5) $(((AB)C)D)$, 数乘次数为 87500.

由上面的例子可知, 不同维数的矩阵相连乘, 相乘的先后顺序不同所得到的数乘次数也是有很大区别的, 该例子中矩阵相乘次序为 $(A(B(CD)))$ 时, 数乘次数最少, 为 10500 次; 相乘次序为 $(((AB)C)D)$ 时, 数乘次数最多, 为 87500 次, 两者差距达 8 倍之多. 所以, 寻找一个矩阵连乘的最优分隔方式就显得尤为重要.

由动态规划算法解决矩阵连乘问题, 有以下几个步骤:

(1) 问题分析: 为了方便起见, 将连乘积 $A_i A_{i+1} \cdots A_j$ 简记为 $A[i:j]$, 其中 A_i 的维度记为 $p_{i-1} \times p_i$. 那么计算目的是求解 $A[1:n]$ 的最优解, 而一个最优策略的子策略也应是最优的, 所以问题可分解为求 $A[i:j]$ 的最优计算次序. 然后考虑计算 $A[i:j]$ 的最优计算次

order of $A[i:j]$. Then consider the optimal calculation order for calculating $A[i:j]$: Assume this calculation order breaks the matrix between the matrix A_k and A_{k+1}, $i \leq k < j$, then the corresponding bracketing method is $(A_i A_{i+1} \ldots A_k)(A_{k+1} A_{k+2} \ldots A_j)$.

Then the calculation cost of $A[i:j]$ is the calculation cost of $A[i:k]$ plus the calculation cost of $A[k+1:j]$, plus the calculation cost of $A[i:k]$ and $A[k+1:j]$ multiplied by each other.

(2) Establish a recursive relationship: The second step of designing a dynamic programming algorithm is to recursively define the optimal value. For the optimal calculation order problem of matrix chain multiplication, assume calculation $A[i:j]$, $1 \leq i \leq j \leq n$ which requires the minimum number of times of multiplication is $dp[i,j]$, the optimal value of the original problem is $dp[1,n]$:

① When $i=j$, $A[i:j] = A_i$, at this time $A[i:j]$ is a single matrix, so there is no need to calculate, $dp[i,i] = 0$, $i=1,2,3,\ldots,n$. ② When $i<j$, the optimal substructure property can be used to calculate $dp[i:j]$. If the optimal order for calculating $A[i:j]$ is disconnected between A_k and A_{k+1}, $i \leq k < j$, Then $dp[i:j] = dp[i:k] + dp[k+1,j] + p_{i-1} \times p_k \times p_j$, where the dimension of A_i is $p_{i-1} \times p_i$. There are $j-i$ possibilities for the position of k, that is, $k \in \{i, i+1, \ldots, j-1\}$. Therefore, k is the position where the calculation cost is the smallest among the $j-i$ positions. We can define $dp[i,j]$ recursively as

$$dp[i,j] = \begin{cases} 0, & i=j \\ \min(dp[i,k] + dp[k+1,j] + p_{i-1} p_k p_j), & i<j \end{cases}$$

$dp[i,j]$ gives the optimal value, that is, the disconnection position k in the optimal order of calculating $A[i:j]$, if the disconnection position k corresponding to $dp[i,j]$ is recorded as $s[i,j]$, then after calculating the optimal value $dp[i,j]$, the corresponding optimal solution can be constructed recursively from $s[i,j]$ through the formula.

次序:设这个计算次序在矩阵 A_k 和 A_{k+1} 之间将矩阵断开,$i \leq k < j$,则其相应加括号的方式为 $(A_i A_{i+1} \cdots A_k)(A_{k+1} A_{k+2} \cdots A_j)$.

那么 $A[i:j]$ 的计算量为 $A[i:k]$ 的计算量加上 $A[k+1:j]$ 的计算量,再加上 $A[i:k]$ 和 $A[k+1:j]$ 相乘的计算量.

(2) 建立递归关系:设计动态规划算法的第二步是递归地定义最优值,对于矩阵连乘的最优计算次序问题,设计算 $A[i:j]$,$1 \leq i \leq j \leq n$ 所需要的最少数乘次数为 $dp[i,j]$,则原问题的最优值为 $dp[1,n]$:

① 当 $i=j$ 时,$A[i:j] = A_i$,此时 $A[i:j]$ 为单一矩阵,因此不需要计算,有 $dp[i,i] = 0$,$i=1,2,3,\cdots,n$. ② 当 $i<j$ 时,可利用最优子结构性质计算 $dp[i:j]$,若计算 $A[i:j]$ 的最优次序在 A_k 和 A_{k+1} 之中断开,$i \leq k < j$,则 $dp[i:j] = dp[i:k] + dp[k+1,j] + p_{i-1} \times p_k \times p_j$,这里 A_i 的维数为 $p_{i-1} \times p_i$. k 的位置有 $j-i$ 种可能,即 $k \in \{i, i+1, \cdots, j-1\}$. 因此,$k$ 是在 $j-i$ 个位置中使得计算量为最小值的位置. 可以递归地定义 $dp[i,j]$ 为

$$dp[i,j] = \begin{cases} 0, & i=j \\ \min(dp[i,k] + dp[k+1,j] + p_{i-1} p_k p_j), & i<j \end{cases}$$

$dp[i,j]$ 给出了最优值,即计算 $A[i:j]$ 的最优次序中断开位置 k,若将对应 $dp[i,j]$ 的断开位置 k 记为 $s[i,j]$,那么计算出最优值 $dp[i,j]$ 之后,可通过公式递归地由 $s[i,j]$ 构造出对应的最优解.

(3) Construct an auxiliary table to calculate the optimal cost: Through the recursive algorithm in the second step, it can be known that many overlapping sub-problems will occur in the solution process, so it is solved by constructing an auxiliary table $s[n][n]$. There are two ways to construct the auxiliary table. One is bottom-up table-filling construction, which requires the solution of sub-problems to be filled incrementally; the other is the top-down table-filling memorandum method, this method initializes each element of the table to a certain special value for calculation, and then fill in the solution of the sub-problem encountered in the recursive process. The following uses the bottom-up method to construct. The main implementation of the algorithm is as follows:

Algorithm 10-2:
```
for(int L=2;L<=n;L++){
    for(int i=1;i<=n;i++){
        int j=i+L-1;
        if(j>n) break;
        for(int k=i;k<j;k++){
            int min_=dp[i][j];
            int temp=dp[i][k]+dp[k+1][j]+p[i-1]*p[k]*p[j];

            if(temp<min_){
                dp[i][j]=temp;
                min_=temp;
                s[i][j]=k;
            }
            dp[i][j]=min(dp[i][j],dp[i][k]+dp[k+1][j]+p[i-1]*p[k]*p[j]);
        }
    }
}
```

Algorithm analysis: At the beginning, multiply two matrices to n matrices to find the free solution of each subproblem. The second for loop is used to traverse each case of matrix multiplication according to the increasing number of matrix multiplications. For example, if two matrices are multiplied, it can be the first matrix and the second matrix, the second and the third, the third and the fourth, and so on. The third for loop is used to traverse the segmentation position of each case when the matrix is multiplied, and find a minimum number of multiplications according to the difference of k value, and record the position of k.

(4) Construct the optimal solution: In the third step, the optimal value is calculated, and the minimum number of multiplications is known, but the calculation order to obtain the minimum number of multiplications is not known yet, so the optimal solution is still not obtained. But the algorithm of the third step has recorded all the information needed to construct the optimal solution. So construct the optimal solution of the problem in the fourth step. The specific algorithm code is as follows:

Algorithm 10-3:
```
void Traceback(int i,int j){
    if(i==j) return;
    int k=s[i][j];
    Traceback(i,k);
    Traceback(k+1,j);
    cout<<"A["<<i<<":"<<k<<"] * A["<<k+1<<":"<<j<<"]"<<endl;
}
int main()
{
    cin>>n;
    MatricChain();
    for(int i=1;i<=n;i++){
        for(int j=1;j<=n;j++)
            cout<<setw(10)<<dp[i][j]<<" ";
        cout<<endl;
    }
    for(int i=1;i<=n;i++){
        for(int j=1;j<=n;j++)
            cout<<setw(5)<<s[i][j]<<" ";
        cout<<endl;
    }
    cout<<"The minimum number of chain multiplications is"
```

第二个矩阵相乘,也可以是第二个和第三个矩阵相乘,也可以是第三个和第四个矩阵相乘,如此等等.第三个for循环用于在矩阵相乘时,遍历每一种情况的分割位置,并且根据 k 值的不同找到一个最小的相乘次数,并记录下 k 的位置.

(4) 构造最优解:第(3)步计算出了最优值,知道了最少数乘法次数,但是还没有知道获得最少数乘法次数所要按照的计算次序是什么,所以仍没有得出最优解.但是第三步的算法已经记录了构造最优解所需要的全部信息.所以第(4)步构造问题的最优解.具体实现的算法代码如下:

算法 10-3:
```
void Traceback(int i,int j){
    if(i==j) return;
    int k=s[i][j];
    Traceback(i,k);
    Traceback(k+1,j);
    cout<<"A["<<i<<":"<<k<<"] * A["<<k+1<<":"<<j<<"]"<<endl;
}
int main()
{
    cin>>n;
    MatricChain();
    for(int i=1;i<=n;i++){
        for(int j=1;j<=n;j++)
            cout<<setw(10)<<dp[i][j]<<" ";
        cout<<endl;
    }
    for(int i=1;i<=n;i++){
        for(int j=1;j<=n;j++)
            cout<<setw(5)<<s[i][j]<<" ";
        cout<<endl;
    }
    cout<<"连乘的最少次数是"<<dp[1]
```

```
<<dp[1][n]<<" times. "<<endl;
    Traceback(1,n);
}
```

Algorithm analysis: The focus of the algorithm in this step is the Traceback( ) function. If there is only one matrix, return directly, otherwise obtain the segmentation position k, divide $A[i,j]$ from the segmentation position into the upper half and the lower half, and then the information of each division is output, you can get the final result.

10.4 The longest common subsequence problem

Before introducing the longest common subsequence problem, first introduce several related concepts. There are two sequences $X = (x_1, x_2, x_3, \ldots, x_m)$ and $Y = (y_1, y_2, y_3, \ldots, y_k)$, if and only if the sequence X has a strictly increasing sequence of subscripts $(i_1, i_2, \ldots, i_k)$, for all $j = 1, 2, \ldots, k, y_j = x_{i_j}$ ($1 \leq i_j \leq m$), then Y is a subsequence of X. For example, for the sequence 1,3,5,4,2,6,8,7, the sequence 3,4,8,7 is a subsequence of it. For a sequence of length n, it has a total of 2^n subsequences and 2^{n-1} non-empty subsequences. It should be noted that the subsequence is not a subset, it is related to the element order of the original sequence.

If sequence C is both a subsequence of sequence A and a subsequence of sequence B, it is called a **common subsequence** of sequence A and sequence B. For example, for sequence 1,3,5,4,2,6,8,7 and sequence 1,4,8,6,7,5, sequence 1,8,7 is a common subsequence of them. It should be noted that the empty sequence is a common subsequence of any two sequences. The longest common subsequence of A and B (the one containing the most elements) is called **the longest common subsequence** of A and B. Therefore, the longest common subsequence problem is: Given two sequences $X = (x_1, x_2, x_3, \ldots, x_m)$ and $Y = (y_1, y_2, y_3, \ldots, y_k)$, find a common subsequence of X and Y, and the sequence is required to be the longest common subsequence of X and Y.

10.4.1 Solving process of the longest common subsequence

The number of the longest common subsequences of the two sequences $A_n = (a_1, a_2, \ldots, a_n)$ and $B_m = (b_1, b_2, \ldots, b_m)$ is not unique, but the length is unique, so the first step to solve the longest common subsequence is to find the length of the longest common subsequence. Use the brute force enumeration method to find the length. Assume sequence A has 2^n subsequences and sequence B has 2^m subsequences. If any two subsequences are compared one by one, the compared subsequences are as high as 2^{n+m}. Brute force enumeration is not suitable for finding the length of the longest common subsequence.

Use the dynamic programming algorithm to solve the longest common subsequence length, let A_x denote the subsequence formed by the continuous first x terms of the sequence A, that is, $A_x = (a_1, a_2, \ldots, a_x)$, $B_y = (b_1, b_2, \ldots, b_y)$, using $LCS(x, y)$ to denote the length of the longest common subsequence of the two sequences, then the original problem is equivalent to finding $LCS(m, n)$. For convenience, we use $L(x, y)$ to denote the longest common subsequence of A_x and B_y. Let x denote the last term considered in the subsequence, we have

(1) When $A_x = B_y$, the last term of the longest common subsequence $L(A_x, B_y)$ of the two sequences A and B must be this element. Use the contradiction method to prove: Let $t = A_x = B_y$, assume the last term of $L(x, y)$ is not t, then either $L(x, y)$ is an empty sequence, or the last term of $L(x, y)$ is $A_a = B_b \neq t$, and obviously $a < x, b < y$. In either case, t can be connected to the back of this $L(x, y)$ to get a longer common subsequence. Contradiction! If you delete the last term a_x from the sequence A_x and delete the last term b_y from the sequence B_y, then delete the last term t

和 Y 的最长公共子序列.

10.4.1 最长公共子序列的求解过程

两个序列 $A_n = (a_1, a_2, \cdots, a_n)$ 和 $B_m = (b_1, b_2, \cdots, b_m)$ 的最长公共子序列的个数不是唯一的,但是长度是唯一的,所以求解最长公共子序列的第一步是先求出最长公共子序列的长度. 使用暴力枚举方法求解长度,设序列 A 有 2^n 个子序列,序列 B 有 2^m 个子序列,如果将任意两个子序列一一比较,被比较的子序列高达 2^{n+m},可知暴力枚举法不适合于求解最长公共子序列长度.

使用动态规划算法求解最长公共子序列的长度,设 A_x 表示序列 A 的连续前 x 项所构成的子序列,即 $A_x = (a_1, a_2, \cdots, a_x)$,$B_y = (b_1, b_2, \cdots, b_y)$,用 $LCS(x, y)$ 表示两个序列的最长公共子序列的长度,那么原问题等价于求 $LCS(m, n)$. 为了方便,我们用 $L(x, y)$ 表示 A_x 和 B_y 的一个最长公共子序列. 令 x 表示在子序列中被考虑的最后一项,有

(1) 当 $A_x = B_y$ 时,这两个序列 A 和 B 的最长公共子序列 $L(A_x, B_y)$ 的最后一项一定是这个元素,使用反证法证明:令 $t = A_x = B_y$,假设 $L(x, y)$ 最后一项不是 t,则要么 $L(x, y)$ 为空序列,要么 $L(x, y)$ 的最后一项是 $A_a = B_b \neq t$,且显然有 $a < x, b < y$. 无论是哪种情况都可以把 t 接到这个 $L(x, y)$ 后面,从而得到一个更长的公共子序列. 矛盾! 如果从序列 A_x 中删掉最后一项 a_x,从序列 B_y 中也删掉最后一项 b_y,则从 $L(x,$

from $L(x,y)$ and the resulting sequence is $L(x-1,y-1)$. Then the subsequence $L(x,y)$ obtained by adding element t to it is also shorter than the subsequence obtained by adding element t to $L(x-1,y-1)$, which contradicts the fact that $L(x,y)$ is the longest common subsequence, the proof is complete. Get $\text{LCS}(A_x,B_y) = \text{LCS}(x-1,y-1)+1$.

(2) When $A_x \neq B_y$, assume $t = L(A_x,B_y)$, or $L(A_x,B_y)$ is an empty sequence, then at least one of $t \neq A_x$ and $t \neq B_y$ is true, if $t \neq A_x$, then there is $L(x,y) = L(x-1,y)$, if $t \neq B_y$, then $L(x,y) = L(x,y-1)$. However, we don't know t in advance, and take the largest one by definition. Therefore, in this case, $\text{LCS}(x,y) = \max(\text{LCS}(x-1,y),\text{LCS}(x,y-1))$.

Summarizing the above situation, we can get the recurrence relation for finding the length of the longest common subsequence:

$$\text{LCS}(x,y) = \begin{cases} \text{LCS}(x-1,y-1)+1, & A_x = B_y \\ \max(\text{LCS}(x-1,y),\text{LCS}(x,y-1)), & A_x \neq B_y \end{cases}$$

In order to obtain the longest common subsequence of the two sequences A_n and B_m, a two-dimensional array $S[n+1][m+1]$ is introduced here, where $S[i][j]$ is used to denote the search state in the process of computing the length of the longest common subsequence. With

$$S[i][j] = \begin{cases} (1) & \text{if } a_i = b_j \\ (2) & \text{if } a_i \neq b_j \text{ and } L(i-1,j) \geq L(i,j-1) \\ (3) & \text{if } a_i \neq b_j \text{ and } L(i-1,j) < L(i,j-1) \end{cases}$$

If $S[i][j] = 1$, it is the first case. This case indicates that $a_i = b_j$, indicating that the next search in the two-dimensional array is to search for $S[i-1][j-1]$ along the diagonal direction; if $S[i][j] = 2$, it is the second case, which means that the search direction in the two-dimensional array is along the horizontal direction to search $S[i][j-1]$; if $S[i][j] = 3$, it is the third case. In this case, the search direction is to search $S[i-1][j]$ along the vertical direction of the two-dimensional array.

y)也删掉最后一项 t 得到的序列是 $L(x-1,y-1)$. 那么它后面接上元素 t 得到的子序列 $L(x,y)$ 也比 $L(x-1,y-1)$ 接上元素 t 得到的子序列短,这与 $L(x,y)$ 是最长公共子序列矛盾,证毕. 得出 $\text{LCS}(A_x,B_y) = \text{LCS}(x-1,y-1)+1$.

(2) 当 $A_x \neq B_y$ 时,设 $t = L(A_x,B_y)$,或者 $L(A_x,B_y)$ 是空序列,则 $t \neq A_x$ 和 $t \neq B_y$ 至少有一个成立,若 $t \neq A_x$,则有 $L(x,y) = L(x-1,y)$,若 $t \neq B_y$,则有 $L(x,y) = L(x,y-1)$. 可是,我们事先并不知道 t,根据定义取最大的一个,因此这种情况下,有 $\text{LCS}(x,y) = \max(\text{LCS}(x-1,y),\text{LCS}(x,y-1))$.

总结上述情况,可以得出求最长公共子序列长度的递推关系式:

$$\text{LCS}(x,y) = \begin{cases} \text{LCS}(x-1,y-1)+1, & A_x = B_y \\ \max(\text{LCS}(x-1,y),\text{LCS}(x,y-1)), & A_x \neq B_y \end{cases}$$

为了得到两个序列 A_n 和 B_m 的最长公共子序列,在此引入一个二维数组 $S[n+1][m+1]$,其中 $S[i][j]$ 用于表示在计算最长公共子序列的长度过程的搜索状态. 有

$$S[i][j] = \begin{cases} (1) & \text{若 } a_i = b_j \\ (2) & \text{若 } a_i \neq b_j \text{ 且 } L(i-1,j) \geq L(i,j-1) \\ (3) & \text{若 } a_i \neq b_j \text{ 且 } L(i-1,j) < L(i,j-1) \end{cases}$$

如果 $S[i][j] = 1$,那么为第一种情况,出现这种情况说明 $a_i = b_j$,表明在该二维数组中的下一次搜索沿着对角线方向搜索 $S[i-1][j-1]$;如果是 $S[i][j] = 2$,那么为第二种情况,说明在二维数组中搜索方向是沿着二维数组 $S[i][j-1]$ 的水平方向;如果 $S[i][j] = 3$,那么为第三种情况,在这种情况下,搜索方向是沿着二维数组的垂直方向上搜索 $S[i-1][j]$.

10.4.2 Algorithm implementation of the longest common subsequence

Calculate the solution from the bottom up using dynamic programming. Take two sequences as input. Save the length of the calculated sequence to a two-dimensional array dp[M][N]. M and N respectively denote the length of the two sequences. The process code is as follows:

Algorithm 10-4:

```
    using namespace std;
    string X,Y;
    int dp[200][200];
int main(){
        cin>>X>>Y;
        int len1=X.length(),len2=Y.length();
        for(int i=1; i<=len1; i++){
          dp[i][0]=0;
          dp[0][i]=0;
        }
        for(int i=1; i<=len1; i++){
            for(int j=1; j<=len2; j++)
}
                if(X[i-1] == Y[j-1])
                    dp[i][j]=dp[i-1][j-1]+1;
                else
                    dp[i][j]=max(dp[i][j-1],dp[i-1][j]);
            }
        }
        cout<<"    ";
        for(int i=1; i<=len2; i++){
            cout<<" "<<Y[i-1];
        }
        cout<<"\n";
        for(int i=0; i<=len1; i++){
            if(i!=0)
                cout<<X[i-1]<<" ";
            else
                cout<<" \n";
```

```
for( int i = 0 ; i <= len1; i++){
if( i! = 0)
cout<<X[i-1]<<" ";
else
cout<<"  ";
for( int j = 0; j <= len2; j++){
    cout<<dp[i][j]<<" ";
}
cout<<"\n";
}
cout<<"\nans = "<<dp[len1][len2];

return 0;
}
```

10.5 0/1 knapsack problem

Given n kinds of items and a backpack with a capacity of c, the weight of item i is W_i and its value is V_i. Solve the problem of how to choose the items loaded into the backpack so that the total value of the items loaded into the backpack is maximized. This kind of problem is called a **knapsack problem**. The 0/1 knapsack problem discussed in this section is a knapsack problem in which the items put are indivisible.

10.5.1 Solution process of 0/1 knapsack problem

In the 0/1 backpack problem, item i is either loaded into the backpack or not loaded into the backpack. Let x_i denote the case that the item i is loaded into the backpack. When $x_i = 0$, it means that the item i is not loaded into the backpack. When $x_i = 1$, it means that the item i is loaded into the backpack. v_i denotes the value of the ith item, w_i denotes the volume (weight) of the ith item; according to the requirements of the problem, there are the following constraint conditions and objective functions:

Constraint conditions: $\begin{cases} \sum_{i=1}^{n} w_i x_i \leq C \\ x_i \in \{0,1\} (1 \leq i \leq n) \end{cases}$

Objective function: $\max \sum_{i=1}^{n} v_i x_i$

With the constraint conditions and objective function, the 0/1 knapsack problem boils down to finding a solution vector $X = (x_1, x_2, x_3, \ldots, x_n)$ that satisfies the above constraint conditions and maximizes objective function.

Using dynamic programming algorithm to solve the 0/1 knapsack problem, first prove whether the problem satisfies the optimality principle. The principle of optimality is the basis of dynamic programming. The principle of optimality refers to "the optimal decision sequence of a multi-stage decision-making process has such a property: Regardless of the initial state and initial decision, for a certain state caused by the previous decision, the decision sequence of each subsequent stage must constitute the optimal strategy". Judge whether the problem satisfies the optimality principle by using the contradiction method to prove:

Assuming $(X_1, X_2, \ldots, X_n)$ is the optimal solution of the 0/1 knapsack problem, then $(X_2, X_3, \ldots, X_n)$ is the optimal solution of the following subproblems:

$$\begin{cases} \sum_{i=2}^{n} w_i x_i \leq C - w_1 x_1 \\ x_i \in \{0,1\} \quad (2 \leq i \leq n) \end{cases}$$
$$\max \sum_{i=2}^{n} v_i x_i$$

Assuming $(Y_2, Y_3, \ldots, Y_n)$ is the optimal solution to the sub-problems of the above problem, it should have

$$v_1 y_1 + \sum_{i=2}^{n} v_i y_i > v_1 x_1 + \sum_{i=2}^{n} v_i x_i$$

But according to the definition of the series:

$$v_1 x_1 + \sum_{i=2}^{n} v_i x_i = \sum_{i=1}^{n} v_i x_i$$

We can draw:

$$v_1 y_1 + \sum_{i=2}^{n} v_i y_i > \sum_{i=1}^{n} v_i x_i$$

目标函数: $\max \sum_{i=1}^{n} v_i x_i$

有了约束条件和目标函数后,0/1背包问题就归结为寻找一个满足上述约束条件,并使目标函数达到最大的解向量$X = (x_1, x_2, x_3, \cdots, x_n)$.

使用动态规划算法解决0/1背包问题,首先证明该问题是否满足最优性原理. 最优性原理是动态规划的基础. 最优性原理是指"多阶段决策过程的最优决策序列具有这样的性质:不论初始状态和初始决策如何,对于前面决策所造成的某一状态而言,其后各阶段的决策序列必须构成最优策略". 判断该问题是否满足最优性原理,可采用反证法:

假设$(X_1, X_2, \cdots, X_n)$是0/1背包问题的最优解,则有$(X_2, X_3, \cdots, X_n)$是下列子问题的最优解:

$$\begin{cases} \sum_{i=2}^{n} w_i x_i \leq C - w_1 x_1 \\ x_i \in \{0,1\} \quad (2 \leq i \leq n) \end{cases}$$
$$\max \sum_{i=2}^{n} v_i x_i$$

假设$(Y_2, Y_3, \cdots, Y_n)$是上述问题的子问题的最优解,则理应有

$$v_1 y_1 + \sum_{i=2}^{n} v_i y_i > v_1 x_1 + \sum_{i=2}^{n} v_i x_i$$

但是根据级数定义可知:

$$v_1 x_1 + \sum_{i=2}^{n} v_i x_i = \sum_{i=1}^{n} v_i x_i$$

可以得出:

$$v_1 y_1 + \sum_{i=2}^{n} v_i y_i > \sum_{i=1}^{n} v_i x_i$$

This formula shows that $(X_1, Y_2, Y_3, \ldots, Y_n)$ is the optimal solution for the 0/1 knapsack problem, which contradicts the initial assumption $(X_1, X_2, \ldots, X_n)$ is the optimal solution for the 0/1 knapsack problem, so the 0/1 knapsack problem satisfies the optimality principle.

Looking for the recurrence relationship, in the 0/1 knapsack problem, there are two possibilities for the current product:

(1) The capacity of the bag is smaller than the volume of the product and cannot fit. The value at this time is the same as the value of the previous $i-1$, that is, $V(i,j) = V(i-1,j)$;

(2) There is still enough capacity to install the product, but it may not reach the current optimal value after it is loaded, so choose the best one between loaded and not loaded, that is, $V(i,j) = \max(V(i-1,j), V(i-1,j-w(i)) + v(i))$, where $V(i-1,j)$ means not to load, $V(i-1,j-w(i)) + v(i)$ means that the ith product is loaded, and the backpack capacity is reduced by $w(i)$ but the value is increased by $v(i)$; from this, the recurrence relation can be obtained as

$$V(i,j) = \begin{cases} V(i-1,j), & j < w_i \\ \max(V(i-1,j), V(i-1,j-w_i) + v_i), & j \geq w_i \end{cases}$$

10.5.2 Algorithm implementation of 0/1 knapsack problem

First define the data structure required by the algorithm:

```
int w[n];       //The volume of the product
int v[n];       //The value of the product
int bagV;       //Backpack size
int dp[n][n];   //Dynamic programming table
int item[n];    //Optimal solution
```

The 0/1 knapsack problem algorithm is implemented as follows:

该式子说明$(X_1, X_2, Y_3, \cdots, Y_n)$才是该0/1背包问题的最优解,这与最开始的假设$(X_1, X_2, \cdots, X_n)$是0/1背包问题的最优解相矛盾,故0/1背包问题满足最优性原理.

寻找递推关系式,在0/1背包问题中,当前商品有两种可能性:

(1) 包的容量比该商品体积小,装不下,此时的价值与前$i-1$个的价值是一样的,即$V(i,j) = V(i-1,j)$;

(2) 还有足够的容量可以装该商品,但装了也达不到当前最优价值,所以在装与不装之间选择最优的一个,即$V(i,j) = \max(V(i-1,j), V(i-1,j-w(i)) + v(i))$,其中$V(i-1,j)$表示不装,$V(i-1,j-w(i)) + v(i)$表示装了第$i$个商品,背包容量减少$w(i)$但价值增加了$v(i)$;由此可以得出递推关系式为

$$V(i,j) = \begin{cases} V(i-1,j), & j < w_i \\ \max(V(i-1,j), V(i-1,j-w_i) + v_i), & j \geq w_i \end{cases}$$

10.5.2 0/1背包问题的算法实现

首先定义好算法所需要的数据结构:

```
int w[n];       //商品的体积
int v[n];       //商品的价值
int bagV;       //背包大小
int dp[n][n];   //动态规划表
int item[n];    //最优解情况
```

0/1背包问题算法实现如下:

Algorithm 10-5:

```
void findMax() {   //Dynamic programming
    for (int i=1; i<=4; i++) {
        for (int j=1; j<=bagV; j++) {
            if (j < w[i])
                dp[i][j]=dp[i-1][j];
            else
                dp[i][j]=max(dp[i-1][j], dp[i-1][j-w[i]] + v[i]);
        }
    }
}
void findWhat(int i, int j) {//Optimal solution
    if (i>=0) {
        if (dp[i][j]==dp[i-1][j]) {
            item[i]=0;
            findWhat(i-1, j);
        }
        else if (j-w[i]>=0 && dp[i][j]==dp[i-1][j-w[i]]+v[i]) {
            item[i]=1;
            findWhat(i-1,j-w[i]);
        }
    }
}
void print() {
    for (int i=0; i<5; i++) {//Dynamic programming table output
        for (int j=0; j<9; j++) {
            cout<<dp[i][j]<<' ';
        }
        cout<<endl;
    }
    cout<<endl;
    for (int i=0;i<5;i++) //Optimal solution output
        cout<<item[i]<<' ';
    cout<<endl;
}
```

Use dynamic programming to solve the 0/1 knapsack problem. By analyzing the time performance of the code, it can be known that the time efficiency of dynamic programming is $O(\text{number} * \text{capacity}) = O(n * c)$. Since a two-dimensional array is used to store the solution of the sub-problem, so the space efficiency of dynamic programming is $O(n * c)$.

利用动态规划解决 0/1 背包问题,通过分析代码的时间性能可知,动态规划的时间效率为 $O(\text{number} * \text{capacity}) = O(n * c)$,因为用到了二维数组存储子问题的解,所以动态规划的空间效率为 $O(n * c)$。

Exercise 10

1. In $O(n^2 2^n)$ time, design a dynamic programming algorithm to solve the traveling salesman problem with a dynamic programming method, and estimate its space complexity.

2. Using the dynamic programming method, find the longest path and the shortest path from vertex 0 to vertex 6 in Fig 10-5.

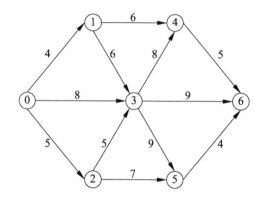

Fig. 10-5　Dynamic programming to find the path

3. There are character sequences $A = xyzzyxzyxxyx$, $B = zyxxyyzyxyzy$, find the solution procedure for the longest common subsequence and its length.

4. There are 6 items whose weights are 5, 3, 7, 2, 3, 4, and their values are 3, 6, 5, 4, 3, 4. There is a backpack with a load capacity of 15, and the items are indivisible. Find the maximum value of the object loaded into the backpack and its solution process.

5. Let $T = \{t_1, t_2, \ldots, t_n\}$ be a collection of n kinds of items, for all $1 \leq i \leq n$, w_i and v_i respectively denote the weight and value of item t_i, and the carrying capacity of the backpack is M. The task is to maximize the total value of items in the backpack while ensuring that the total weight of the items does

not exceed the capacity of the backpack. Write a dynamic programming algorithm to solve this problem.

算法来求解这个问题.

习题 10 答案
Key to Exercise 10

Chapter 11 Greedy Algorithm

11.1 Basic idea of greedy algorithm

Greedy algorithms are usually used to solve optimization problems with maximum or minimum values. It is like climbing a mountain, moving forward step by step, starting from a certain initial state, according to the current local optimal decisions rather than the global ones, with the condition that satisfies the constraint equation, with the criterion that makes the value of the objective function increase fastest or slowest, select an input element that can meet the requirements as quickly as possible, so as to form a feasible solution to the problem as soon as possible.

The design method of the greedy algorithm is described as follows:

Algorithm 11-1:
(1) greedy (A,n)
(2) {
(3)   Solution = φ;
(4)   for (i=1;i<n;i++) {
(5)   x = select(A);
(6)   if (feasible (solution, x))
(7)   Solution = union (solution, x);
(8)   }
(9)   return solution;
(10) }

At the beginning, make the initial solution vector solution empty; then, use "select" to select an input x from A according to a certain decision criterion, and use "feasible" to judge: after the solution vector "solution" is added to x, whether it is feasible, if feasible, merge x to the solution vector "solution" and delete it from A; otherwise, discard x, re-select another input from A, and repeat the above steps until a solution that satisfies the problem is found.

In general, the greedy algorithm consists of an iterative loop, in each loop, a small amount of local calculations are performed to try to find a local optimal solution without considering the future. Therefore, it builds the solution of the problem step by step. Each step of the work increases the scale of the partial solution, and the choice of each step greatly increases the objective function it hopes to achieve. Because each step is composed of a small amount of work based on a small amount of information, the resulting algorithm is particularly effective. However, because of this, in many instances, the local optimal solution produced by it can be transformed into the global optimal solution; but in some instances, the global optimal solution cannot be given. Therefore, when designing a greedy algorithm, the difficulty lies in proving that the designed algorithm is the optimal algorithm that really solves this problem.

Problems suitable for solution by greedy algorithms, generally have the following two important properties: greedy selection properties and optimal substructure properties.

The so-called greedy selection property refers to the global optimal solution of the problem sought, which can be achieved through a series of local optimal selections. Every time a selection is made, a partial solution is obtained and the problem to be solved is simplified into a similar sub-problem on a smaller scale.

For example, in the cashier's currency payment problem, the set $P = \{p_1, p_2, \ldots, p_{60}\}$ is used to denote the cash in the hands of the cashier, and the elements in the set denote, in order, the ten 10 yuan, ten 5 yuan, ten 1 yuan, ten 50 cents, ten 20 cents and ten 10 cents of currency in the cashier's hand; the vector $X = (x_1, x_2, \ldots, x_{60})$ denotes the currency that the cashier pays to the customer. In order to pay off 57 yuan and 80 cents as quickly as possible, and to minimize the number of currency paid, in the current state, choosing $p_1 = 10$ yuan can achieve this goal. Therefore, in the first step, the selected currency set is $S_1 = \{p_1\}$, a local solution $Y_1 = (1, 0, \ldots)$ is obtained, and the problem is simplified as a subproblem of choosing currency in the set $P = \{p_1, p_2, \ldots, p_{60}\}$, and paying off 47 yuan and 80 cents to the customer. In the subsequent steps, you can use the same method for selection, and the global optimal solution of the problem can be obtained.

The so-called optimal substructure refers to the optimal solution of a problem that contains the optimal solutions of its subproblems. In the above currency payment problem, the optimal solution for the set of currency paid to the customer is $S_n = \{p_1, p_2, p_3, p_4, p_5, p_{11}, p_{21}, p_{22}, p_{31}, p_{41}, p_{51}\}$. The optimal solution of the sub-problem simplified in the first step is $S_{n-1} = \{p_2, p_3, p_4, p_5, p_{11}, p_{21}, p_{22}, p_{31}, p_{41}, p_{51}\}$. Obviously, $S_{n-1} \subset S_n$, and $S_{n-1} \cup \{p_1\} = S_n$. Therefore, the cashier payment problem has an optimal substructure property.

Example 11.1.1 Assuming there are 5 cities, the cost matrix is shown in Fig. 11-1. If the salesperson starts from the first city and uses the greedy algorithm to solve the problem, then the solution process is shown in Fig. 11-2.

下,挑选 $p_1 = 10$ 元可以达到这个目的. 于是,在第一步,所挑出的货币集合是 $S_1 = \{p_1\}$,得到了一个局部解 $Y_1 = (1, 0, \cdots)$,并把问题简化为在集合 $P = \{p_1, p_2, \cdots, p_{60}\}$ 中挑选货币,付出 47 元 8 角给客户这样一个子问题. 在以后的步骤中,可以用同样的方法进行挑选,并能得到问题的全局最优解.

所谓最优子结构,是指一个问题的最优解中包含它的子问题的最优解. 在上述货币兑付问题中,付给客户的货币集合的最优解是 $S_n = \{p_1, p_2, p_3, p_4, p_5, p_{11}, p_{21}, p_{22}, p_{31}, p_{41}, p_{51}\}$. 第一步所简化了的子问题的最优解是 $S_{n-1} = \{p_2, p_3, p_4, p_5, p_{11}, p_{21}, p_{22}, p_{31}, p_{41}, p_{51}\}$. 显然,$S_{n-1} \subset S_n$,并且 $S_{n-1} \cup \{p_1\} = S_n$. 所以,出纳员付钱问题具有最优子结构性质.

例 11.1.1 假定有 5 个城市,费用矩阵如图 11-1 所示. 如果售货员从第一个城市出发,采用贪心算法求解,那么选择过程如图 11-2 所示.

	1	2	3	4	5
1	∞	3	3	2	6
2	3	∞	7	3	2
3	3	7	∞	2	5
4	2	3	2	∞	3
5	6	2	5	3	∞

图 11-1 5 个城市的费用矩阵

Fig. 11-1 Cost matrix of 5 cities

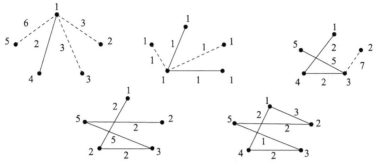

图 11-2 货郎担问题的求解过程
Fig. 11-2 Solution process of the travelling salesman problem

Since the route with the least cost is always selected to move forward, the route selected is 1→4→3→5→2→1, and the total cost is 14. It is easy to see that when using the greedy method, only one city is selected as the starting city, and the required running time is $O(n^2)$. If all n cities can be used as starting cities, then n routes can be obtained, and then the shortest route is selected from the n routes, the required running time is $O(n^3)$. If the exhaustive method is used, when $n>20$, it needs to run for tens of millions of years, while the greedy method can be completed in a short time. Compared with the exhaustive method, the efficiency is greatly improved. But the result is not the optimal route. The optimal route from city 1 is 1→2→5→4→3→1, and the total cost is only 13. If the city is taken as the vertex of the graph and the road between the cities is taken as the edge between the vertices, then the greedy method starts from city 1, and the selected edge set is $\{e_{14},e_{43},e_{35},e_{52},e_{21}\}$, but the edge set of the optimal solution is $\{e_{12},e_{25},e_{54},e_{43},e_{31}\}$, which shows that when the greedy method is used to solve the problem of traveling salesman, it does not have the property of optimal substructure and the property of greedy selection. Because the greedy method chooses the edge e_{14} as the element of the partial solution in the local state, it does not consider whether the future choice can still reach the optimum. Therefore, it cannot guarantee that edge e_{14} is

由于总是选择费用最少的路线前进,选择的路线是 1→4→3→5→2→1,总费用是 14. 容易看到,采用贪心法求解时,只选择一个城市作为出发城市,所需要的运行时间是 $O(n^2)$. 如果 n 个城市都可以作为出发城市,那么可以得到 n 条路线,然后从 n 条路线中选取一条最短的路线,所需运行时间是 $O(n^3)$. 如果采用穷举法,当 $n>20$ 时,需要运行数千万年,而采用贪心法,在很短的时间里就可完成. 与穷举法比较起来,效率大大提高了. 但所得结果不是最优的路线. 从城市 1 出发的最优的路线是 1→2→5→4→3→1, 总费用只有 13. 如果把城市作为图的顶点,把城市之间的道路作为顶点之间的边,那么利用贪心法从城市 1 出发,所选择的边集是 $\{e_{14},e_{43},e_{35},e_{52},e_{21}\}$, 而最优解的边集是 $\{e_{12},e_{25},e_{54},e_{43},e_{31}\}$, 这说明用贪心法来解货郎担问题时,不具有最优子结构性质,也不具有贪心选择性质. 因为贪心法在局部状态下,选择边 e_{14},作为部分解的元素时,它并没有考虑到将来的选择是否仍可以达到最优. 因此,它无法保证边 e_{14} 是全局最优解的元素.

an element of the global optimal solution.

11.2 The single-source shortest path problem

Given a directed weighted graph $G = (V, E)$, each edge in the graph has a non-negative length, and one vertex u is called the source vertex. The so-called single-source shortest path problem is to determine the distance from the source vertex u to all other vertices. Here, the distance from vertex u to vertex x is defined as the length of the shortest path from u to x. This problem can be implemented using the Dijkstra algorithm, which is based on the greedy method.

11.2.1 Dijkstra algorithm for solving the shortest path

Assume that (u, v) is the edge in E, and $c_{u,v}$ are the length of the edge. If the vertex set V is divided into two sets S and T: the distance between the vertices contained in S and u has been determined; the distance between the vertices contained in T and u has not yet been determined. At the same time, the distance $d_{u,x}$ from the source vertex u to the vertex x in T is defined as the length of the shortest path starting from u, passing through the vertices in S, but not passing other vertices in T, and directly reaching the vertex x in T, then the thinking method of the Dijkstra algorithm is as follows: At the beginning, $S = \{u\}$, $T = V - \{u\}$. For all vertices x in T, if there is an edge from u to x, set $d_{u,x} = c_{u,x}$; otherwise, set $d_{u,x} = \infty$. Then, for all vertices x in T, find the vertex t with the smallest $d_{u,x}$, namely

$$d_{u,t} = \min(d_{u,x} \mid x \in T)$$

Then $d_{u,t}$ is the shortest distance from vertex t to vertex u. At the same time, the vertex t is also the vertex closest to u among all the vertices in the set T. Delete the vertex t from T and merge it into S. Then, for all vertices x adjacent to t in T, use the following formula to update the value of $d_{u,x}$:

$$d_{u,x} = \min(d_{u,x}, d_{u,t} + c_{t,x})$$

11.2 单源最短路径问题

给定有向赋权图 $G = (V, E)$，图中每一条边都具有非负长度，其中有一个顶点 u 称为源顶点. 所谓单源最短路径问题，是确定由源顶点 u 到其他所有顶点的距离. 在这里，将顶点 u 到顶点 x 的距离定义为由 u 到 x 的最短路径的长度. 这个问题可以用狄斯奎诺算法来实现，它是基于贪心法的.

11.2.1 解最短路径的狄斯奎诺算法

假定 (u, v) 是 E 中的边，$c_{u,v}$ 是边的长度. 如果把顶点集合 V 划分为两个集合 S 和 T：S 中所包含的顶点到 u 的距离已经确定；T 中所包含的顶点到 u 的距离尚未确定. 同时，把源顶点 u 到 T 中顶点 x 的距离 $d_{u,x}$ 定义为从 u 出发，经过 S 中的顶点，但不经过 T 中其他顶点，而直接到达 T 中顶点 x 的最短路径的长度，则狄斯奎诺算法的思想方法如下：开始时，$S = \{u\}$，$T = V - \{u\}$. 对 T 中的所有顶点 x，如果存在从 u 到 x 的边，则设 $d_{u,x} = c_{u,x}$；否则，设 $d_{u,x} = \infty$. 然后，对 T 中的所有顶点 x，寻找 $d_{u,x}$ 最小的顶点 t，即

$$d_{u,t} = \min(d_{u,x} \mid x \in T)$$

则 $d_{u,t}$ 就是顶点 t 到顶点 u 的最短距离. 同时，顶点 t 也是集合 T 中的所有顶点中距离 u 最近的顶点. 从 T 中删去顶点 t，把它并入 S. 然后，对 T 中与 t 相邻的所有顶点 x，用下面的公式更新 $d_{u,x}$ 的值：

$$d_{u,x} = \min(d_{u,x}, d_{u,t} + c_{t,x})$$

Continue the above steps until T is empty.

Thus, if $p(x)$ is the previous vertex of x in the shortest path from vertex u to vertex x, then the Dijkstra algorithm can be described by the following steps:

(1) Set $S=\{u\}$, $T=V-\{u\}$.

(2) For $\forall x \in T$, if $(u,x) \in E$, then $d_{u,x}=c_{u,x}$, $p(x)=u$; otherwise, $d_{u,x}=\infty$, $p(x)=-1$.

(3) Find $t \in T$, so that $d_{u,t}=\min(d_{u,x} | x \in T)$, then $d_{u,t}$ is the distance from t to u.

(4) $S=S \cup \{t\}$, $T=T-\{t\}$.

(5) If $T=\varphi$, the algorithm ends; otherwise, go to step (6).

(6) For all vertices x adjacent to t, if $d_{u,x} \leq d_{u,t}+c_{t,x}$, go directly to step (3); otherwise, let $d_{u,x}=d_{u,t}+c_{t,x}$, $p(x)=t$, and go to step (3).

Example 11.2.1 In the directed weighted graph shown in Fig. 11-3, find the distance from vertex a to all other vertices. If the adjacency list is used to store the distance between vertices, the adjacency list is shown in Fig. 11-4.

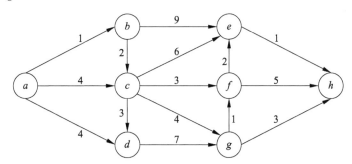

图 11-3 顶点 a 到其他所有顶点的最短距离的有向赋权图
Fig. 11-3 A directed weighted graph with the shortest distance from vertex a to all other vertices

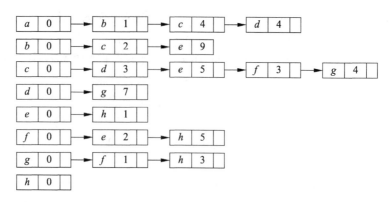

图 11-4 图 11-3 中所表示的赋权图的邻接表
Fig. 11-4 The adjacency list of the weighted graph shown in Fig. 11-3

Tab. 11-1 shows the execution process of each cycle when the Dijkstra algorithm is executed on the above directed weighted graph. The path from each vertex to vertice a is shown in Fig. 11-5.

表 11-1 表示对上面的有向赋权图执行狄斯奎诺算法时,每一轮循环的执行过程. 各个顶点到顶点 a 的路径如图 11-5 所示.

表 11-1 狄斯奎诺算法的执行过程
Tab. 11-1 The execution process of the Dijkstra algorithm

	S	$d_{a,b}$	$d_{a,c}$	$d_{a,d}$	$d_{a,e}$	$d_{a,f}$	$d_{a,g}$	$d_{a,h}$	$d_{a,t}$	t
1	a	1	4	4	∞	∞	∞	∞	1	b
2	a,b		3	4	10	∞	∞	∞	3	c
3	a,b,c			4	9	6	7	∞	4	d
4	a,b,c,d				9	6	7	∞	6	f
5	a,b,c,d,f				8		7	11	7	g
6	a,b,c,d,f,g				8			10	8	e
7	a,b,c,d,f,g,e							9	9	h
8	a,b,c,d,f,g,e,h									

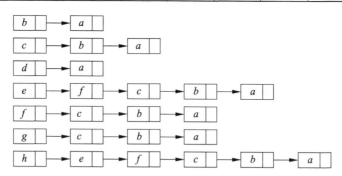

图 11-5 各个顶点到顶点 a 的路径
Fig. 11-5 The path from each vertex to vertex a

11.2.2 Implementation of Dijkstra algorithm

For simplicity, in the directed weighted graph $G=(V, E)$, the vertices are numbered with numbers, and the set of vertices is $V=\{0, 1, \ldots, n-1\}$; put the length of the edges (i, j) in the edge set E stored in the adjacency list of the graph; use the Boolean array s to denote the vertices in S, $s[i]$ is true, it means that the vertex i is in S, otherwise it is not in S; use array element $d[i]$ to denote the distance from vertex i to the source vertex; use array element $p[i]$ to store the number of the front vertex on the shortest path from vertex i to the source vertex; and assume that the source vertex is given by the variable u. The data structure of the graph adjacency list is defined as follows:

(1) struct adj_list {/* Data structure of adjacency list node */
(2)   int v_num;   /* Number of the adjacency vertices */
(3)   float len;   /* Distance between the adjacency vertex and this vertex */
(4)   struct adj_list *next;   /* Next adjacency vertex */
(5) };

Then the description of the Dijkstra algorithm is as follows:

Algorithm 11-2: Dijkstra Algorithm

Input: the number of vertices n, the head node of the adjacency list of the directed graph, node[ ], the source vertex u

Output: distance $d[\ ]$ between other vertices and the source vertex u, the number of the front vertex on the shortest path to the source vertex $p[\ ]$

(1) #define MAX_FLOT_NUM 3.14E38
           /* Largest floating point */
(2) void dijkstra (NODE node[ ], int n, int u, float d[ ], int p[ ])
(3) {
(4)   float temp;
(5)   int i, j, t;
(6)   BOOL *s=new BOOL(n);
(7)   NODE *pnode;

(8)　　for (i=0;i<n;i++){
(9)　　　　d[i]=MAX_FLOT_NUM;　s[i]=FALSE;
　　　　　p[i]=-1;　　　　　/* Initialization */
(10)　　}
(11)　　if (!(pnode=node[u].next))
　　/* Source vertex are not adjacent to other vertices */
(12)　　　　return;
(13)　　while (pnode) {
　　　　　/* Preset the distance of the vertex adjacent
　　　　　　to the source vertex */
(14)　　　　d[pnode->v_num]=pnode->len;
(15)　　　　p[pnode->v_num]=u;
(16)　　　　pnode=pnode->next;
(17)　　}
　/* At the beginning, the set s contains only the vertex u */
(18)　　d[u]=0;　s[u]=TRUE;
(19)　　for (i=1;i<n;i++) {
(20)　　　　temp=MAX_FLOT_NUM; t=u;
　　　　　　/* Find the closest vertex t to u */
(21)　　　　for (j=0;j<n;j++)
(22)　　　　　　if(!s[j]&&d[j]<temp) {
(23)　　　　　　　　t=j;　temp=d[j];
　　　　　　　/* If not found, break out of the loop */
(24)　　　　　　}
　　　　　　/* Otherwise, merge t into the set s */
(25)　　　　if (t==u) break;
　/* Update the distance from the vertex adjacent t to u */
(26)　　　　s[t]=TRUE;
(27)　　　　pnode=node[t].next;
(28)　　　　while (pnode) {
(29)　if(!s[pnode->v_num]&&d[pnode->v_num]>d[t]+pnode->len) {
(30)　　　　　　　d[pnode->v_num]=d[t]+pnode->len;
(31)　　　　　　　p[pnode->v_num]=t;
(32)　　　　　　}
(33)　　　　　　pnode=pnode->next;
(34)　　　　}
(35)　　}
(36)　　delete s;}

(8)　　for (i=0;i<n;i++){
(9)　　　　d[i]=MAX_FLOT_NUM;
　　　　　s[i]=FALSE;p[i]=-1;
　　　　　　　　　　/* 初始化 */
(10)　　}
(11)　　if (!(pnode=node[u].next))
　　　　　/* 源顶点不与其他顶点相邻 */
(12)　　　　return;
(13)　　while (pnode) {
　　　　　/* 设置与源顶点相邻的顶点的距离 */
(14)　　　　d[pnode->v_num]=pnode->len;
(15)　　　　p[pnode->v_num]=u;
(16)　　　　pnode=pnode->next;
(17)　　}
　　　/* 开始时,集合 s 仅包含顶点 u */
(18)　　d[u]=0;　s[u]=TRUE;
(19)　　for (i=1;i<n;i++) {
(20)　　　　temp=MAX_FLOT_NUM; t=u;
　　　　　　/* 找到距离顶点 u 最近的顶点 t */
(21)　　　　for (j=0;j<n;j++)
(22)　　　　　　if(!s[j]&&d[j]<temp) {
(23)　　　　　　　　t=j;　temp=d[j];
　　　　　　　　/* 如果没有找到,跳出循环 */
(24)　　　　　　}
　　　　　　/* 否则,将 t 合并到集合 s 中 */
(25)　　　　if (t==u) break;
　　　/* 更新与顶点 t 相邻的顶点到顶点 u 的距离 */
(26)　　　　s[t]=TRUE;
(27)　　　　pnode=node[t].next;
(28)　　　　while (pnode) {
(29)　if(!s[pnode->v_num]&&d[pnode->v_num]>d[t]+pnode->len) {
(30)　　　　　　　d[pnode->v_num]=d[t]+pnode->len;
(31)　　　　　　　p[pnode->v_num]=t;
(32)　　　　　　}
(33)　　　　　　pnode=pnode->next;
(34)　　　　}
(35)　　}
(36)　　delete s;}

At the beginning, the head node of the adjacency list of the directed graph is stored in the array node[ ]. Therefore, the length of all the out edges associated with vertex i and the numbers of all the vertices adjacent to vertex i are stored in the linked list pointed to by node[i]. The algorithm is divided into two stages: the initialization stage and the selection of the vertex with the shortest distance stage. In the initialization stage, lines 8~10 of the algorithm set the distance from the source vertex to all other vertices to infinity, set the set S to be empty, and set the number of the front vertex on the shortest path from all vertices to the source vertex to −1; lines 11~12 judge whether the source vertex has adjacent vertices. If not, it means that the source vertex and other vertices are not reachable, then the algorithm ends. Otherwise, lines 13~17 preset the distance from the source vertex to the adjacent vertices. At this time, there are only these adjacent vertices x, their distance to the source vertex $d[x]$ is assigned, and the distance from other vertices to the source vertex is still infinite; line 18 merges the source vertex u into the set S, ending the initialization stage.

In the stage of selecting the vertex with the shortest distance, because there are n vertices, the algorithm executes a loop with $n-1$ rounds. In lines 20~24, look for the vertex t closest to u in T. If it cannot be found, the vertex u to the vertex in T is unreachable, and the algorithm ends. Otherwise, it is the vertex we are looking for and merge it into S; lines 27~34 update the distance from the vertex adjacent to t to u, and then enter a new loop. Finally, either $n-1$ vertices are processed, or several vertices are unreachable.

11.2.3 Analysis of Dijkstra's algorithm
The time complexity of this algorithm is estimated as follows: lines 8~10 cost $O(n)$ time; lines 13~17 cost $O(n)$ time; lines 19~35 are a double loop, and the loop body of the outer loop executes at most $n-1$ rounds,

in the inner loop of lines 21~24, it takes $O(n)$ time at most to find the vertex t closest to u in T. Lines 27~34 update the distance from the vertex adjacent to t to u, it costs at most $O(n)$ time. These two inner loops need to execute at most $n-1$ rounds, therefore, it takes $O(n^2)$ time for lines 19~35. Therefore, the time complexity of the algorithm is $O(n^2)$. In addition, this algorithm requires $O(n)$ workspace.

The correctness of the algorithm is proved below. First of all, in the formula $d_{u,t} = \min(d_{u,x} | x \in T)$, the path length $d_{u,x}$ from vertex u to vertex x in the set $\{d_{u,x} | x \in T\}$, it only denotes the path length that in the current search process, the detected vertex u passes through the vertices in S, but does not pass through the vertices in T to reach x, it is not necessarily the shortest path length from vertex u to vertex x. Because the shortest path from u to x may contain other vertices in T except vertex x. For example, in Fig. 11-6, the path length $d_{u,t}$ from u to t via v may be 8, and the path length $d_{u,x}$ from u to x via w may be 12. At this time, it cannot be considered that $d_{u,x}$ is the length of the shortest path from u to x, because there may be a path of length 2 from t to x. If so, the length of the path from u to x via t is only 10. However, as long as the equation $d_{u,t} = \min(d_{u,x} | x \in T)$ is satisfied, $d_{u,t}$ must be the shortest distance from vertex u to vertex t. This is confirmed by the following theorem.

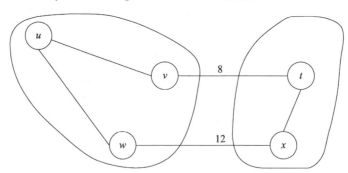

图 11-6 说明顶点距离的例子

Fig. 11-6 Example illustrating vertex distance

Theorem 11.2.1 Let $G=(V,E)$ be a directed weighted graph, $S\subseteq V, u\in S, T=V-S$, if $t\in T, d_{u,t}=\min(d_{u,x}|x\in T)$, then $d_{u,t}$ is the distance from vertex u to vertex t.

Proof Use proof by contradiction. If there is a path from u to t. If the length is less than $d_{u,t}$, the path must contain the vertices in $T-\{t\}$. If x is such a vertex, then $d_{u,x}<d_{u,t}$, and $d_{u,t}=\min(d_{u,x}|x\in T)$ is contradictory. So $d_{u,t}$ is the distance from vertex u to t.

In this way, as long as it is proved that the $d_{u,x}$ obtained in the formula $d_{u,x}=\min(d_{u,x}, d_{u,t}+c_{t,x})$ is the length of the shortest path that passes through the vertices in $S\cup\{t\}$, but does not pass through other vertices in $T-\{t\}$, and directly reaches x, then the $d_{u,x}$ after processing by the formula $d_{u,x}=\min(d_{u,x}, d_{u,t}+c_{t,x})$, then after the processing of $d_{u,t}=\min(d_{u,x}|x\in T)$, the new $d_{u,t}$ obtained will be the distance from the vertex u to the new vertex t. In this way, it proved the correctness of the algorithm. The following theorem proves the correctness of the formula $d_{u,x}=\min(d_{u,x}, d_{u,t}+c_{t,x})$.

Theorem 11.2.2 Let $G=(V,E)$ be a directed weighted graph, $S\subseteq V$, $u\in S$, $T=V-S$, $t\in T$, $d_{u,t}=\min(d_{u,x}|x\in T)$; let $\overline{S}=S\cup\{t\}$, $\overline{T}=T-\{t\}$, for any $x\in\overline{T}$, there is

$$d_{u,x}=\min(d_{u,x}, d_{u,t}+c_{t,x})$$

Proof The distance $d_{u,x}$ from the source vertex u to the vertex x in $\overline{T}$ is defined as the length of the shortest path starting from u, passing through the vertices in S, but not passing other vertices in $\overline{T}$, and directly reaching the vertex x in $\overline{T}$. The shortest path from u to x without passing through the vertices in $\overline{T}$ has the following two cases:

(1) The vertices on the path are all in S except for x, so that it does not pass through the vertex t, then

the $d_{u,x}$ in the right side of the above formula is the length of the shortest path directly to x through S without passing through T. Therefore, the formula $d_{u,x} = \min(d_{u,x}, d_{u,t}+c_{t,x})$ holds.

(2) The vertices on the path, in addition to the vertices in S, also pass the vertex t, then directly get $d_{u,x} = d_{u,t} + c_{t,x}$. Therefore, the formula $d_{u,x} = \min(d_{u,x}, d_{u,t}+c_{t,x})$ still holds.

In summary, Dijkstra's algorithm is correct.

11.3 The minimum spanning tree problem

In real life, the minimum cost spanning tree problem of graphs has a wide range of applications. For example, if the vertices of the graph denote cities, the edges between vertices denote roads or communication lines between cities, and the weights of edges denote the length of roads or the cost of communication lines, the minimum cost spanning tree problem is denoted as the problems of the shortest roads or the least costly communication lines between cities.

11.3.1 Introduction to minimum spanning tree

Definition 11.3.1 Assume graph $G = (V, E)$ and graph $G' = (V', E')$. If $G' \subseteq G$ and $V' = V$, then G' is said to be a spanning subgraph of G.

For example, Fig. 11-7a is an undirected complete graph, Fig. 11-7b ~ e are its several spanning subgraphs.

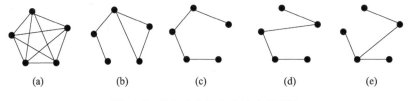

Fig. 11-7 Undirected complete graph and its spanning subgraphs

Definition 11.3.2 If the spanning subgraph T of the undirected graph G is a tree, then T is called the spanning tree or supporting tree of G. The edges in the spanning tree T are called branches.

For example, Fig. 11-7b is not the spanning tree of Fig. 11-7a, but Fig. 11-7c~e are all the spanning trees of Fig. 11-7a.

If the connected graph $G=(V,E)$, T is the spanning tree of G, then the spanning tree T has the following properties:

Property 1 T is a connected graph without simple loops.

Property 2 Every pair of vertices u and v in T has exactly one basic path from u to v.

Property 3 If $|V|=n$ and $|E|=m$, then $m=n-1$.

Property 4 Adding an edge between any two non-adjacent vertices in T will get the only basic path in T.

Definition 11.3.3 If the graph $G=(V, E, W)$ is a weighted graph, and T is the spanning tree of G. The sum of the weights on each branch of T is called T's weight. The spanning tree with the smallest weight in G is called the minimum cost spanning tree or minimum spanning tree of G.

In the following discussion, it is assumed that the graphs are all connected. If the graph is not connected, the algorithm can be applied to each connected branch of the graph.

11.3.2 Kruskal's algorithm

There are many algorithms for finding the minimum cost spanning tree. Among them, Kruskal's algorithm and Prim's algorithm are typical algorithms designed using greedy algorithm strategy.

1. Thinking method of Kruskal's algorithm

Kruskal's algorithm is commonly known as the loop avoidance method. Its thinking method is as follows: At the beginning, all the vertices of the graph are regarded as isolated vertices, and each vertex constitutes a tree with only root nodes, and these trees constitute a forest T; then, all edges are sorted by weight in non-descending order to form a non-descending sequence of the edge set; take the edge with the smallest weight from the edge set. If this edge is added to the forest T, it will not make T form a loop, then add it to the forest (or connect some two trees in the forest to form a tree); otherwise, give it up. In the both cases, delete it from the edge set; repeat this process until all $n-1$ edges are placed in the forest, after that, the process ends, and all the trees in the forest are connected to a tree T, which is the minimum spanning tree of the required graph.

When the edge e is added to T, if the vertices u and v associated with the edge e are on two trees respectively, with the addition of the edge e, the two trees will be merged into one tree; if the vertices u and v associated with edge e are both in the same tree, with the newly addition of the edge e, it will connect these two nodes, so that the original tree forms a loop. In order to judge whether adding edge e to T will form a loop, the operations of find(u), find(v) and union(u,v) can be used. The first two operations look for the root node of the tree where u and v are located. If the find(u) and find(v) operations indicate that the root nodes of u and v are not the same, continue the union(u,v) operation that will add edge e to T, and merge the two trees where u and v are into one tree; if the find(u) and find(v) operations show that the positions of root nodes of u and v are the same, then u and v are on a same tree, the union(u,v) operation is not executed at this time, and edge e is discarded.

Therefore, for an undirected connected weighted graph $G=(V, E, W)$, the steps of Kruskal's algorithm for finding the minimum cost spanning tree of the graph can be described as follows:

(1) Sort the edges in E in non-descending order of weight.

(2) Let the edge set of the minimum cost spanning tree be T, and T is initialized to $T=\varphi$.

小的一条边,如果把这条边加入森林 T 中,不会使 T 构成回路,那么就把它加入森林中(或者是把森林中某两棵树连接成一棵树),否则就放弃它,在这两种情况下,都把它从边集中删去;重复这个过程,直到把 $n-1$ 条边都放到森林中,结束这个过程,这时这个森林中所有的树就被连接成一棵树 T,它就是所要求取的图的最小生成树.

在把边 e 加入 T 中时,如果与边 e 相关联的顶点 u 和 v 分别在两棵树上,随着边 e 的加入,这两棵树合并成一棵树;如果与边 e 相关联的顶点 u 和 v 都在同一棵树上,则新加入的边 e 将把这两个结点连接起来,使原来的树构成回路. 为了判断把边 e 加入 T 中是否会构成回路,可以使用 find(u)、find(v) 操作及 union(u,v) 操作. 前两个操作寻找 u 和 v 所在树的根结点,如果 find(u)、find(v) 操作表明 u 和 v 的根结点不相同,那么继续 union(u,v) 操作,将把边 e 加入 T 中,并使 u 和 v 所在的两棵树合并成一棵树;如果 find(u)、find(v) 操作表明 u 和 v 的根结点位置相同,那么 u 和 v 同在一棵树上,这时就不执行 union(u,v) 操作,并丢弃边 e.

于是,对无向连通赋权图 $G=(V, E, W)$,求该图的最小花费生成树的克鲁斯卡尔算法的步骤,可叙述如下:

(1) 按权的非降顺序排序 E 中的边.

(2) 令最小花费生成树的边集为 T,T 初始化为 $T=\varphi$.

(3) Initialize each vertex as the root node of the tree.

(4) Let $e=(u,v)$ be the edge with the smallest weight in E, and $E=E-\{e\}$.

(5) If find$(u) \neq$ find(v), perform union(u,v) operation, $T=T\cup\{e\}$.

(6) If $|T|<n-1$, go to step (4); otherwise, the algorithm ends.

Example 11.3.1 Fig. 11-8 shows the execution process of Kruskal's algorithm. Fig. 11-8a shows an undirected weighting graph; Steps ① and ② add edges with weights 1 and 2 to T respectively, as shown in Fig. 11-8b~c; in step ③, the edge with weight 3 and the edge in T form a loop and are discarded; in steps ④ and ⑤, the edges with weights 4 and 5 are added to T, as shown in Fig. 11-8d~e; in steps ⑥, ⑦ and ⑧, the edges with weights 6, 7, and 8 form a loop with the edges in T and are discarded; in step ⑨, add the edge with weight 9 to T, as shown in Fig. 11-8f; so far, 5 edges have been added to T, and the number of vertices is 6, so the algorithm ends.

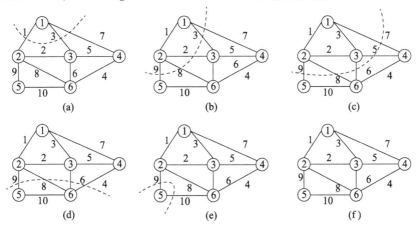

Fig. 11-8 Execution process of Kruskal's algorithm

2. Realization of Kruskal's algorithm

Assume that the undirected weighted graph $G=(V, E, W)$ has n vertices and m edges. For simplicity, vertices are numbered numerically. Define the following data structure:

```
(1)   typedef struct {      /* Data structure of the edge */
(2)       float key;        /* Weight of edge */
(3)       int u;            /* Vertex number associated with
                                the edge */
(4)       int v;            /* Vertex number associated with
                                the edge */
(5)   } EDGE;
(6)   struct node {         /* Data structure of the vertex */
(7)       struct node *p;
                            /* Point to parent node */
(8)       int rank;         /* Rank of the node */
(9)       int u;            /* Vertex number */
(10)  };
(11)  typedef struct node NODE;
(12)  EDGE E[m+1], T[n];
(13)  NODE V[n];
```

Among them, use the array E to store the edge set to form a minimum heap; use the array V to store the vertex set for "find" and "union" operations, and easily determine whether the added edges will form a loop; use the array T to store the edges of the generated minimum cost spanning tree. Make some necessary modifications to "find" and "union" operations and heap operations to adapt them to the above data structure, Kruskal's algorithm can be described as follows.

Algorithm 11-3: Kruskal's Algorithm

Input: array $V[\ ]$ storing n vertices, array $E[\ ]$ storing m edges

Output: array $T[\ ]$ storing the edge set of the minimum cost spanning tree

```
(1)  void kruskal (NODE V[], EDGE E[], EDGE T[],
         int n, int m)
(2)  {
(3)      int i,j,k;
(4)      EDGE e;
(5)      NODE *u, *v;
(6)      make_heap(E,m);  /* Minimal heap with edge set */
(7)      for (i=0;i<n;i++) {
```
/* Each vertex acts as the root node of the tree, forming a forest */

```
(8)         V[i].rank=0;    V[i].p=NULL;
(9)     }
(10)    i=j=0;
(11)    while((i<n-1)&&(m>0)){
(12)        e=delete_min(E,&m);
/* Remove the edge with the smallest weight from the min-heap */
(13)        u=find(&V[e.u]);
/* Retrieve the root node of the tree where the vertex adjacent
   to the edge is located */
(14)        v=find(&V[e.v]);
(15)        if(u!=v){
            /* Two root nodes are not in the same tree */
(16)            union(u,v);  /* Connect them */
(17)            T[j++]=e;
            /* Add edges to the minimum cost spanning tree */
(18)            i++;
(19)        }
(20)    }
(21) }
```

Line 6 composes the array E into a minimum heap according to the weight of the edge. Lines 7~8 regard each vertex as the root node of the tree to form a forest. Lines 11~20 execute a loop to remove the edge e with the smallest weight from the minimum heap, and reduce the number of edges m by 1; use the "find" operation to obtain the root node of the tree where the two vertices adjacent to the edge are located, if these two root nodes are not the root nodes of the same tree, use the "union" operation to merge the two trees into one tree, and add edge e to the minimum cost spanning tree T. This loop continues to execute until the minimum cost spanning tree that produces $n-1$ edges or all m edges have been processed.

3. Analysis of Kruskal's algorithm

Line 6 of the algorithm uses m edges to form the minimum heap, which takes $O(m\log m)$ time. Lines 7~8 initialize n root nodes, which requires $O(n)$ time. The loop in lines 11~20 is executed at most $n-1$ times. In the body of the loop, the edge with the smallest weight is deleted from the smallest heap in line 12. Each execution takes $O(\log m)$ time, and a total of $O(n\log m)$ time; the "find" operation in the loop body can be executed at most $2m$ times, and the total cost is at most $O(m\log n)$ time. Therefore, the running time of the algorithm is determined by line 6, and the time taken is $O(m\log m)$. If the graph being processed is a complete graph, then there will be $m = n(n-1)/2$, which is measured by the number of vertices, and the time taken is $O(n^2\log n)$; if the graph being processed is a flat graph, then, there will be $m = O(n)$, and the time taken at this time is $O(n\log n)$. In addition, the space required by the algorithm to store the edge set of the minimum cost spanning tree is $O(n)$, and the remaining work unit required is $O(1)$.

The following proves the correctness of the algorithm, with the following theorem:

Theorem 11.3.1 Kruskal's algorithm correctly obtains the minimum spanning tree of an undirected weighted graph.

Proof Let G be an undirected connected graph, T^* is the edge set of G's minimum spanning tree; T is the spanning tree edge set generated by Kruskal's algorithm, then the vertices in O are both the vertices in T^* and the vertices in T. If the number of vertices of G is n, then $|T^*| = |T| = n - 1$. Let's use induction to prove that $T = T^*$.

(1) Let e_1 be the edge with the smallest weight in G. According to Kruskal's algorithm, $e_1 \in T$. At this time, if $e_1 \notin T^*$, because T^* is the minimum cost spanning tree of G, the vertices associated with e_1 must be two non-adjacent vertices in T^*. According to property 4 of spanning tree, adding e_1 to T^* will make T^* the only loop. Assume this loop is $e_1, e_{a2}, \cdots, e_{ak}$, and e_1 is the edge with the smallest weight in this loop. Let $T^{**} = T \cup \{e_1\} - \{e_{ai}\}$ and e_{ai} be any edge in the loop $e_1, e_{a2}, \cdots, e_{ak}$ except e_1, then the edge set T^{**} is still the spanning tree of G, and the weight of T^{**} is smaller than or equal to the weight of T^*. If the weight of T^{**} is greater than the weight of T^*, it contradicts that T^{**} is the edge set of the mini-

时间;循环体中的 find 操作,最多执行 $2m$ 次,总花费最多为 $O(m\log n)$ 时间. 因此,算法的运行时间由第 6 行所决定,所花费的时间为 $O(m\log m)$. 如果所处理的图是一个完全图,那么,将有 $m = n(n-1)/2$,这时用顶点个数来衡量,所花费的时间为 $O(n^2\log n)$;如果所处理的图是一个平面图,那么将有 $m = O(n)$,这时所花费的时间为 $O(n\log n)$. 此外,算法用来存放最小花费生成树的边集所需要的空间为 $O(n)$,其余需要的工作单元为 $O(1)$.

下面证明算法的正确性,有下面的定理:

定理 11.3.1 克鲁斯卡尔算法正确地得到无向赋权图的最小生成树.

证明 G 是无向连通图,T^* 是 G 的最小生成树的边集;T 是由克鲁斯卡尔算法所产生的生成树边集,则 O 中的顶点,既是 T^* 中的顶点,也是 T 中的顶点. 若 G 的顶点数为 n,则 $|T^*| = |T| = n-1$. 下面用归纳法证明 $T = T^*$.

(1)设 e_1 是 G 中权最小的边,根据克鲁斯卡尔算法,有 $e_1 \in T$. 此时,若 $e_1 \notin T^*$,因为 T^* 是 G 的最小花费生成树,所以,与 e_1 关联的顶点必是 T^* 中的两个不相邻的顶点,根据生成树的性质4,把 e_1 加入 T^*,将使 T^* 构成唯一的一条回路. 假定这条回路是 $e_1, e_{a2}, \cdots, e_{ak}$,且 e_1 是这条回路中权值最小的边. 令 $T^{**} = T \cup \{e_1\} - \{e_{ai}\}$,$e_{ai}$ 是回路 $e_1, e_{a2}, \cdots, e_{ak}$ 中除 e_1 外的任意一条边,则边集 T^{**} 仍然是 G 的生成树,且 T^{**} 的权小于或等于 T^* 的权. 如果 T^{**} 的权值大于 T^* 的权,则与 T^{**} 是 G 的最小花费生成树的边集相矛盾,所

mum cost spanning tree of G, so $e_1 \in T^*$; if the weight of $e_1 \in T$ is equal to the weight of T^*, then T^{**} is also the edge set of G's minimum cost spanning tree, and $e_1 \in T^{**}$. At this time, a new T^* can be used to mark T^{**}. In both cases, there is $e_1 \in T^*$.

(2) If e_2 is the second smallest edge in G, the same can be proved $e_2 \in T$, and $e_2 \in T^*$.

(3) Let $e_1, \cdots, e_k$ be the first k edges with the smallest weight in G, and they all belong to T, and also belong to T^*, let e_{k+1} be the $(k+1)$-th edge with the smallest weight in G, and $e_{k+1} \in T$, but $e_{k+1} \notin T^*$. Similarly, the vertices associated with e_{k+1} are also two non-adjacent vertices in T^*. Adding e_{k+1} to T^* will make T^* the only loop. Assuming that this loop is $e_{k+1}, e_{a1}, \ldots, e_{am}$, in $e_{a1}, \ldots, e_{am}$, there must be an edge $e_{ai} \in T^*$, but $e_{ai} \notin T$, otherwise there will be a loop in T. Because $e_1, \ldots, e_{k+1}$ is the first $k+1$ edges with the smallest weight in G, and all $e_1, \ldots, e_k$ belong to T, the weight of e_{ai} is greater than or equal to the weight of e_{k+1}. Let $T^{**} = T \cup \{e_{k+1}\} - \{e_{ai}\}$, then T^{**} is still the spanning tree of G, and the weight of T^{**} is less than or equal to the weight of T^*. Same as the above reason, there must be $e_{k+1} \notin T^*$.

(4) $e_1, \ldots, e_k$ is the first k edges with the smallest weight in G, and they all belong to T, and also belong to T^*, let e_{k+1} be the $(k+1)$-th edge with the smallest weight in G, and $e_{k+1} \notin T$, but $e_{k+1} \in T^*$. Because $e_1, \ldots, e_k$ all belong to T, and $e_{k+1} \notin T$, according to Kruskal's algorithm, there must be $e_1, \ldots, e_k$, e_{k+1} to form a loop. Because $e_1, \ldots, e_k, e_{k+1}$ also belongs to T^*, if $e_{k+1} \in T^*$, T^* will have a loop. Therefore only $e_{k+1} \notin T^*$.

In summary, there is $T = T^*$. Therefore, Kruskal's algorithm correctly obtains the minimum spanning tree of the undirected weighted graph.

11.3.3 Prim's algorithm

Prim's algorithm is also an algorithm designed with a greedy strategy, but it is completely different from the Kruskal's algorithm, and is somewhat similar to the

Dijkstra algorithm for finding the shortest path. Here, it is also assumed that the graph G is connected.

1. Thinking method of Prim's algorithm

Let $G=(V, E, W)$, for simplicity, let the vertex set be $V=\{0,1,\ldots,n-1\}$. Assume that the edges associated with vertices i, j are $e_{i,j}$, and the weights of $e_{i,j}$ are denoted by $c[i][j]$, and T is the edge set of the minimum spanning tree. This algorithm maintains two sets of vertices S and N, at the beginning: let $T=\varphi$, $S=\{0\}$, $N=V-S$. Then, make a greedy choice. Choose $i \in S, j \in N$, and i and j that minimize $c[i][j]$; and set $S=S \cup \{j\}$, $N=N-\{j\}$, $T=T \cup \{e_{i,j}\}$. Repeat the above steps until N is empty or $n-1$ edges are found. At this time, the edge set in T is the minimum cost spanning tree in G that is required to be taken. Therefore, the steps of Prim's algorithm can be described as follows:

(1) $T=\varphi, S=\{0\}, N=V-S$.

(2) If N is empty, the algorithm ends; otherwise, go to step (3).

(3) Find i and j that make $i \in S, j \in N$, and $c[i][j]$ the smallest.

(4) $S=S \cup \{j\}, N=N-\{j\}, T=T \cup \{e_{i,j}\}$; go to step (2).

Example 11.3.2 Fig. 11-9 shows the working process of Prim's algorithm. One side of the dashed line denotes the set of vertices S, the other side denotes the set of vertices N, the thin lines denote the edges associated with the vertices, and the thick lines denote the minimum spanning tree generated. At the beginning, as shown in Fig. 11-9a, at this time, $S=\{1\}, N=\{2,3,4,5,6\}$. In the set N, there are 3 vertices 2, 3, 4 adjacent to the set S, and the weight of the edge $e_{1,2}$ is the smallest; in Fig. 11-9b, the vertex 2 is merged into the set S, and the edge $e_{1,2}$ is merged into T. At this time, vertices 3, 4, 5, and 6 are adjacent to set S, and edge $e_{2,3}$ has the smallest weight; in Fig. 11-9c, vertex 3 is merged into set S, and edge $e_{2,3}$ is merged into T. At this time, vertices 4, 5, and 6 are adjacent to set S, and the weight of

最短路径的狄斯奎诺算法. 在这里, 也假定图 G 是连通的.

1. 普里姆算法的思想方法

令 $G=(V, E, W)$, 为简单起见, 令顶点集为 $V=\{0,1,\cdots,n-1\}$. 假定与顶点 i,j 相关联的边为 $e_{i,j}$, $e_{i,j}$ 的权值用 $c[i][j]$ 表示, T 是最小生成树的边集. 这个算法维护两个顶点集合 S 和 N, 开始时: 令 $T=\varphi, S=\{0\}, N=V-S$. 然后, 进行贪心选择. 选取 $i \in S, j \in N$, 并且 $c[i][j]$ 最小的 i 和 j; 并使 $S=S \cup \{j\}, N=N-\{j\}, T=T \cup \{e_{i,j}\}$. 重复上述步骤, 直到 N 为空, 或找到 $n-1$ 条边为止. 此时, T 中的边集, 就是所要求取的 G 中的最小花费生成树. 因此, 普里姆算法的步骤可描述如下:

(1) $T=\varphi, S=\{0\}, N=V-S$.

(2) 如果 N 为空, 算法结束; 否则, 转步骤(3).

(3) 寻找使 $i \in S, j \in N$, 且 $c[i][j]$ 最小的 i 和 j.

(4) $S=S \cup \{j\}, N=N-\{j\}, T=T \cup \{e_{i,j}\}$; 转步骤(2).

例 11.3.2 图 11-9 表示普里姆算法的工作过程. 虚线一侧表示顶点集合 S, 另一侧表示顶点集合 N, 细线表示与顶点关联的边, 粗线表示所产生的最小生成树. 开始时, 如图 11-9a 所示, $S=\{1\}, N=\{2,3,4,5,6\}$. 在集合 N 中, 有 3 个顶点 2, 3, 4 与集合 S 邻接, 边 $e_{1,2}$ 的权最小; 在图 11-9b 中, 把顶点 2 并入集合 S, 把边 $e_{1,2}$ 并入 T, 此时, 顶点 3, 4, 5, 6 都与集合 S 邻接, 而边 $e_{2,3}$ 的权最小; 在图 11-9c 中, 把顶点 3 并入集合 S, 把边 $e_{2,3}$ 并入 T, 此时, 顶点 4, 5, 6 与集合 S 邻接, 而边 $e_{3,4}$ 的权最小; 在图 11-9d 中, 把顶点 4 并入集合 S, 把边 $e_{3,4}$ 并入 T, 此时, 剩下顶点 5 和 6 与集合 S 邻接, 而边 $e_{4,6}$

edge $e_{3,4}$ is the smallest; in Fig. 11-9d, vertex 4 is merged into set S, and edge $e_{3,4}$ is merged into T. At this time, vertices 5 and 6 are left adjacent to set S, and edge $e_{4,6}$ has the smallest weights; in Fig. 11-9e, vertex 6 is merged into set S, and edge $e_{4,6}$ is merged into T, at this time, the last vertex 5 is left adjacent to the set S, and the weight of the edge $e_{2,5}$ is the smallest; in Fig. 11-9f, the vertex 5 is merged into the set S, and the edge $e_{2,5}$ is merged into T, the resulting minimum spanning tree as shown in the thick line in Fig. 11-9f.

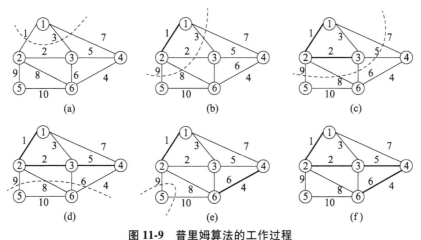

Fig. 11-9 Working process of Prim's algorithm

2. Implementation of Prim's algorithm

Similarly, in the directed weighted graph $G=(V,E)$, the vertices are numbered by numbers, so that the set of vertices is $V=\{0, 1, \ldots, n-1\}$; use the adjacency matrix $c[i][j]$ to represent the adjacency relationship between vertices i and j and the weight of edges $e_{i,j}$ in the graph $G=(V,E)$; if i and j are not adjacent, set $c[i][j]$ to MAX_FLOAT_NUM; use a Boolean array s to denote the vertices in S, $s[i]$ is true, indicating that the vertex i is in S, otherwise it is not in S; use the array $T[n]$ of the data structure type EDGE described in Section 11.3.2 to store the edge set of the generated minimum spanning tree.

In order to effectively find i and j such that $i \in S$, $j \in N$, and $c[i][j]$ are the smallest, consider the following facts: if edge $e_{i,j}$ is such an edge, such that vertex $i \in S$, and vertex $j \in N$, vertex j is called boundary point. The boundary point is a candidate for transfer from the set N to the set S. If j is a boundary point, then there is at least one vertex i in S that is adjacent to j. For the sake of simplicity, the vertex i adjacent to j and has the smallest weight $c[i][j]$ in S is referred to as the nearest neighbor of vertex j. Use the array neig[j] to store the nearest neighbors of vertex j; use the array w[j] to store the weights of the edges associated with j and its nearest neighbors. These two arrays are called the nearest neighbor information table. In this way, there is the following data structure:

(1) float c[n][n]; /* Adjacency matrix of the graph */
(2) BOOL S[n]; /* Set S */
(3) EDGE T[n];
    /* The set of edges of the minimum cost spanning tree */
(4) int neig[n]; /* The nearest neighbors of vertex j */
(5) float w[n];
/* The weight of the edge associated with vertex j and its nearest neighbors */

For the sake of simplicity, it is assumed that two-dimensional arrays can be passed through parameters and can be directly referenced in functions. Therefore, Prim's algorithm is described as follows:

Algorithm 11-4 Prim's Algorithm

Input: the adjacency matrix $c[\ ][\ ]$ of the undirected connected weighted graph, the number of vertices is n

Output: the minimum cost spanning tree $T[\ ]$ of the graph, the number k of edges in T

(1) #define MAX_FLOAT_NUM 3.14E38
(2) void prim (float c[ ][ ], int nm EDGE T[ ], int &k)
(3) {
(4)   int i, j, u;
(5)   BOOL *s=new BOOL[n];
(6)   int *neig=new int [n];
(7)   float min, *w=new float[n];
(8)   s[0]=TRUE;   /* S={0} */

实:如果边 $e_{i,j}$ 是这样的一条边,使得顶点 $i \in S$,且顶点 $j \in N$,就把顶点 j 称为边界点.边界点是由集合 N 转移到集合 S 的候选者.如果 j 是一个边界点,那么在 S 中至少有一个顶点 i 与 j 相邻接.为简单起见,把 S 中与 j 相邻接并且权 $c[i][j]$ 最小的顶点 i,称为顶点 j 的近邻.用数组 neig[j] 来存放顶点 j 的近邻;用数组 w[j] 来存放与 j 及其近邻相关联的边的权值.把这两个数组称为近邻信息表.这样,有如下的数据结构:

(1) float c[n][n];
        /* 图的邻接矩阵 */
(2) BOOL S[n]; /* 集合 S */
(3) EDGE T[n];
        /* 最小花费生成树的边集 */
(4) int neig[n]; /* 顶点 j 的近邻 */
(5) float w[n];
    /* 顶点 j 与近邻相关联的边的权 */

为了简明起见,假定二维数组可以通过参数传递,并可在函数中直接引用.于是,普里姆算法描述如下:

算法 11-4:普里姆算法

输入:无向连通赋权图的邻接矩阵 $c[\ ][\ ]$,顶点个数 n

输出:图的最小花费生成树 $T[\ ]$,T 中边的数目 k

(1) #define MAX_FLOAT_NUM 3.14E38
(2) void prim (float c[ ][ ], int nm EDGE T[ ], int &k)
(3) {
(4)   int i, j, u;
(5)   BOOL *s=new BOOL[n];
(6)   int *neig=new int [n];
(7)   float min, *w=new float[n];
(8)   s[0]=TRUE;   /* S={0} */

```
(9)     for (i=1;i<n;i++) {
/* Initialize the initial state of each vertex in the set N */
(10)        w[i]=c[0][1];
/* The weight of the incident edge of vertex i and its nearest
   neighbors */
(11)        neig[i]=0;
                    /* The nearest neighbor of vertex i */
(12)        s[i]=FALSE;  /* N={1,2,...,n-1} */
(13)    }
(14)    k=0;
/* The set T of edges of the minimum spanning tree is empty */
(15)    for (j=1; j<n; j++) {
(16)        u=0;
(17)        min=MAX_FLOAT_NUM;
(18)        for (j=1; j<n; j++)
/* Retrieve the closest vertex u to S in N */
(19)            if(!s[j]&&w[j]<min) {
(20)                u=j; min=w[j];
(21)            }
(22)        if (u==0) break;
/* The graph is not connected, if it is connected then jump
   out of the loop */
(23)        T[k].u=neig[u];
/* Record the edges of the minimum spanning tree */
(24)        T[k].v=u;
(25)        T[k++].key=w[u];
                            /* S=S∪{u} */
(26)        s[u]=TRUE;  /* S=S∪{u} */
(27)        for (j=1; j<n; j++) {
/* Update the nearest neighbor information of the vertices in N
   */
(28)            if (!s[j]&&c[u][j]<w[j]) {
(29)                w[j]=c[u][j];
(30)                neig[j]=u;
(31)            }
(32)        }
(33)    }
(34)    delete s; delete w; delete neig;
(35) }
```

To simplify the description, the set of vertices processed by this algorithm is $V=\{0, 1, \ldots, n-1\}$. Use the Boolean array S to denote the vertex set, and the corresponding element of the array denotes the vertex of the corresponding number. If the array element is true, it means that the corresponding vertex is in the set S, otherwise the corresponding vertex is in the set N. Lines 8~14 of the algorithm is the initialization part: Line 8 sets the initial element S of the set $S=\{0\}$. Lines 9~13 set the nearest neighbor information of all vertices in N, and initialize the nearest neighbor information table: set the nearest neighbors of all vertices i in the set N to vertex 0; the weights of the edges associated with the nearest neighbors are set to $c[0][i]$. In this way, in the subsequent processing, as long as the information of the nearest neighbors is retrieved, i and j that make $i \in S, j \in N$, and $c[i][j]$ the smallest can be found. Line 14 sets the initial storage location of the minimum cost spanning tree edge set.

Lines 15~33 are the second part of the algorithm, which is also the core part. This is a loop. The body of the loop is executed $n-1$ times, each time a minimum cost spanning tree edge is generated, and a vertex in the set N is merged into the set S. Lines 16~17 prepare for retrieving the closest vertex to S in N; lines 18~21 perform the retrieval. At this time, just retrieve the nearest neighbor information table and find the j that minimizes the weight $w[j]$ from the set N. Line 22 further determines whether such j is found. If it cannot be found, then all $w[j]$ in the set N have the value MAX_FLOAT_NUM, indicating that all the vertices in N are not connected with the vertices in S, so the algorithm ends. If found, the edge associated with its nearest neighbor is an edge in the minimum cost spanning tree. Lines 23~25 register the information of this edge in the edge set T of the minimum cost spanning tree. Line 26 merges the vertex into the set S. Lines 27~32 update the nearest neighbor information of the vertices in N, go to the beginning of the loop, and continue the next round of loop.

的相应元素表示对应编号的顶点. 如数组元素为真, 则表示对应顶点在集合 S 中, 否则对应顶点在集合 N 中. 算法的第 8~14 行是初始化部分: 第 8 行设置集合 $S=\{0\}$ 的初始元素 S. 第 9~13 行设置 N 中所有顶点的近邻信息, 初始化近邻信息表: 把集合 N 中所有顶点 i 的近邻都置为顶点 0; 与近邻相关联的边的权都置为 $c[0][i]$. 这样, 在以后的处理中, 只要检索近邻的信息, 就可以找到使 $i \in S, j \in N$, 并且 $c[i][j]$ 最小的 i 和 j. 第 14 行设置最小花费生成树边集的初始存放位置.

第 15~33 行是算法的第二部分, 也是核心部分. 这是一个循环, 循环体共执行 $n-1$ 次, 每一次产生一条最小花费生成树的边, 并把集合 N 中的一个顶点并入集合 S. 第 16~17 行为在 N 中检索与 S 最接近的顶点做准备; 第 18~21 行进行检索, 这时只要检索近邻信息表, 从集合 N 中找出使权 $w[j]$ 最小的 j 即可. 第 22 行进一步判断, 是否找到这样的 j. 如果找不到, 这时集合 N 中所有的 $w[j]$ 的值都为 MAX_FLOAT_NUM, 说明 N 中的所有顶点与 S 中的顶点不连通, 于是结束算法. 如果找到, 它与它的近邻所关联的边, 就是最小花费生成树中的一条边, 第 23~25 行把这条边的信息登记在最小花费生成树的边集 T 中. 第 26 行把该顶点并入集合 S. 第 27~32 行更新 N 中顶点的近邻信息, 转到循环的开始部分, 继续下一轮的循环.

3. Analysis of Prim's algorithm

The time complexity of this algorithm is estimated as follows: lines 8~14 initialize the nearest neighbor information table and vertex set, which takes $O(n)$ time; the loop body in lines 15~33 is executed $n-1$ times in total. Lines 16, 17, and 22~26, each cycle takes $O(1)$ time, a total of $n-1$ times are executed, and the total $O(n)$ time is taken. Lines 18~20, retrieve the vertex closest to S in N, using an internal loop to complete, the loop body needs to be executed $n-1$ times, so it takes $O(n^2)$ time; lines 27~32 update the nearest neighbor information table is also completed with an internal loop, and the loop body needs to be executed $n-1$ times, so it takes a total of $O(n^2)$ time. It follows that the time complexity of this algorithm is $O(n^2)$. At the same time, it can be seen from the algorithm that the space used for the work unit is $O(n)$.

The correctness of the algorithm is given by the following theorem.

Theorem 11.3.2 Prim's algorithm for finding the minimum spanning tree in an undirected weighted graph is correct.

Proof Assuming that the edge set of the minimum spanning tree generated by Prim's algorithm is T, and the edge set of the minimum spanning tree of the undirected weighted graph G is T^*, the following uses induction to prove that $T = T^*$.

(1) At the beginning, $T = \varphi$, the above argument is true.

(2) Assuming that the algorithm adds edge $e = (i,j)$ to T before line 14, the argument is true, $\overline{G} = (S, T)$ is the subtree of G's minimum spanning tree. According to Prim's algorithm, when $e = (i,j)$ is selected to be added to T, $i \in S$ and $j \in N$ are satisfied, and i and j that minimize $c[i][j]$ are used as the associated vertices of edge e. And let $S' = S \cup \{j\}$, $T' = T \cup \{e\}$, $G' = (S', T)$. At this time, there are

① G' is a tree. Because e is only associated with one vertex in S, adding e will not make G' form a loop, and G' is still connected. ② G' is the subtree of G's minimum spanning tree. Because, if $e \in T^*$, this conclusion holds; if $e \notin T^*$, then the vertices associated with e must be two non-adjacent vertices in T^*. According to the property 4 of spanning tree, $T^* \cup \{e\}$ will contain a loop. e is an edge in this loop, and $e=(i,j)$, $i \in S, j \in N$. Then there must be another edge $e'=(x,y)$ in the loop, $x \in S, y \in N$. According to the choice of Prim's algorithm, the weight of e is less than or equal to the weight of e'. Let $T^{**} = T^* \cup \{e\} - \{e'\}$, then the weight of T^{**} is less than or equal to the weight of T^*. If the weight of T^{**} is less than the weight of T^*, it contradicts that T^* is the edge set of the minimum spanning tree, so $e \in T^*$; if the weight of T^{**} is equal to the weight of T^*, then use the new T^* to mark T^{**}. In both cases, there is $e_1 \in T^*$.

In summary, $T = T^*$, the spanning tree produced by Prim's algorithm is the minimum spanning tree of G.

① G'是树. 因为e只和S中的一个顶点关联,加入e后不会使G'构成回路,并且G'仍然连通. ② G'是G的最小生成树的子树. 因为,如果$e \in T^*$,以下结论成立:如果$e \notin T^*$,那么与e关联的顶点必是T^*中两个不相邻的顶点,根据生成树的性质4,$T^* \cup \{e\}$将包含一个回路. e是这个回路中的一条边,并且$e=(i,j)$, $i \in S, j \in N$. 则回路中必存在另一条边$e'=(x,y)$, $x \in S, y \in N$. 按照普里姆算法的选择,e的权小于或等于e'的权. 令$T^{**} = T^* \cup \{e\} - \{e'\}$,则$T^{**}$的权小于或等于$T^*$的权. 若$T^{**}$的权小于$T^*$的权,则与$T^*$是最小生成树的边集相矛盾,所以,$e \in T^*$;若$T^{**}$的权等于$T^*$的权,这时用新的$T^*$来标记$T^{**}$. 在这两种情况下,都有$e_1 \in T^*$.

综上所述,$T = T^*$,普里姆算法所产生的生成树是G的最小生成树.

Exercise 11

1. Use a greedy algorithm to design a solution to the traveling salesman problem. Start from 5 cities and choose the route with the shortest cost. Verify that this algorithm cannot get the optimal solution.

2. Use the Dijkstra algorithm to solve the single-source shortest path problem shown in Fig. 11-10.

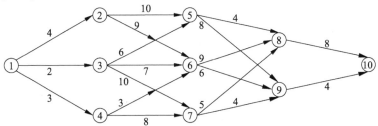

Fig. 11-10 The shortest path problem of a directed weighted graph

3. Change the graph of question 2 to an undirected weighted graph, use Kruskal's algorithm to find the minimum spanning tree of the graph, and draw the process of generating the minimum spanning tree.

4. Change the graph of question 2 to an undirected weighted graph, use Prim's algorithm to find the minimum spanning tree of the graph, and draw the process of generating the minimum spanning tree.

5. Find the shortest path between each point in the network shown in Fig. 11-11 from point A.

6. Use Kruskal's algorithm to find the minimum cost spanning tree shown in Fig. 11-11, and draw the minimum cost spanning tree generation process.

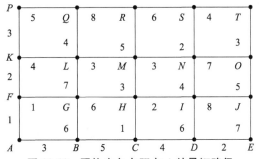

Fig. 11-11 The shortest path from each point in the network to point A

7. Use Prim's algorithm to find the minimum cost spanning tree shown in Fig. 11-11, and draw the minimum cost spanning tree generation process.

8. Assume that coins with value of 25 cents, 10 cents, 5 cents, and 1 cent are used to pay n cents. Design an algorithm to minimize the number of coins paid out.

9. In question 8, assume that the currency value of the coin is 1, 2, 4, 8, 16, …, 2^k, and k is a positive integer. If the paid money $n < 2^{k+1}$, design an $O(\log n)$ algorithm to solve this problem.

10. Let $G = (V, E)$ be an undirected graph. The vertex cover set S of G is a subset of G, such that $S \subseteq V$, and every edge in E is associated with at least one vertex in S. Consider the following vertex covering algorithm for finding G: first, sort the vertices in V in descending order of vertex degree; then perform the following steps until all edges are covered: pick the vertex with the highest degree and that is associated with at least one edge in the rest of the graph, this vertex is added to the vertex cover set, and all edges associated with this vertex are deleted. Design this algorithm and show that this algorithm does not always get the minimum vertex cover set.

11. Let $G = (V, E)$ be an undirected graph, and the clique C in G is a complete subgraph of G. If in G, there is no other clique C' with more vertices than C, then C is called the largest clique of G. At the beginning, let $C = G$, and then repeatedly delete vertices that are not adjacent to other vertices from C until C becomes a clique. Design this algorithm and show that this algorithm cannot always get the largest clique of G.

7. 用普里姆算法求图 11-11 所示的最小花费生成树,画出最小花费生成树的生成过程.

8. 假定用面值为 25 美分、10 美分、5 美分、1 美分的硬币来支付 n 美分. 设计一个算法,使付出硬币的枚数最少.

9. 在第 8 题中,假定硬币的币值是 1, 2, 4, 8, 16, …, 2^k, k 为正整数. 如果所支付的钱 $n < 2^{k+1}$, 设计一个 $O(\log n)$ 的算法来解这个问题.

10. 令 $G = (V, E)$ 是一个无向图, G 的顶点覆盖集 S 是 G 的一个子集, 使得 $S \subseteq V$, 并且 E 中的每一条边至少与 S 中的一个顶点相关联. 考虑下面寻找 G 的顶点覆盖算法: 首先, 按顶点度的递减顺序给 V 中的顶点排序; 接着执行下面的步骤, 直到所有的边全被覆盖: 挑出最高度的顶点, 且至少与其余图中的一条边相关联, 把这个顶点加入到顶点覆盖集, 并删去和这个顶点相关联的所有的边. 设计这个算法, 并说明这个算法不总能得到最小顶点覆盖集.

11. 令 $G = (V, E)$ 是一个无向图, G 中的团 C 是 G 的一个完全子图. 如果在 G 中, 不存在另一个顶点个数多于 C 中顶点数的团 C', 就称 C 为 G 的最大团. 开始时, 令 $C = G$, 然后, 反复地从 C 中删去与其他顶点不相邻的顶点, 直到 C 成为一个团. 设计这个算法, 并说明这个算法不是总能得到 G 的最大团.

习题 11 答案
Key to Exercise 11

Chapter 12　Lower Bounds

This chapter discusses whether a more efficient algorithm may still exist for a given problem or how to prove that an algorithm is the most efficient algorithm for solving a given problem, which involves a lower bound on the time necessary to solve all algorithms for that problem. Thus, there are two concepts here: a lower bound on the time necessary to solve a given algorithm for a given problem and a lower bound on the time necessary to solve all algorithms for a given problem. This was covered in the previous chapters, and the problem discussed in this chapter deals with the later one: if a function $g(n)$ can be found and it can be determined and proved that it is a lower bound on the time of all algorithms for solving a given problem, then it is impossible to find an algorithm of lower order than the function $g(n)$, and if an algorithm A is designed with a computation time of the same order as $g(n)$, it is considered to be the optimal algorithm for solving the problem.

12.1　Trivial lower bounds

Determining and proving a lower bound on the time required by an algorithm to solve a given problem is generally difficult. This is because it involves all algorithms for solving the problem, and enumerating all possible algorithms and analyzing them is usually not possible. Therefore, some kind of computational model must be used. However, some problems have obvious lower bounds that can be deduced in an intuitive way. For example, reading n input elements of a problem takes $\Omega(n)$ time, so $\Omega(n)$ is a time lower bound for that problem, and such a lower bound is called a trivial lower bound. Here are two examples of trivial lower bounds.

Example 12.1.1　Check the number of elements in an integer array of n elements whose value is even. It is clear that this requires judging and accumulating each element in the array. Because it takes $\Omega(1)$ time to judge each element, it takes $\Omega(n)$ time to judge n elements. Therefore, $\Omega(n)$ is a lower bound on all algorithms for solving this problem.

Example 12.1.2 The problem of checking the reachability matrix of a directed graph with n vertices. The reachability matrix of a directed graph with n vertices is a matrix of $n \times n$. Obviously, n^2 elements need to be checked, and each element checked takes at least $\Omega(1)$ time, then checking n^2 elements takes at least $\Omega(n^2)$ time. Therefore, $\Omega(n^2)$ is a lower bound on all algorithms for solving this problem.

12.2 Decision tree model

When considering the time complexity of certain problems in previous chapters, the comparison operation is often used as the basic operation. With the result of the comparison operation, the execution of the algorithm is divided into two parts: for the part larger than the number of operations being compared, the execution of the algorithm proceeds along a branch node; for the part smaller than the number of operations being compared, the execution of the algorithm proceeds along another branch node. Thus, a binary tree can be used to depict the execution process of the algorithm, and this number is called the decision tree. Thus, a decision tree is a binary tree such that each of its internal nodes corresponds to a comparison of the form $x \leq y$. If the relation holds, control is transferred to the left son node; otherwise, control is transferred to the right son node. Each of its leaf nodes, denotes an outcome of the problem. When using the decision tree model to build the lower bound of a problem, it is common to ignore all arithmetic operations in solving the problem and concentrate only on the number of transfers in the branch execution.

The execution of the decision tree starts at the root node, and then control is transferred to their son nodes based on the results of comparison operations. This process is carried out until the leaf nodes are reached. The time complexity of the problem, therefore, is related to the height of the decision tree. The use of the decision tree model is illustrated below for the retrieval problem and the sorting problem.

12.2.1 Retrieval problems

Retrieval problem: Let an array A be an ordered array with n elements, and given an element x, determine whether x is in the array A. When using the decision tree model to determine the lower bound of a comparison-based retrieval problem, a binary tree is used to denote the retrieval process of the array, and each node in the tree denotes a comparison between element x and some element $A[i]$ of the array. Each comparison has three possible outcomes: $x<A[i]$, $x=A[i]$ and $x>A[i]$. It is assumed that if $x=A[i]$, the algorithm retrieve succeeds and terminates; if $x<A[i]$, the execution of the algorithm transfers to the left branch of the binary tree; if $x>A[i]$, the execution of the algorithm transfers to the right branch of the binary tree. If the execution of the algorithm proceeds along the left and right branches until none of the leaf nodes can find an i such that $x=A[i]$, the algorithm retrieval fails and terminates. This is because the retrieval process starts from the root node and continues until the leaf node. Therefore, the number of comparisons and determinations is the height of the tree plus 1. When the number of retrieved elements is n, because the elements are ordered, the internal nodes of the decision tree are at most 2^i-1, where $k=\lceil \log n \rceil$. If all nodes are concentrated on the k-th level of the tree and its lower levels, then the height of the tree is at least $\lceil \log n \rceil$. It follows that retrieving n elements requires at least $\lceil \log n \rceil +1$ comparisons in the worst case, which is obviously a lower bound on the retrieval problem.

判定树的执行是从根结点开始的,然后根据比较操作的结果,将控制转移到它们的儿子结点.这个过程一直进行,直到叶子结点为止.因此,问题的时间复杂度就与判定树的高度有关.下面就检索问题和排序问题来说明判定树模型的使用.

12.2.1 检索问题

检索问题:令数组 A 是一个具有 n 个元素的有序数组,给定元素 x,确定 x 是否在数组 A 中.用判定树模型来确定基于比较的检索问题的下界时,用一棵二叉树来表示数组的检索过程,树中的每一个结点表示元素 x 和数组中某个元素 $A[i]$ 的一次比较.每次比较有3种可能的结果:$x<A[i]$,$x=A[i]$ 及 $x>A[i]$.假定:若 $x=A[i]$,则算法检索成功并终止;若 $x<A[i]$,则算法的执行转移到二叉树的左分支;若 $x>A[i]$,则算法的执行转移到二叉树的右分支.若算法的执行沿着左、右分支前进,直到叶子结点都找不到一个 i,使得 $x=A[i]$,则算法以检索失败而终止.因为检索过程是从根结点开始,到叶结点为止的,所以比较与判定的次数是树的高度加1.当被检索元素的个数为 n 时,因为元素是有序的,判定树的内部结点最多为 2^i-1 个,其中 $k=\lceil \log n \rceil$.如果所有结点都集中在树的第 k 层及其较低的层上,那么树的高度至少为 $\lceil \log n \rceil$.由此得到,检索 n 个元素,在最坏情况下,至少需要进行 $\lceil \log n \rceil +1$ 次比较,显然,这也是检索问题的下界.由此可以得到下面的定理:

This leads to the following theorem:

Theorem 12.2.1 The number of comparisons to retrieve an ordered array with n elements is $\lceil \log n \rceil + 1$ in the worst case.

Therefore, the lower bound of the retrieval problem is $\Omega(\log n)$, and the binary retrieval algorithm is the optimal algorithm in the retrieval problem.

For example, if $A = \{3, 4, 7, 10, 15, 18, 26, 30, 31, 38\}$, the decision tree of the binary retrieval problem is shown in Fig. 12-1.

定理 12.2.1 检索具有 n 个元素的有序数组,在最坏情况下的比较次数是 $\lceil \log n \rceil + 1$.

因此,检索问题的下界是 $\Omega(\log n)$,而二叉检索算法是检索问题中的最优算法.

例如,$A = \{3, 4, 7, 10, 15, 18, 26, 30, 31, 38\}$,则二叉检索问题的判定树如图 12-1 所示.

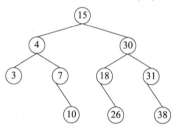

图 12-1 检索有序数组的判定树

Fig. 12-1 Retrieving the decision tree of an ordered array

Fig. 12-1 draws the internal nodes of a decision tree that retrieves 10 ordered elements. As seen in the figure, the number of comparisons in the worst-case for retrieving an ordered array with 10 elements is 4. The number of comparisons in the worst-case, for any retrieval algorithm, is never less than this number. So, 4 is a lower bound for all algorithms that retrieve 10 elements.

The decision tree model is similar to a binary sorted tree with the following characteristics:

(1) If the left sub-tree is non-empty, the values of all nodes on the left sub-tree are less than the value of the root node.

(2) If the right sub-tree is non-empty, the values of all the nodes on the right sub-tree are greater than the value of the root node.

(3) The left and right sub-trees are also a binary sorted tree respectively.

According to its characteristic definition, the left sub-tree node value<root node value<the right sub-tree

图 12-1 画出了检索 10 个有序元素的判定树的内部结点.从图中看到,检索一个具有 10 个元素的有序数组,在最坏情况下的比较次数是 4.任何检索算法,在最坏情况下的比较次数,都不会小于这个数目.所以,4 是检索 10 个元素的所有算法的下界.

该判定树模型与二叉排序树相似,其特征有:

(1) 若左子树非空,则左子树上的所有结点的值均小于根结点的值.

(2) 若右子树非空,则右子树上的所有结点的值均大于根结点的值.

(3) 左、右子树也分别是一棵二叉排序树.

根据其特征定义,左子树结点值<根结点值<右子树结点值,然后对二叉

node value, and then the binary sorted tree is traversed in the middle order, which can obtain an ascending ordered sequence.

A description of the algorithm for constructing a binary sorted tree is given below:

```
void Ctreat_BST(BiTree &T, KeyType str[], int n){
    T=NULL;     //Initialize T to an empty tree
    int i=0;
    While(i<n){     //Insert each keyword into the binary tree in turn
        BST_Insert(T, str[i]);
        i++;
    }
}
```

Algorithm description of insertion operation of binary sorted tree:

```
int BST_Insert(BiTree &T, KeyType k){
    if (T==NULL)    //The original tree is empty and the newly inserted record is the root node
    {
        T=(BiTree)malloc(sizeof(BSTNode));
        T->key=k;
        T->lchild=T->rchild=NULL;
        return 1;     //Return 1, inserted successfully
    }
    else if (k==T->key)    //If there are nodes with the same keyword in the tree, insertion failed
        return 0;
    else if (k<T->key)    //Insert into the left subtree of T
        return BST_Insert(T->lchild,k);
    else    //Insert into the right subtree of T
        return BST_Insert(T->rchild,k);
}
```

There are also descriptions of algorithms such as searching and deleting of binary sorted trees, which are not repeated here.

12.2.2 Sorting problems

Now consider the use of comparative sorting. Let the array A be an unordered array with n elements, and

排序树进行中序遍历,可以得到一个递增的有序序列.

下面给出构造二叉排序树的算法描述:

```
void Ctreat_BST(BiTree &T, KeyType str[], int n){
    T=NULL;     //初始化 T 为空树
    int i=0;
    While(i<n){     //依次将每个关键字插入二叉树中
        BST_Insert(T, str[i]);
        i++;
    }
}
```

二叉排序树的插入操作的算法描述:

```
int BST_Insert(BiTree &T, KeyType k){
    if (T==NULL)    //原树为空,新插入的记录为根结点
    {
        T=(BiTree)malloc(sizeof(BSTNode));
        T->key=k;
        T->lchild=T->rchild=NULL;
        return 1;     //返回1,插入成功
    }
    else if (k==T->key)    //树中存在相同关键字的结点,插入失败
        return 0;
    else if (k<T->key)    //插入 T 的左子树中
        return BST_Insert(T->lchild,k);
    else    //插入 T 的右子树中
        return BST_Insert(T->rchild,k);
}
```

还有二叉排序树的查找和删除等算法描述,此处不再赘述.

12.2.2 排序问题

现在考虑使用比较排序的方式. 令数组 A 是一个具有 n 个元素的无序

sort the array A in non-increasing or non-decreasing order. In the sorted case, each internal node of the decision tree denotes a decision, and each leaf node denotes an output. In each decision, two elements of the array $A[i]$ and $A[j]$ are compared, control is transferred to the left branch node if $A[i] \leq A[j]$; otherwise, control is transferred to the right branch node. Starting from the root node, two certain elements of the array are determined, and so on until the leaf node, an ordered array is obtained. Fig. 12-2 shows a kind of decision tree for sorting an array with 3 elements.

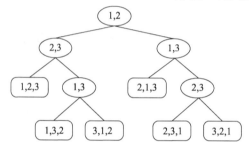

Fig. 12-2　A decision tree for sorting 3 elements

There are 6 leaf nodes in the graph, and each leaf node denotes one possible output. If the number of elements being sorted is n, the number of leaf nodes in the decision tree is $n!$ because there are $n!$ sorts of n elements.

Using different sorting algorithms, the order in which the elements are determined will be different, and the elements that are compared in each decision will be different. Therefore, the corresponding decision trees are different. However, no matter how different their decision trees are, for sorting n elements, the number of leaf nodes of their decision trees is $n!$. Since the worst-case time complexity is the longest path from the root node of the decision tree to the leaf nodes, i.e., the height of the decision tree, so their time complexity depends on the height of the decision tree. Regarding the height of the decision tree, the following theorem is given.

Theorem 12.2.2 If T is a binary tree with at least $n!$ leaf nodes, then the height of T is at least
$$n\log n - 1.5n = \Omega(n\log n)$$

Proof Let h be the height of a binary tree. By the property of binary trees, the number of leaf nodes at level h of a binary tree is at most 2^h. Since T has at least $n!$ leaf nodes, there is $n! \leq 2^h$, i.e., $h \geq \log n!$. Since
$$\log n! = \sum_{i=1}^{n} \log i = \sum_{i=2}^{n} \log i$$
From the integral property
$$\sum_{i=2}^{n} \log i \geq \int_{1}^{n} \log x \, dx$$
$$= n\log n - n\log e + \log e$$
$$\geq n\log n - 1.5n$$
$$= \Omega(n\log n)$$

Theorem 12.2.2 is often called the information-theoretic lower bound because each comparison yields at most one bit of information, and $\log n!$ information will be obtained through the sorting process. It also shows that for comparison-based sorting algorithms, no matter which algorithm is used to sort n elements, although their decision trees are different, the height of the decision trees will not be smaller than $\Omega(n\log n)$. Therefore $\Omega(n\log n)$ is a lower bound on the time of these algorithms. From this, we have the following theorem:

Theorem 12.2.3 Any comparison-based sorting algorithm that sorts n elements has a worst-case time lower bound of $\Omega(n\log n)$.

12.3 Algebraic decision tree model

The above decision tree model allows comparison between only two elements in each node of the decision tree, and its function is relatively simple. If the decision function of the decision tree node is expanded to allow calculation and comparison of polynomials with n

input variables, and then branching based on the result of the decision, the function is much more powerful than that of the decision tree model. The decision tree that is generated by calculating and deciding in this way is called an algebraic decision tree.

An algebraic decision tree with n input variables $x_1, x_2, \ldots, x_n$ is a binary tree, and each of its nodes can be labeled with a statement. The statement used to label the nodes is the following statement: if $f(x_1, x_2, \ldots, x_n) \sigma 0$ holds, then it is transferred to the left son node; otherwise, it is transferred to the right son node. Where σ is any of the 3 relational operators $=, >, \geq$, and the statement used to label the leaf node is the answer.

In an algebraic decision tree, an algebraic decision tree is said to be a linear algebraic decision tree, or simply a linear decision tree, if for some $d \geq 1$, the statements that mark the internal nodes, and their associated polynomials $f(x_1, x_2, \ldots, x_n)$, are linear.

Let Π be a decision problem and $x_1, x_2, \ldots, x_n$ be the input instances of this decision problem. When each input instance $(x_1, x_2, \ldots, x_n)$ of Π is considered as a point in the n-dimensional space E^n, then for a subset $W \subseteq E^n$ in the n-dimensional space E^n, W is said to be a member point set of the decision problem Π, if $(x_1, x_2, \ldots, x_n) \in W$ when and only when the answer of the problem Π to the input instance $(x_1, x_2, \ldots, x_n)$ is yes. If the point $p = (x_1, x_2, \ldots, x_n)$ is regarded as input parameters, the computation starts at the root node of the algebraic decision tree T and the leaf node which eventually reaches the algebraic decision tree T is regarded as yes, when and only when $(x_1, x_2, \ldots, x_n) \in W$, then the algebraic decision tree T is said to decide the members in W.

This can be analogous to the decision tree model, where a lower bound on the time complexity of the problem Π can be inferred in the worst case by estimating the height of the algebraic decision tree T at the time of solving the problem Π.

比较判定,然后根据判定的结果进行分支,其功能就比判定树模型的功能强大得多. 用这种方式进行计算和判断所产生的判定树,称为代数判定树.

n 个输入变量 $x_1, x_2, \cdots, x_n$ 的代数判定树是一棵二叉树,其每一个结点都可以用一个语句来标记. 用来标记结点的语句,是如下的语句:若 $f(x_1, x_2, \cdots, x_n) \sigma 0$ 成立,则转移到左儿子结点;否则,转移到右儿子结点. 其中,σ 是3个关系运算符 $=, >, \geq$ 中的任何一个,而用来标记叶子结点的语句则是答案.

在代数判定树中,对某个 $d \geq 1$,如果标记内部结点的语句,与其相关的多项式 $f(x_1, x_2, \cdots, x_n)$ 都是线性的,则称该代数判定树是线性代数判定树,或简称线性判定树.

令 Π 是一个判定问题,$x_1, x_2, \cdots, x_n$ 是该判定问题的输入实例. 当把 Π 的每一个输入实例 $(x_1, x_2, \cdots, x_n)$ 看成 n 维空间 E^n 中的一点时,对 n 维空间 E^n 中的子集 $W \subseteq E^n$,如果 $(x_1, x_2, \cdots, x_n) \in W$,当且仅当问题 Π 对输入实例 $(x_1, x_2, \cdots, x_n)$ 的答案为 yes 时,那么称 W 为判定问题 Π 的成员点集. 如果把点 $p = (x_1, x_2, \cdots, x_n)$ 作为输入参数,在代数判定树 T 的根结点上开始进行计算,最终到达代数判定树 T 的叶子结点为 yes,当且仅当 $(x_1, x_2, \cdots, x_n) \in W$,就称代数判定树 T 判定 W 中的成员.

这就可以类似于判定树模型,只要估计在解问题 Π 时代数判定树 T 的高度,即可以在最坏的情况下,推断出问题 Π 的时间复杂度的下界.

Theorem 12.3.1 Let $V \subseteq E^n$ be the set of points defined by the following polynomial equation of order d with $d \geq 1$, then

$$\begin{cases} p_i(x_1, x_2, \cdots, x_n) = 0, & 1 \leq i \leq m \\ q_j(x_1, x_2, \cdots, x_n) > 0, & 1 \leq j \leq s \\ r_k(x_1, x_2, \cdots, x_n) \geq 0, & s \leq k \leq t \end{cases}$$

Let $\#V$ be the number of connected branches that V has in E^n, then $\#V \leq d(2d-1)^{n+t-1}$.

The theorem states that the point set V, which consists of all points in the n-dimensional space E^n that satisfy the above equation, and the number of connected branches in E^n is a function of the parameters d, n, and t. Where d is the order of the highest order polynomial in the polynomial equation; n is the number of independent variables in the polynomial equation, i.e., the dimension in the n-dimensional space; and t is the number of inequality equations in the polynomial equation. Theorem 12.3.1 also shows that the number of connected branches of the point set V in E^n is independent of the number of equations in the polynomial equation.

12.4 Linear time reductions

In previous chapters, reduction was used to solve problems by reducing one problem to another, e.g., by reducing the computational complexity of one problem to that of another. Similarly when discussing the lower bounds of a problem, the lower bounds of two problems can be related by reduction. If the lower bound of problem A is known and it is possible to reduce problem A to problem B by reduction, then it is possible to use the lower bound of problem A to build the lower bound of problem B. The basic steps can be divided into three steps: First, the input of problem A is transformed into the compatible input of problem B. Then solve the problem B. Finally, the output of problem B is converted into the correct solution of problem A. If the first and third steps can be done in $O(\tau(n))$ time, where n is the input size of the problem and $\tau(n)$ is a polynomial in n, then problem A is said to be reduced to problem B in polynomial time, denoted $A \propto_{\tau(n)} B$, and problem A and problem B are said to be $\tau(n)$ time equivalent; in particular, if $\tau(n)$ is a linear function of n, then problem A is said to be reduced to problem B in linear time, denoted as $A \propto_n B$, and say that problem A and problem B are equivalent. In this sense, problem A and problem B are said to have the same computational complexity.

The sorting problem SORTING is known to have a time lower bound of $\Omega(n\log n)$ in the worst case. The sorting problem can be reduced to the convex hull problem CONVEX HULL in linear time, such that the convex hull problem also has a time lower bound of $\Omega(n\log n)$ in the worst case.

Let the set of positive real numbers $\{x_1, x_2, \ldots, x_n\}$ be an input of the sorting problem. First, for all $i, 1 \leq i \leq n$, transform each real number x_i in the set, into a point $p_i = (x_i, x_i^2)$ on the two-dimensional plane, while taking the latter as an input to the convex hull problem. Obviously, this transformation process can be done in linear time; meanwhile, all these n points constructed up, are on the parabola $y = x^2$.

Next, solve this input instance with any of the algorithms for solving the convex hull problem, and the result will output an ordered table of the coordinates of the poles on the convex hull of Fig. 12-3.

时间内完成,其中 n 为问题的输入规模,$\tau(n)$ 为 n 的多项式,则称问题 A 以多项式时间归约于问题 B,记为 $A \propto_{\tau(n)} B$,并称问题 A 和问题 B 是 $\tau(n)$ 时间等价的;特别地,如果 $\tau(n)$ 为 n 的线性函数,则称问题 A 以线性时间归约于问题 B,记为 $A \propto_n B$,并称问题 A 和问题 B 是等价的. 在这种意义下,称问题 A 和问题 B 具有相同的计算复杂性.

已知排序问题 SORTING 在最坏情况下的时间下界为 $\Omega(n\log n)$. 排序问题可以用线性时间归约于凸壳问题 CONVEX HULL,从而,凸壳问题在最坏的情况下也有时间下界 $\Omega(n\log n)$.

令正实数集合 $\{x_1, x_2, \cdots, x_n\}$ 是排序问题的一个输入. 首先,对所有的 $i, 1 \leq i \leq n$,把集合中的每一个实数 x_i,变换为二维平面上的点 $p_i = (x_i, x_i^2)$,而把后者作为凸壳问题的输入. 显然,这一变换过程可以用线性时间来完成;同时,所构造出来的这 n 个点,都在抛物线 $y = x^2$ 上.

其次,用求解凸壳问题的任何一个算法,对这个输入实例进行求解,其结果将输出图 12-3 中凸壳上极点坐标的一个有序表.

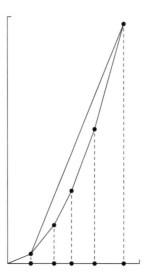

图 12-3 把排序问题归约为凸壳问题
Fig. 12-3 Reducing the sorting problem to a convex hull problem

Finally, reading the x-coordinate values of each point in the polar coordinates sequentially yields the ordered set of the original real numbers. This process is also done in linear time, and then the output of the convex hull problem is also transformed in linear time to the output of the sorting problem. These yields:

$$SORTING \propto_n CONVEX\ HULL$$

Thus the lower bound on the time of the convex hull problem CONVEX HULL in the worst case is $\Omega(n\log n)$.

The interpolation problem for polynomials is: given n pairs of real numbers $(x_1,y_1),(x_2,y_2),\ldots,(x_n,y_n)$, where $x_i \neq y_i$ when $i \neq j$. Find the $(n-1)$st order polynomial $f(x)$, such that $f(x_i)=y_i, 1 \leq i \leq n$.

To determine a lower bound for the polynomial interpolation problem, consider a special $(n-1)$st order polynomial $P(x)$ with $P(x_i)=(-1)^{i+1}$ when $x_i=i, 1 \leq i \leq n$, as shown in Fig. 12-4. Since the polynomials are continuous, for all $i, 1 \leq i \leq n-1$, there exist s_i and t_i satisfying $i < s_i < t_i < i+1$ such that $P(s_i)=1$ and $P(t_i)=-1$. Therefore, s_i and t_i are respectively the $n-1$ roots of the equation:

$$P(x)=1, P(x)=-1$$

最后,按顺序读取极点坐标中每一点的 x 坐标值,则得到将原实数集排序后的集合. 这一过程也用线性时间完成,则凸壳问题的输出也以线性时间变换为排序问题的输出. 由此得到:

$$SORTING \propto_n CONVEX\ HULL$$

因此凸壳问题 CONVEX HULL 在最坏情况下的时间下界为 $\Omega(n\log n)$.

多项式的插值问题是:给定 n 对实数 $(x_1,y_1),(x_2,y_2),\cdots,(x_n,y_n)$,其中,当 $i \neq j$ 时,$x_i \neq y_i$. 求 $n-1$ 阶多项式 $f(x)$,使得 $f(x_i)=y_i, 1 \leq i \leq n$.

为了确定多项式插值问题的下界,考虑一个特殊的 $n-1$ 阶多项式 $P(x)$,当 $x_i=i, 1 \leq i \leq n$ 时,$P(x_i)=(-1)^{i+1}$,如图 12-4 所示. 因为多项式是连续的,所以对所有的 $i, 1 \leq i \leq n-1$,存在 s_i 和 t_i,满足 $i<s_i<t_i<i+1$,使得 $P(s_i)=1, P(t_i)=-1$. 所以,s_i 和 t_i 分别是方程的 $n-1$ 个根:

$$P(x)=1, P(x)=-1$$

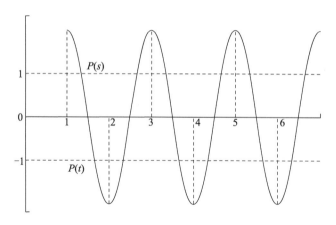

图 12-4 一个特殊的 $n-1$ 阶多项式

Fig. 12-4 A special polynomial of order $n-1$

In this way, a decision problem can be constructed such that its member point set W is

$$W = \{(x_1, y_1, x_2, y_2, \ldots, x_n, y_n) \mid y_i^2 = 1 \land P(x_i) = y_i \land 1 \leq i \leq n\}$$

Now, the input $(x_1, y_1, x_2, y_2, \ldots, x_n, y_n)$ of the above decision problem can be transformed in linear time to the input of the interpolation problem. Then, using any algorithm for solving the interpolation problem, the polynomial $f(x)$ is computed such that $f(x_i) = y_i$. Finally, then for all i, $1 \leq i \leq n$, it is checked whether $y_i^2 = 1$ and $f(x_i) = P(x_i)$ hold, and the last checking step can be done in linear time. Therefore, the linear time of the above decision problem is reduced to the polynomial interpolation problem, and then the lower bound of the above decision problem is also the lower bound of the polynomial interpolation problem.

Each point in the above member point set W is an isolated point, if $(x_1, y_1, x_2, y_2, \ldots, x_n, y_n) \in W$, then we have $y_i^2 = 1$ and:

$x_i \in \{s_k \mid 1 \leq k \leq n-1\}$, when $y_i = 1$;

$x_i \in \{t_k \mid 1 \leq k \leq n-1\}$, when $y_i = -1$.

Therefore, the number of points contained in W is

这样,可以构造一个判定问题,使得它的成员点集 W 为

$$W = \{(x_1, y_1, x_2, y_2, \cdots, x_n, y_n) \mid y_i^2 = 1 \land P(x_i) = y_i \land 1 \leq i \leq n\}$$

现在,可以把上述判定问题的输入 $(x_1, y_1, x_2, y_2, \cdots, x_n, y_n)$ 用线性时间变换为插值问题的输入. 然后, 用求解插值问题的任何算法, 计算多项式 $f(x)$, 使得 $f(x_i) = y_i$. 最后, 再对所有的 $i, 1 \leq i \leq n$, 检查 $y_i^2 = 1$ 及 $f(x_i) = P(x_i)$ 是否成立, 而最后的检查步骤, 可以用线性时间完成. 因此, 上述判定问题的线性时间归约于多项式插值问题, 上述判定问题的下界也是多项式插值问题的下界.

在上述成员点集 W 中的每一点都是孤立点, 如果 $(x_1, y_1, x_2, y_2, \cdots, x_n, y_n) \in W$, 则有 $y_i^2 = 1$, 并且

$x_i \in \{s_k \mid 1 \leq k \leq n-1\}$, 当 $y_i = 1$ 时;

$x_i \in \{t_k \mid 1 \leq k \leq n-1\}$, 当 $y_i = -1$ 时.

所以, W 中所包含的点数为

$$\#(W) = \binom{2n-2}{n} n!$$
$$= \frac{(2n-2)!}{(n-2)!}$$

By Theorem 12.2.3, the lower bound for the above decision problem is

$$\log\left[\frac{(2n-2)!}{(n-2)!} - n\right] = \Omega(n\log n)$$

Therefore, the lower bound is $\Omega(n\log n)$ for the polynomial interpolation problem.

Exercise 12

1. Draw the decision tree of the 4 element linear search algorithm.

2. Let A and B be two unordered tables with n elements, and determine whether the elements in A are the same as those in B, i.e., whether the elements in A are of a certain ordering of the elements in B. Use Ω notation to denote the number of comparisons required to solve this problem.

3. When an array A with n elements is a heap, state the minimum number of comparisons needed to test this array.

4. The problem of equality of sets can be described as follows: given a set $A = \{x_1, x_2, \ldots, x_n\}$, $B = \{y_1, y_2, \ldots, y_n\}$, determine whether $A = B$ holds. Use the lower bound theorem to prove that the problem is lower bounded in time by $\Omega(n\log n)$ under the algebraic decision tree model.

5. Let A be an integer array of n elements, the size of each element is in the range 1 to m, where $m > n$. Find an element x in the range 1 to m that is not in A. What is the minimum number of comparisons needed to solve this problem?

习题 12 答案
Key to Exercise 12

Chapter 13 Backtracking Method

13.1 Basic idea of backtracking method

13.1.1 Backtracking method

Backtracking method is called "general problem solving method". It can be used to systematically search all or any solutions of a problem. Backtracking is a systematic and opportunistic search algorithm. In the solution space tree of the problem, it searches the solution space tree from the root node according to the depth first strategy. When the algorithm searches any node of the solution space tree, it first judges whether the node contains the solution of the problem. If not, skip the search for the subtree with this node as the root and trace back to other ancestor nodes layer by layer. Otherwise, enter the subtree and continue to search according to the depth first strategy. When finding all solutions of the problem by backtracking method, it is necessary to backtrack to the root, and it ends when all subtrees of the root node have been searched. When the backtracking method is used to find a solution of the problem, it can end as long as a solution of the problem is searched. The algorithm for systematically searching for a problem in a depth-first method is called backtracking method, which is used to solve the problem with large combinatorial number. Generally speaking, some solution spaces are very large and can be considered as a very large tree. At this time, the time complexity of complete traversal is unbearable. Thus, you can check some conditions while traversing. When traversing a branch, if you find that the conditions are not met, you will return to the root node and enter the traversal of the next branch. This is the origin of the word "backtracking". Selective traversal based on conditions is called pruning or branch and bound.

13.1.2 Solution space of problem

When using backtracking method to solve a problem, the solution space of the problem should be clearly defined. The solution space of the problem contains at least one (optimal) solution of the problem. For example, for the 0/1 knapsack problem with n selectable items, the solution space is composed of 0/1 vectors with length n. The solution space contains all possible 0/1 assignments to variables. For example, when $n=3$, the solution space is {(0,0,0),(0,0,1),(0,1,0),(0,1,1),(1,0,0),(1,0,1),(1,1,0),(1,1,1)}.

After defining the solution space of the problem, the solution space should also be well organized, so that the whole solution space can be easily searched by backtracking method. The solution space is usually organized in the form of tree or graph.

For example, for the 0/1 knapsack problem with $n=3$, the solution space can be denoted by a complete binary tree, as shown in Fig. 13-1.

13.1.2 问题的解空间

用回溯法解问题时,应明确定义问题的解空间.问题的解空间至少包含问题的一个(最优)解.例如对于有 n 种可选择物品的 0/1 背包问题,其解空间由长度为 n 的 0/1 向量组成.该解空间包含对变量的所有可能的 0/1 赋值.例如 $n=3$ 时,其解空间是{(0,0,0),(0,0,1),(0,1,0),(0,1,1),(1,0,0),(1,0,1),(1,1,0),(1,1,1)}.

定义了问题的解空间后,还应该将解空间很好地组织起来,使得能用回溯法方便地搜索整个解空间.通常用树或图的形式组织解空间.

例如,对于 $n=3$ 时的 0/1 背包问题,可用一棵完全的二叉树表示其解空间,如图 13-1.

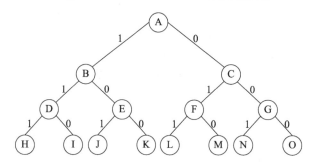

图 13-1 0/1 背包问题,$n=3$ 时的解空间

Fig. 13-1 The solution space of the 0/1 knapsack problem when $n=3$

The labels on the edges of layers i to $i+1$ of the solution space tree give the values of the variables. Any path from the tree root to the leaf denotes an element in the solution space. For example, the path from the root node to node H is equivalent to the element $(1,1,1)$ in the solution space.

13.1.3 Basic idea of backtracking method

After determining the organizational structure of the

解空间树的第 i 层到第 $i+1$ 层边上的标号给出了变量的值.从树根到叶子的任一路径表示解空间中的一个元素.例如,从根结点到结点 H 的路径相当于解空间中的元素 $(1,1,1)$.

13.1.3 回溯法的基本思想

在确定了解空间的组织结构后,回

solution space, the backtracking method starts from the root node and searches the whole solution space in the way of depth first search. The backtracking method searches recursively in the solution space in this way until the required solution is found or all solutions in the solution space are traversed.

When the backtracking method searches the solution space tree, two strategies are usually used to avoid invalid search and improve the search efficiency of the backtracking method. Firstly, the constraint function is used to cut out the subtree that does not meet the constraint at the current node (extension node); the second is to cut the subtree that can not get the optimal solution with the bound function. These two kinds of functions are collectively referred to as pruning functions.

Backtracking usually involves the following three steps:

(1) Define the solution space of the given problem;

(2) Determine the solution space structure that is easy to search;

(3) Search the solution space by depth first, and use pruning function to avoid invalid search in the search process.

When applying backtracking method to solve a problem, we should first clarify the solution space of the problem. The solution of a complex problem is often composed of many parts, that is, a large solution can be regarded as composed of several small decisions. Many times they form a decision sequence. All possible decision sequences for solving a problem constitute the solution space of the problem. The decision sequence satisfying the constraints in the solution space is called feasible solution. Generally speaking, solving any problem has a goal. The feasible solution that makes the goal reach the optimal under constraint conditions is called the optimal solution of the problem.

溯法从根结点出发,以深度优先搜索方式搜索整个解空间.回溯法以这种工作方式递归地在解空间中搜索,直到找到所要求的解或解空间所有解都被遍历为止.

回溯法搜索解空间树时,通常采用两种策略避免无效搜索,提高回溯法的搜索效率.其一是用约束函数在当前结点(扩展结点)处剪去不满足约束的子树;其二是用限界函数剪去得不到最优解的子树.这两类函数统称为剪枝函数.

用回溯法解题通常包含以下三个步骤:

(1) 定义所给问题的解空间;

(2) 确定易于搜索的解空间结构;

(3) 以深度优先方式搜索解空间,并在搜索过程中用剪枝函数避免无效搜索.

在应用回溯法解问题时,首先应该明确问题的解空间.一个复杂问题的解往往由多个部分构成,即一个大的解决方案可以看作是由若干个小的决策组成的.很多时候它们构成一个决策序列.解决一个问题的所有可能的决策序列构成该问题的解空间.解空间中满足约束条件的决策序列称为可行解.一般说来,解任何问题都有一个目标,在约束条件下使目标达到最优的可行解称为该问题的最优解.

13.2 *n*-queens problem

13.2.1 Algorithm description

Place *n* queens on a chessboard of *n*×*n* grids, so that they can't attack each other, that is, any two queens can't be in the same row, column or slash. Ask how many ways there are to place them.

Fig. 13-2 *n*-queens problem

N-queens problem evolved from the eight queens problem. The problem was put forward by international chess player Max Bethel in 1848: place eight queens on a chessboard of 8×8 grids, so that they can't attack each other, that is, any two queens can't be in the same row, column or slash. Ask how many kinds of placement methods there are. Gauss thought there are 76 options. In 1854, different authors published 40 different solutions in the chess magazine in Berlin. Later, 92 results were solved by the method of graph theory.

13.2.2 Algorithm analysis

With the popularization and development of computers, computers can simply calculate the problems that people could not solve before. And the idea is very clear, that is, brute-force search, traverse all cases, and then calculate the number of solutions.

13.2.3 Backtracking method

Search forward according to the optimization conditions to achieve the goal. However, when a certain step is explored and it is found that the original choice is not excellent or can not achieve the goal, it will go back and choose again. This technology of going back and going again if it fails is the backtracking method.

13.2.4 Idea of backtracking method

Use the array to simulate the chessboard. Start from the first line and select the position in turn. If the current position meets the conditions, select the position downward. If the conditions are not met, move the current position backward by one bit(Tab. 13-1).

选择并不优或达不到目标,就退回一步重新选择,这种走不通就退回再走的技术为回溯法.

13.2.4 回溯法的思路

用数组模拟棋盘,从第一行开始,依次选择位置,如果当前位置满足条件,则向下选位置,如果不满足条件,则当前位置后移一位(表 13-1).

表 13-1 回溯法
Tab. 13-1 Backtracking method

*							
unsatisfied 不满足	unsatisfied 不满足	*					
unsatisfied 不满足	unsatisfied 不满足	unsatisfied 不满足	unsatisfied 不满足	*			
unsatisfied 不满足	unsatisfied 不满足	unsatisfied 不满足	unsatisfied 不满足	*			
unsatisfied 不满足	unsatisfied 不满足	unsatisfied 不满足	unsatisfied 不满足	unsatisfied 不满足	unsatisfied 不满足	unsatisfied 不满足	unsatisfied 不满足

The last one is not satisfied. Go back to the previous row, select the next position and continue to explore.

In fact, we don't need an $n \times n$ array. We only need an n-length array to store the location.

Representation: arr[i] = k; it denotes that a queen is placed at position k on row i. Such an array of arr[n] can denote a feasible solution. Due to backtracking, we can find all solutions.

13.2.5 *n*-queens backtracking solution

The eight queens can't be in the same row, column or slash, so:

(1) Putting a queen in each row solves the problem of not being in the same row.

(2) In row i, traverse column n to test the position. Compare with the position of all previous rows.

(3) Compare column: the current column col is not equal to all previous columns. That is, col!=arr[i].

最后一个不满足,回溯到上一行,选下一个位置,继续试探.

其实并不需要一个 $n \times n$ 的数组,我们只需要一个长度为 n 的数组来存放位置.

表示方式:arr[i]=k;它表示在第 i 行的第 k 个位置放一个皇后. 这样,一个形为 arr[n]的数组就可以表示一个可行解. 利用回溯法,我们就可以求所有解.

13.2.5 *n* 个皇后回溯求解

因为 8 个皇后不能在同一行、同一列、同一斜线上,所以:

(1) 每一行放一个皇后,就解决了不在同一行的问题.

(2) 在第 i 行的时候,遍历 n 列,试探位置. 与之前所有行存放的位置进行比较.

(3) 比较列:当前列 col 不等于之前所有的列. 即 col!=arr[i].

(4) Compare slash: because the queens are not in the same slashes with a slope of 1 or −1, that is, (row−i)/(col−arr[i])! =1 or −1.

(5) You can skillfully use the absolute value function:abs(row−i)= abs(col−arr[i]).

13.2.6 Time complexity

Worst case: each row has n cases and there are n rows, so the time complexity is $O(n^n)$.

However, because the backtracking method will judge and abandon some situations in advance, the time complexity is not as high as expected. But $O(n!)$ isn't quite right either, and it's not quite clear how to calculate it exactly for the time being.

13.2.7 Space complexity

Only the array of arr[n] is used, that is, $O(n)$.

13.3 m-coloring problem of graphs

13.3.1 Problem description

Give undirected connected graph G and m different colors. These colors are used to color the vertices of graph G, and each vertex has a color. Is there a coloring method that makes the 2 vertices of each edge in G have different colors? This problem is the decision problem of m-coloring problem of graphs. If a graph needs at least m colors to make the two vertices connected by each edge in the graph have different colors, the number m is called the chromatic number of the graph. The problem of finding the chromatic number m of a graph is called the m colorable optimization problem of a graph. If all vertices and edges of a graph can be drawn on the plane in some way and no two sides intersect, the graph is said to be planar graph. The famous four-color conjecture of planar graphs is a special case of the problem of determining the m colorability of graphs.

The colorability of general connected graphs is discussed, which is not limited to plane graphs. Given a graph $G=(V,E)$ and m colors, if the graph is not m

colorable, give a negative answer; if the graph is m colorable, find all different coloring methods (Fig. 13-3).

着色的,给出否定回答;如果这个图是 m 可着色的,找出所有不同的着色法(图13-3).

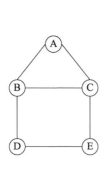
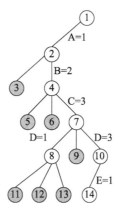

图 13-3 图的 m 着色问题
Fig. 13-3 The m-coloring problem of graphs

13.3.2 Algorithm idea

Backtracking method, the solution vector $\{c_1, c_2, c_3, \ldots, c_n\}$ of the problem can construct the solution space tree as a subset tree (complete m-ary tree).

Constraint condition: adjacent nodes cannot be of the same color.

1. Time complexity

The upper bound of computing time of backtracking algorithm for m-coloring problem of graphs can be estimated by calculating the number of internal nodes in the solution space tree. The number of inner nodes in the solution space tree of m-coloring problem of graphs is m^n. For each inner node, in the worst case, it takes $O(n)$ to check the availability of the color corresponding to each son of the current extension node with the constraint function. Therefore, the total time cost of backtracking method is $O(n \times m^n)$.

2. Space complexity

Use adjacency matrix to store graph, space complexity $S(n) = O(n^2)$.

13.3.2 算法思想

回溯法,问题的解向量 $\{c_1, c_2, c_3, \cdots, c_n\}$ 可以构造解空间树为子集树(完全 m 叉树).

约束条件:相邻结点不能同色.

1. 时间复杂度

图 m 可着色问题的回溯算法的计算时间上界可以通过计算解空间树中内结点个数来估计.图 m 可着色问题的解空间树中内结点个数是 m^n. 对于每个内结点,在最坏情况下,用约束函数检查当前扩展结点的每个儿子所对应的颜色的可用性需耗时 $O(n)$. 因此,回溯法总的时间耗费是 $O(n \times m^n)$.

2. 空间复杂度

用邻接矩阵来存储图,空间复杂度 $S(n) = O(n^2)$.

13.4 0/1 knapsack problem

13.4.1 Problem description

There are n items, each of which has its own volume and value. For the existing backpack with a given capacity, how to make the items loaded in the backpack have the maximum total value?

In order to facilitate explanation and understanding, the examples described below are first substituted with specific numbers, i.e., number=4, capacity=8.

Insert Tab. 13-2 here:

Tab. 13-2 0/1 knapsack problem

i(Item Number)	1	2	3	4
w(Volume)	2	3	4	5
V(Value)	3	4	5	6

13.4.2 General idea

Find out the optimal solution and solution composition of 0/1 knapsack problem according to the problem-solving steps of dynamic programming (problem abstraction, establishing model, looking for constraint conditions, judging whether it meets the optimality principle, finding the recursive relationship between large and small problems, filling in tables, looking for solution composition), and then write code to realize it.

13.4.3 Principle of dynamic programming

Dynamic programming is similar to divide and conquer. It divides big problems into small problems, and solves small problems one by one by looking for the recursive relationship between big problems and small problems, so as to finally achieve the effect of solving the original problems. The difference is that the divide and conquer method has been repeatedly calculated many times on subproblems and subproblems, while the dynamic programming has memory. All the answers to the solved subproblems are recorded by filling in the table, the subproblems needed in the new problem can be extracted directly, avoiding repeated calculation, thus saving time. Therefore, after the problem meets the optimality principle, the core of solving the problem with dynamic programming is to fill in the table. After filling in the table, the optimal solution will be found.

The optimality principle is the basis of dynamic programming. The optimality principle means that "the optimal decision sequence of multi-stage decision-making process has such a property: regardless of the initial state and initial decision-making, for a certain state caused by the previous decision-making, the decision-making sequence of subsequent stages must constitute the optimal strategy".

13.4.4 Solution process of knapsack problem

Before solving the problem, for the convenience of description, first define some variables: V_i denotes the value of the i-th item, W_i denotes the volume of the i-th item, $V(i,j)$ denotes when the current knapsack capacity is j, the value corresponding to the best combination of the first i items. Abstract the knapsack problem $(X_1, X_2, \ldots, X_n$, where X_i takes 0 or 1, denoting whether the i-th item is selected or not).

(1) **Establish the model**, that is, find max $(V_1X_1+V_2X_2+\ldots+V_nX_n)$;

(2) **Find constraint conditions**, $W_1X_1+W_2X_2+\ldots+W_nX_n<$capacity;

(3) **Find a recurrence relation**, there are two possibilities in the face of current commodities:

The capacity of the knapsack is smaller than that of the commodity and the commodity cannot be loaded. At this time, the value is the same as that of the first $i-1$, that is, $V(i,j)=V(i-1,j)$;

There is still enough capacity to load the commodity, but it may not reach the current optimal value, so choose the optimal one between loading and non-loading, that is, $V(i,j)=\max(V(i-1,j),V(i-1,j-w(i))+v(i))$.

Where $V(i-1,j)$ means non-loading, $V(i-1, j-w(i))+v(i)$ means the i-th commodity is loaded, the knapsack capacity decreases $w(i)$, but the value increases $v(i)$;

Thus, the recurrence relation can be obtained:
$V(i,j) = V(i-1,j)$, $j<w(i)$;
$V(i,j) = \max(V(i-1,j), V(i-1,j-w(i)) + v(i))$, $j \geq w(i)$.

Here, we need to explain why it needs to be solved in this way when it can be loaded (this is the key to this problem!).

It can be understood that, how many ways are there to reach the state of $V(i,j)$? There must be two kinds. The first is that the i-th commodity is not loaded in, and the second is that the i-th commodity is loaded in. It's easy to understand that it's not loaded in, which is $V(i-1,j)$; how to understand it is loaded in? If the i-th commodity is loaded, what is the state before loading? It must be $V(i-1,j-w(i))$. Due to the optimality principle (mentioned above), $V(i-1, j-w(i))$ is a state caused by the previous decision, and the latter decision must constitute the optimal strategy. The two cases are compared to obtain the optimal solution.

(4) **Fill in the table**, first initialize the boundary conditions, $V(0,j) = V(i, 0) = 0$;

Insert Tab. 13-3 here:

其中 $V(i-1,j)$ 表示不装，$V(i-1, j-w(i))+v(i)$ 表示装了第 i 个商品，背包容量减少 $w(i)$，但价值增加了 $v(i)$;

由此可以得出递推关系式:
$V(i,j) = V(i-1,j)$, $j<w(i)$;
$V(i,j) = \max(V(i-1,j), V(i-1,j-w(i)) + v(i))$, $j \geq w(i)$.

这里需要解释一下，为什么能装的情况下，需要这样求解（这才是本问题的关键所在!）。

可以这么理解，如果要到达 $V(i,j)$ 这个状态有几种方式？肯定是两种，第一种是第 i 件商品没有装进去，第二种是第 i 件商品装进去了。没有装进去很好理解，就是 $V(i-1,j)$；装进去了怎么理解呢？如果装进去第 i 件商品，那么装入之前是什么状态，肯定是 $V(i-1,j-w(i))$。由最优性原理（上文已讲到），$V(i-1,j-w(i))$ 就是之前决策造成的一种状态，后面的决策就要构成最优策略。比较两种情况，可得出最优解。

(4) **填表**，首先初始化边界条件，$V(0,j) = V(i, 0) = 0$;

这里插入表 13-3:

表 13-3 动态规划解背包问题
Tab. 13-3 Dynamic programming solution for the knapsack problem

i/j	0	1	2	3	4	5	6	7	8
0	0	0	0	0	0	0	0	0	0
1	0								
2	0								
3	0								
4	0								

Then fill in the form line by line:

For example, $i=1, j=1, w(1)=2, v(1)=3$, there is $j<w(1)$, so $V(1,1)=V(1-1,1)=0$;

If $i=1, j=2, w(1)=2, v(1)=3$, there is $j=w(1)$, so $V(1,2)=\max(V(1-1,2), V(1-1, 2-w(1))+v(1))=\max(0, 0+3)=3$;

In this way, fill into the last one, $i=4, j=8$, $w(4)=5, v(4)=6$, there is $j>w(4)$, so $V(4,8)=\max(V(4-1,8), V(4-1, 8-w(4))+v(4))=\max(9, 4+6)=10\ldots$

Therefore, the completed table is shown in Tab. 13-4:

Tab. 13-4 The solution of the knapsack problem

i/j	0	1	2	3	4	5	6	7	8
0	0	0	0	0	0	0	0	0	0
1	0	0	3	3	3	3	3	3	3
2	0	0	3	4	4	7	7	7	7
3	0	0	3	4	5	7	8	9	9
4	0	0	3	4	5	7	8	9	10

$V(\text{number}, \text{capacity}) = V(4,8) = 10$.

13.4.5 Backtracking of optimal solution of knapsack problem

The optimal solution of the knapsack problem can be obtained by the above method, but we don't know what commodities the optimal solution consists of, so we should find out the composition of the solution by backtracking according to the optimal solution. According to the principle of filling in the table, there can be the following solution finding methods:

When $V(i,j) = V(i-1,j)$, it means that the i-th commodity is not selected, then return to $V(i-1,j)$;

When $V(i,j) = V(i-1, j-w(i)) + v(i)$, it means that the i-th commodity is loaded, which is part of the optimal solution. Then we have to go back to before loading that commodity, i.e. to $V(i-1, j-w(i))$;

Keep traversing until $i=0$, the composition of all solutions will be found.

Take the example above:

The optimal solution is $V(4, 8) = 10$, and $V(4, 8)! = V(3, 8)$ has $V(4,8) = V(3, 8-w(4)) + v(4) = V(3,3) + 6 = 4 + 6 = 10$, so the fourth commodity is selected, return to $V(3, 8-w(4)) = V(3, 3)$;

There is $V(3, 3) = V(2, 3) = 4$, so the third commodity is not selected. Go back to $V(2, 3)$;

And $V(2, 3)! = V(1, 3)$, there is $V(2, 3) = V(1, 3-w(2)) + v(2) = V(1, 0) + 4 = 0 + 4 = 4$, so the second commodity is selected, return to $V(1, 3-w(2)) = V(1, 0)$;

There is $V(1, 0) = V(0, 0) = 0$, so the first commodity is not selected. The results are shown in Tab. 13-5:

一直遍历到 $i=0$ 结束为止,所有解的组成都被找到.

以上面的例子来说:

最优解为 $V(4, 8) = 10$,而 $V(4, 8)! = V(3,8)$ 却有 $V(4,8) = V(3, 8-w(4)) + v(4) = V(3, 3) + 6 = 4 + 6 = 10$,所以第 4 件商品被选中,并且回到 $V(3, 8-w(4)) = V(3, 3)$;

有 $V(3, 3) = V(2, 3) = 4$,所以第 3 件商品没被选择,回到 $V(2, 3)$;

而 $V(2, 3)! = V(1,3)$,有 $V(2, 3) = V(1, 3-w(2)) + v(2) = V(1, 0) + 4 = 0 + 4 = 4$,所以第 2 件商品被选中,并且回到 $V(1, 3-w(2)) = V(1, 0)$;

有 $V(1, 0) = V(0, 0) = 0$,所以第 1 件商品没被选择. 结果如表 13-5:

表 13-5 背包问题最优解回溯
Tab. 13-5 Backtracking of optimal solution of knapsack problem

i/j	0	1	2	3	4	5	6	7	8
0	0	0	0	0	0	0	0	0	0
1	0	0	3	3	3	3	3	3	3
2	0	0	3	4	4	7	7	7	7
3	0	0	3	4	5	7	8	9	9
4	0	0	3	4	5	7	8	9	10

Exercise 13

1. Design a backtracking algorithm to solve n-queens problem with recursive function.

2. Design a backtracking algorithm to solve m-coloring problem of graphs with recursive function.

3. Design a backtracking algorithm to solve Hamiltonian loop with recursive function.

4. Design an algorithm to solve the traveling salesman problem with backtracking method.

5. Design a backtracking algorithm to solve the problem of horse traveling in chess: given an 8×8 chessboard, the horse starts from a certain position on the chessboard, passes through each square in the chessboard exactly once, and finally returns to the starting position.

6. Given the carrying capacity of the backpack $M = 20$, there are 6 objects with values of 11, 8, 15, 18, 12 and 6 and weights of 5, 3, 2, 10, 4 and 2 respectively. Explain the process of solving the above 0/1 knapsack problem by backtracking method. Draw the search tree, number the nodes according to the generation order, and mark the results of the actions performed when generating the node next to the node.

7. There is an $n \times m$ grid of the maze closed on all sides. Only the grid in the upper left corner has an entrance and the grid in the lower right corner has an exit. There may or may not be entrances and exits in the east, west, south and north of the grid inside the maze. Design a backtracking algorithm to get out of the maze.

习题 13

1. 用递归函数设计一个解 n 个皇后问题的回溯算法.

2. 用递归函数设计一个解图的 m 着色问题的回溯算法.

3. 用递归函数设计一个解哈密尔顿回路的回溯算法.

4. 用回溯法设计一个解货郎担问题的算法.

5. 设计一个回溯算法,求解国际象棋中马的周游问题:给定一个 8×8 的棋盘,马从棋盘的某一个位置出发,只经过棋盘中的每一个方格一次,最后回到它出发的位置.

6. 给定背包的载重量 $M = 20$,有 6 个物体,价值分别为 11,8,15,18,12,6,质量分别为 5,3,2,10,4,2. 说明用回溯法求解上述 0/1 背包问题的过程. 画出搜索树,给按照生成顺序结点编号,并在结点旁边标出生成该结点时所执行动作的结果.

7. 设有一个 $n \times m$ 格的迷宫,四面封闭,仅在左上角的格子有一个入口,在右下角的格子有一个出口. 迷宫内部的格子,其东、西、南、北四面可能有出入口,也可能没有出入口. 设计一个回溯算法通过迷宫.

习题 13 答案
Key to Exercise 13

Part VI Modelling and Reasoning about Systems

Chapter 14 Model-based Verification

In the second chapter of the text book, we have learned the related concepts of finite automata, including the formal definition and formal description of finite automata. In this chapter, we will examine a system attribute verification method based on a finite state machine model. Firstly, we need to look at the concepts surrounding formal verification.

14.1 Formal verification and model checking

The verification of design correctness is an important research topic concerned by academia and industry at present. We can feel **the importance of design correctness verification** from the following two events.

In 1994, the Pentium processor was found to be performing a particular floating-point operation with an error that occurs only once every 27000 years. In response, Intel paid a whopping \$475 million to recycle defective Pentium processors.

On 4 June 1996, the Ariana V rocket developed by the European Space Agency exploded less than 40 seconds after launch. An investigation later revealed that the error is an exception occurs when a large 64-bit floating point was converted to a 16-bit signed integer. A decade of hard work was undone by a tiny mistake.

It can be seen from the above events, it is important to ensure that the design of digital systems is correct, whether they are used in high-risk areas or in the ordinary household. Software, hardware and protocol are the three most basic forms of digital system design, and any complex digital system is roughly composed of these three parts. If mistakes are inevitable

第Ⅵ部分 系统建模与推理

第14章 基于模型的验证

在本书第2章我们学到了有穷自动机的相关概念,包括有穷自动机的形式定义和形式化的描述方法.在这一章,我们将研究基于有穷状态机模型的系统属性验证方法.首先,我们需要了解一下形式化验证的相关概念.

14.1 形式化验证和模型检测

设计正确性的验证问题是目前学术界和工业界均予以关注的重要研究课题,我们可从以下两个事件切实感受**设计正确性验证的重要性**.

1994年,奔腾处理器被发现在执行某个特定的浮点运算时出现错误,这种错误27000年才可能出现一次.对此,因特尔付出4.75亿美元的巨额代价回收有缺陷的奔腾处理器.

1996年6月4日,欧洲航天局研制的阿里亚娜五型火箭在发射后不到40秒爆炸.事后调查发现,爆炸发生的原因是当一个很大的64位浮点数转换为16位带符号整数时出现异常.十年的努力毁于这个细微错误.

从以上事件可以看出,无论是用于高风险性领域还是普通家用,保证数字系统的设计正确性都是至关重要的.软件、硬件和协议是数字系统设计所包含的三种最基本的形式,任何复杂的数字系统大致都由这三个部分组成.如果说在复杂系统设计过程中错误

in the design of complex systems, the only reason for accidents is that a complete validation of the product has not been done before it is put into use, that is, to ensure the system has fulfilled the designer's intent.

So far, the verification methods can be divided into **analog vertification**, **emulation vertification** and **formal verification**. Analog verification is a traditional verification method, and is still the mainstream verification method. Analog verification is to apply the excitation signal to the design, calculate and observe the output results, and judge whether the results are consistent with the expected. **The main drawback of analog verification is imperfection, that is, it can only prove error but not error free.** Thus, analog verification are generally suitable for finding large and obvious design errors in the early stages of verification, but not for complex and subtle errors.

Emulation verification is similar in principle to analog verification, it integrates the three main parts of analog verification, namely excitation generation, monitor and coverage measure into **testbench** to construct FPGA. Emulation is much faster than analog verification, and its disadvantages are expensive and poor flexibility.

Since analog vertification and emulation verification have such great limitations, it is natural to explore a more perfect verification method, namely **formal verification. Formal verification is to prove mathematically whether the system realizes the designer's intention or not.** This means first constructing a mathematical model of the system using some language and logic, and then applying rigorous mathematical reasoning to prove the correctness of the design. **The main advantage of formal verification is completeness, which can completely determine the correctness of the design. Its disadvantage is that the original design has to be model extracted first, which requires mathematical skills and experience**

on the user. Moreover, some tools require manual guidance (such as theorem proving), and some tools have **state space explosion problems** (such as model testing).

While formal validation is not yet a mainstream tool for validation, researchers in the industry are continuing to work and making progress. According to the content or need of verification, formal verification can be divided into two types: **equivalence test** (Fig. 14-1) and **property test** (Fig. 14-2).

虽然形式验证目前还不能成为验证的主流工具,但业界的研究人员仍在继续努力,并不断取得进展. 按照验证的内容或需要,形式验证可分为两类:**等价性检验**(图 14-1)和**性质检验**(图 14-2).

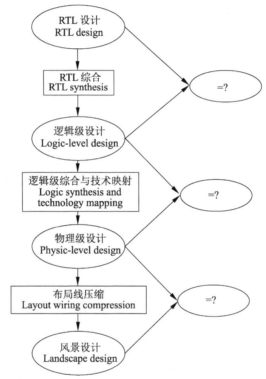

图 14-1 数字系统设计流程中的等价性问题
Fig. 14-1 Equivalence problem in digital system design process

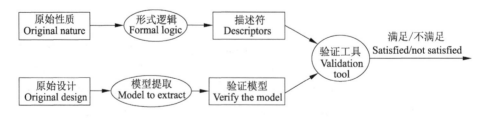

图 14-2 性质检验的框架
Fig. 14-2 A framework for property testing

(1) The question to be answered by the equivalence test is: are the two designs functionally equivalent or mutually inclusive? Seen from the design process of the whole digital system, equivalence problem almost exists in each adjacent design level, as shown in Fig. 14-1. There are also many equivalence tests between designs at the same level. Equivalence is divided into two kinds: combinatorial equivalence and sequential equivalence. For two combined circuits, the equivalence test determines whether the two circuits have the same output for all possible inputs; For two sequential circuits, the equivalence test is to prove whether the two circuits have the same output for all possible input sequences.

(2) Property test refers to the verification of a design whether it satisfies certain properties. Property testing is mostly used at high levels, such as behavior level verification. Property testing requires a formal description of both the system model and the properties to be verified, as shown in Fig. 14-2. The system model is generally denoted by finite state machine, and the properties are denoted by formal logic. Based on the system model and properties, the verification tool gives the judgment that it is satisfied or not satisfied.

Any formal verification system can be summarized into three parts: specification, implementation and verification algorithm. The specification is the design correctness requirement, the implementation is the design to be verified, and the verification algorithm gives the verification process.

According to the content of the description, specifications can be divided into:

(1) **System specification**: description of the design system at various levels of abstraction (behavior level, RTL, gate level, switch level). The description tool can be a hardware description language such as RTL, Verilog, or a high-level language such as C,

(1)等价性检验要回答的问题是：两个设计在功能上是否等价或互相包含？从整个数字系统的设计流程看，等价性问题几乎存在于每个上下相邻的设计层次中，见图 14-1. 在相同层次设计间做等价性检验的也很多. 等价性又分为组合等价性和时序等价性两种. 对于两个组合电路，等价性检验要确定的是对于所有可能输入，两个电路是否有相同输出；对于两个时序电路，等价性检验要证明的则是对于所有可能的输入序列，两个电路是否有相同输出.

(2)性质检验是指对一个设计，验证它是否满足某些性质. 性质检验多用于高层，如行为级验证. 性质检验需要对系统模型和要验证的性质都进行形式化的描述，见图 14-2. 系统模型一般用有穷状态机表示，性质多用形式逻辑表示. 验证工具根据系统模型和性质，给出满足或不满足的判断.

任何形式的验证系统都可概括为三个部分：规范、实现及验证算法. 规范是设计正确性要求，实现是待验证的设计，验证算法给出了验证的过程.

规范根据其所描述的内容可分为：

(1) **系统规范**：在各种不同抽象层次(行为级、RTL、门级、开关级)上对设计系统进行的描述. 描述工具可以是某种硬件描述语言如 RTL、Verilog，或者是某种高级语言 C、C++、System C,

C++, System C, or a form of a finite state automaton accepted by a verification tool.

(2) **Property specification**: denotes the temporal property or function that the system to be verified should have during operation. The most appropriate tool for describing temporal properties is **temporal logic**. Temporal logic is an expression that describes the property of the state in which the system is located and the sequence of transitions between states. Temporal logic is an extension of propositional logic by adding operators to denote temporal relations. According to the structure of time expansion, temporal logic can be divided into **linear temporal logic** (**LTL**) and **computation tree logic** (**CTL**).

The three properties that are most often tested are **safety**, **liveness**, and **fairness**. Safety describes that a dangerous event will never occur, liveness describes that a necessary event will eventually occur, and fairness describes that an event must occur indefinitely and often.

Formal verification methods can be classified into three parts: theorem proving, model checking and equivalence testing. This chapter focuses on the verification method of model checking.

Model checking is an important automatic verification technology. It was first proposed by Clarke, Emerson, Quielle and Sifakis respectively in 1981, mainly through explicit state search or implicit fixed point calculation to verify the modal/propositional properties of finite state concurrent systems.

Model checking is more highly regarded in industry than deductive proof because it can be performed automatically and provides a counterexample path if the system does not satisfy the properties. Although limitation is a disadvantage on finite systems, model checking can be applied to many vital systems, such as hardware controllers and communication protocols that are finite state systems. In many cases, model

（2）性质规范：表示待验证系统运行过程中应该具有的时态性质或功能.描述时态性质的最合适工具是**时态逻辑**.时态逻辑是描述系统所处状态及在状态间迁移序列的性质的表达式.时态逻辑是对命题逻辑的扩充,增加了表示时间关系的运算符.按时间展开的结构,时态逻辑分为**线性时态逻辑**（LTL）和**分支时态逻辑**（CTL）两种.

三种最经常验证的性质是：**安全性、活性和公平性**.安全性描述的是某危险事件永不发生,活性描述的是某必需事件终将发生,公平性描述的是某事件必须无限经常地发生.

可将形式验证方法分为三类：定理证明、模型检测、等价性检验三个部分.本章关注模型检测的验证方法.

模型检测是一种很重要的自动验证技术.它最早由 Clarke、Emerson、Quielle 和 Sifakis 分别在1981年提出,主要通过显式状态搜索或隐式不动点计算来验证有穷状态并发系统的模态/命题性质.

由于模型检测可以自动执行,并能在系统不满足性质时提供反例路径,因此在工业界比演绎证明更受推崇.尽管限制在有穷系统上是一个缺点,但模型检测可以应用于许多非常重要的系统,如硬件控制器和通信协议都是有穷状态系统.很多情况下,可以把模型检测和各种抽象与归纳原则

checking can be combined with various abstract and inductive principles to verify non-finite state systems (such as real-time systems).

Model checking is based on **temporal logic**. The idea of temporal logic is that in a model the true and false of formulas are not static, as they are in propositional or predicate logic. Instead, the model of temporal logic consists of several states, and a formula can be true in some states and false in others. Thus, the concept of static truth is replaced by the concept of dynamic truth, regarding this concept of dynamic truth, the formula can change its truth value as the state of the system evolves. In model checking, model M is a **transition systems**, and property φ is a temporal logic formula. To verify that a system satisfies a property, we must do three things:

(1) Use the descriptive language of the model-checker to model the system and get a model M;

(2) The properties are encoded by the specification language of the model checker, and a temporal logical formula φ is generated;

(3) Run the model checker with M and φ as inputs.

If $M \models \varphi$, the model checker output answers "yes", otherwise outputs "no"; in the latter case, most model checkers also generate a trajectory of system behavior that leads to failure. The automatic generation of "reverse tracing" is an important tool in system design and debugging.

Since model checking is a model-based processing, this chapter only studies the concept of satisfaction, that is, the satisfaction relationship between a model and a formula ($M \models \varphi$).

There is a large family of temporal logics that have been proposed and used in the study of all sorts of things. These vast number of formal systems can be classified and organized according to their particular view of "time". Linear temporal logic views time as a set of paths, where a path is an instantaneous sequence of time. Computing tree logic views time as a tree, branching into the future with the current time as the root. The branching time seems to make the uncertainty of the future clearer. Another property of time is to think of time as continuous or discrete. If analog computers are studied, the former is recommended, and for synchronous networks, the latter is preferred.

Temporal logic has a dynamic aspect, because in a model, the truth values of formulas are not fixed as they are in predicate or propositional logic, but depend on points in time within the model. In this chapter, we examine one type of logic where time is linear, called linear temporal logic, and another type of logic where time is branching, called computation tree logic. These logics have been proven to be very fruitful in verifying hardware and communication protocols and people are already using them for software verification. Model checking is the process of calculating the answer to the questions whether M and $s \models \varphi$ hold, where φ is a formula in LTL or CTL, M is an appropriate model of the system under consideration, s is a state of the model, and $\models$ is a satisfying relationship.

14.2 Syntax and semantics of linear temporal logic

14.2.1 Syntax of linear temporal logic

Linear temporal logic is a special kind of temporal logic that can express the concept of time. It regards the time axis as a linear sequence, which is often used to accurately denote the dynamic semantics of the model. The role of LTL: describes the logic of LT characteristics, but does not give a specific moment, only the relative characteristics of time.

The syntax of LTL is described as follows.

Propositional logic: describe the static behavior of the system at a certain moment. The most commonly used propositional logic in LTL are the following three:

(1) Atomic proposition: $a \in AP$;
(2) Negation law: $\neg \varphi$;
(3) Associative law: $\varphi \wedge \psi$;

Temporal operator: describe the properties of the system under the trajectory, the most basic are the following two:

(1) The next moment is satisfied φ: $\bigcirc \varphi$, pronounced "next";
(2) Every moment is satisfied φ until ψ: $\varphi \cup \psi$, pronounced "until".

An operator derived from propositional logic and temporal operators:
(1) $\varphi \vee \psi \equiv \neg(\neg \varphi \wedge \neg \psi)$;
(2) $\varphi \Rightarrow \psi \equiv \neg \varphi \vee \psi$;
(3) $\varphi \Leftrightarrow \psi \equiv (\varphi \Rightarrow \psi) \wedge (\psi \Rightarrow \varphi)$;
(4) $\varphi \oplus \psi \equiv (\varphi \wedge \neg \psi) \wedge (\neg \varphi \wedge \psi)$;
(5) true $\equiv \varphi \vee \psi$;
(6) false $\equiv \neg$ true;
(7) $\Diamond \varphi \equiv$ true $\cup \varphi$ ($\Diamond$ pronounced "eventually", means that φ will be satisfied at some point in the future);
(8) $\Box \varphi \equiv \neg \Diamond \neg \varphi$ ($\Box$ pronounced "always", it means that φ will always be satisfied at this time).

These operators have priority, $\neg$, $\bigcirc$ execute first, followed by $\cup$, then $\vee$, $\wedge$ and finally $\rightarrow$. Intuitive explanation of the operator:

(1) atomic prop. A

(2) next step $\bigcirc a$

(3) until $a \cup b$

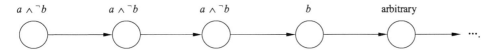

(4) eventually ◇a

(5) always □a

On a trajectory, if from a certain moment, each moment only satisfies b, even if a has not occurred, then $a \cup b$ is also satisfied at that moment.

Mutual exclusion: Two processes cannot enter critical resources at the same time.

Starvation-free: When the process keeps waiting to enter the critical resource, it can finally keep entering the critical resource.

We use LTL language to denote the logic of mutual exclusion and starvation-free at each moment:

(1) Mutual exclusion: $\Box \neg (c_1 \wedge c_2)$;

(2) Starvation-free:
$(\Box \Diamond w_1 \Rightarrow \Box \Diamond c_1) \wedge (\Box \Diamond w_2 \Rightarrow \Box \Diamond c_2)$.

$\Box \Diamond w_1$ understanding method: look at the outermost layer first, and treat $\Diamond w_1$ as one φ, then $\Box \varphi$ denotes that φ will always be satisfied from this time; then look at the inner layer $\Diamond w_1$, which denotes that infinitely always meets w_1; therefore, add it up to denote that from this point on, it will start to satisfy w_1 all the time infinitely.

14.2.2 Semantics of linear temporal logic

The property of LT caused by LTL formula φ on AP is: $Words(\varphi) = \{\sigma \in (2^{AP})^\omega | \sigma \models \varphi\}$, where σ is a locus, $\sigma = A_0 A_1 A_2 \ldots$. From this LT property, the following properties can be deduced:

(1) $\sigma \models true$, denotes that there must be a correct path in σ.

(2) $\sigma \models a$, iff $a \in A_0$ (that is $A_0 \models a$), denotes that path σ contains a if and only if a is a possibility

在一条轨迹上,如果从某时刻起, 每个时刻只满足 b, 即便没出现过 a, 那么该时刻也满足 $a \cup b$。

互斥现象: 两个进程不可以同时进入临界资源。

饥饿自由: 当进程不停地等待进入临界资源的时候, 它最终可以不停地进入临界资源。

我们用 LTL 语言表示互斥现象和饥饿自由在每个时刻的逻辑:

(1) 互斥现象: $\Box \neg (c_1 \wedge c_2)$;

(2) 饥饿自由:
$(\Box \Diamond w_1 \Rightarrow \Box \Diamond c_1) \wedge (\Box \Diamond w_2 \Rightarrow \Box \Diamond c_2)$。

$\Box \Diamond w_1$ 理解方法: 先看最外层, 把 $\Diamond w_1$ 看作一个 φ, 则 $\Box \varphi$ 表示从此时起将会一直满足 φ; 再看内层, $\Diamond w_1$ 表示无限经常地满足 w_1; 因此, $\Box \Diamond w_1$ 表示从此时起开始会一直无限次满足 w_1。

14.2.2 线性时态逻辑的语义

由 LTL 公式 φ 在 AP 上引起的 LT 性质为: $Words(\varphi) = \{\sigma \in (2^{AP})^\omega | \sigma \models \varphi\}$, 其中 σ 是一个轨迹, $\sigma = A_0 A_1 A_2 \cdots$。由这个 LT 性质可以推出如下性质:

(1) $\sigma \models true$, 表示 σ 上一定有一条正确的路径。

(2) $\sigma \models a$, iff $a \in A_0$ (也就是 $A_0 \models a$), 表示当且仅当 a 是状态集合 A_0 当

in the set of states A_0, A_0 includes all the possibilities for the initial state, and a is just one of them.

(3) $\sigma \models \varphi_1 \wedge \varphi_2, iff \ \sigma \models \varphi_1$ and $\sigma \models \varphi_2$, denotes that if the path σ contains the states φ_1 and φ_2, then σ contains $\varphi_1 \wedge \varphi_2$.

(4) $\sigma \models \neg \varphi, iff \ \sigma \models \varphi$.

(5) $\sigma \models \bigcirc \varphi, iff \ \sigma[i..] = A_1 A_2 A_3 ... \models \varphi$.

Note: $\sigma[i..] = A_i A_{i+1} A_{i+2} ...$, denotes the suffix of σ starting with index i.

(6) $\sigma \models \Diamond \varphi, iff \ \exists j \geq 0. \ \sigma[j..] \models \varphi$.

(7) $\sigma \models \Box \varphi, iff \ \forall j \geq 0. \ \sigma[j..] \models \varphi$.

(8) $\sigma \models \Box \Diamond \varphi, iff \ \forall j \geq 0. \ \exists \geq \sigma[i..] \models \varphi$.

(9) $\sigma \models \Diamond \Box \varphi, iff \ \exists j \geq 0. \ \forall \geq \sigma[i..] \models \varphi$.

Example 14.2.1 From Fig. 14-3, we can derive four properties of model TS:

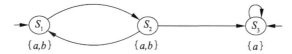

图 14-3 S_1, S_2, S_3 的转换图

Fig. 14-3 Transformation diagram of S_1, S_2, and S_3

(1) Start with the initial state, and every state satisfies a, that is $TS \models \Box a$.

(2) From the initial state, each state is satisfied, and if the state does not satisfy b, then the state satisfies $a \wedge b$, that is $TS \models \Box(\neg b \Rightarrow \Box(a \wedge \neg b))$.

(3) If starting from the initial state S_1 on the left, the next state can satisfy both a and b, if starting from the initial state S_3 on the right, the next state of S_3 is still S_3, which does not satisfy both a and b, so $TS \models \bigcirc(a \wedge b)$.

(4) Starting with the initial state S_1 on the left, each state satisfies b, until $a \wedge \neg b$ is satisfied, b is no longer satisfied, but starting with the initial state on the right, it does not satisfy, so $TS \models b \cup (a \wedge \neg b)$.

Temporal operators are limited to describing events along a path from a given state.

<>p: p will be true at some moment in the future

| | | | | | | p | |

[ ]p: p is always true from the present moment

| p | p | p | p | p | p | p | p |

Xp: p is true at the next moment

| | p | | | | | | |

14.3 Syntax and semantics of computation tree logic

14.3.1 Syntax of computation tree logic

Computation tree logic is a branching temporal logic, that is, its temporal model is a tree-like structure in which the future is uncertain. There are different paths to the future, and any one of them could be the "actual" path of reality.

As before, we fix a set of atomic formulas/descriptions (e. g. $p, q, r, \ldots$ or $p_1, p_2, \ldots$).

Definition 14.3.1 Like LTL, the CTL formula is summarized and defined by Backus-Naur normal form:

$\varphi ::= \bot \mid T \mid p \mid (\neg \varphi) \mid (\varphi \wedge \varphi) \mid (\varphi \vee \varphi) \mid (\varphi \rightarrow \varphi) \mid AX\varphi \mid EX\varphi \mid AF\varphi \mid EF\varphi \mid AG\varphi \mid EG\varphi \mid A[\varphi \cup \varphi] \mid E[\varphi \cup \varphi] \mid$

Here p takes the set of atomic formulas.

Notice that each CTL temporal conjunction is a pair of symbols. The first pair is either A or E. A is "along all paths" (without exception), and E means "along at least one path (exists)" (possible). The second symbol in the symbol pair is X, F, G, or U, the meaning are "next state", "some future state", "all future states (global)", and "until" respectively. For example, the symbol pair in $E[\varphi_1 \cup \varphi_1]$ is EU. In CTL, pairs of symbols like EU are not separable. Notice that AU and EU are binary. The symbols X, F, G, and U cannot appear alone without an A or E preceding them. Similarly, each A or E must be accompanied by one of X, F, G or U.

Weak until (W) and release (R) are not normally included in CTL, but they can be derived.

Convention 14.3.1 Assume binding precedence similar to that in propositional and predicate logic for CTL connectives. Unary connectives (including ¬ and temporal connectives AG, EG, AF, EF, EF, AX and EX) have the closest binding; then $\wedge$ and $\vee$; and finally $\rightarrow$, AU and EU.

Naturally, we can ignore these priorities using parentheses. To understand the syntax, let's look at some examples of the well-formed CTL formula and the non-well-formed CTL formula. Assume p, q and r are atomic formulas. Here are some well-formed CTL formulas:

- $AG(q \rightarrow EG\ r)$, note that it is different from $AG\ q \rightarrow EG\ r$, because according to Convention 14.3.1, the latter formula refers to $(AG\ q) \rightarrow (EG\ r)$.
- $EF\ E[r \cup q]$
- $A[p \cup EF\ r]$
- $EF\ EG\ p \rightarrow AF\ r$, also note that the parenthesis binding is $(EF\ EG\ p) \rightarrow AF\ r$, not $EF(EG\ p \rightarrow AF\ r)$ or $EF\ EG(p \rightarrow AF\ r)$
- $A[p_1 \cup A[p_2 \cup p_3]]$
- $E[A[p_1 \cup p_2] \cup p_3]$
- $AG(p \rightarrow A[p \cup (\neg p \wedge A[\neg p \cup q])])$

It's worth spending some time seeing how the syntax rules allow these formulas to be constructed. The following formula is not a well-formed formula:

- $EF\ G\ r$
- $A \neg G \neg p$
- $F[r \cup q]$
- $EF(r \cup q)$
- $AEF\ r$
- $A[(r \cup q) \wedge (p \cup r)]$

It is especially worth understanding why the rules of syntax do not allow the construction of these formulas. Take $EF(r \cup q)$ as an example, the problem with this string is that U can only appear in pairs with A and E, and E can be paired with F. To change this string into a well-formed CTL formula, you can only write $EF\ E[r \cup q]$ or $EF\ A[r \cup q]$.

Note that square brackets are used when the operator paired after A or E is U. There is no particular reason to do this. You can use the usual parentheses instead. However, it is often helpful to read the formula (because we can more easily find the position of corresponding closing parentheses). Another reason to use square brackets is that SMV always insists on doing so.

$A[(r \cup q) \wedge (p \cup r)]$ is not a well-formed formula because syntax does not allow Boolean connectors to be placed directly in $A[\ ]$ or $E[\ ]$. The appearance of A or E must be followed by one of G, F, X or U. If U follows them, it must be of the form $A[\varphi \cup \psi]$. Now φ and ψ can contain $\wedge$ because they are arbitrary formulas. Therefore, $A[(p \wedge q) \cup (\neg r \rightarrow q)]$ is a well-formed formula.

Note that AU and EU are binary connectives using a mixture of infix and prefix notation. The strict infix form would be $\varphi_1 AU \varphi_2$, and the strict prefix form would be $AU(\varphi_1, \varphi_2)$.

For any formal language, as in the previous two chapters, it is useful to draw a well-formed syntax parse tree. Fig. 14-4 shows the syntax parse tree for $A[AX \neg p \cup E[EX\ (P \wedge q) \cup \neg q]]$.

- $EF\ G\ r$
- $A \neg G \neg p$
- $F[r \cup q]$
- $EF(r \cup q)$
- $AEF\ r$
- $A[(r \cup q) \wedge (p \cup r)]$

特别值得理解的是,为什么语法规则不允许构造这些公式.以$EF(r \cup q)$为例,这个字符串的问题在于U只能与A和E配对出现,E又是与F成对出现的.为了将这个串改成一个合式CTL公式,只能写$EF\ E[r \cup q]$或$EF\ A[r \cup q]$.

注意,当在A或E后面配对的算子是U时,使用方括号.这样做并没有什么特别的理由.可以用通常的圆括号代替.然而,这样通常有助于人们阅读公式(因为我们能更容易地找到相应的关闭括号的位置).使用方括号的另一个原因是SMV一直坚持这样做.

$A[(r \cup q) \wedge (p \cup r)]$不是合式公式的原因是语法不允许将布尔连接词直接放在$A[\ ]$或$E[\ ]$中.A或E的出现必须紧跟着G,F,X或U之一出现.如果U跟着它们出现,一定是$A[\varphi \cup \psi]$的形式.现在φ和ψ可以包含$\wedge$,因为它们是任意公式.因此,$A[(p \wedge q) \cup (\neg r \rightarrow q)]$是一个合式公式.

注意AU和EU是混合使用中缀和前缀记号的二元连接词.按严格的中缀形式,应该写成$\varphi_1 AU \varphi_2$,而按严格的前缀形式,应该写成$AU(\varphi_1, \varphi_2)$.

就任何形式语言而言,如前两章,画出合式的语法分析树是有用的.$A[AX \neg p \cup E[EX(p \wedge q) \cup \neg q]]$的语法分析树如图14-4所示.

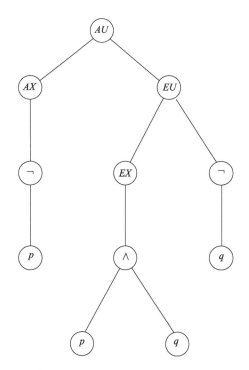

图 14-4　$A[AX\neg p\cup E[EX(P\wedge q)\cup\neg q]]$ 的语法分析树
Fig. 14-4　Syntax parse tree of $A[AX\neg p\cup E[EX(P\wedge q)\cup\neg q]]$

Definition 14.3.2　The subformula of a CTL formula φ is the formula ψ, whose syntax parse tree is the subtree of the syntax parse tree of φ.

14.3.2　Semantics of computation tree logic

Given model $M=(S,\rightarrow,L)$, $s\in S$, ϕ is the CTL formula. The definition of whether $M,s\models\phi$ holds is to recurse the structure of ϕ. The following content can be used to help understand:

(1) If ϕ is atomic, the satisfaction relationship is determined by L.

(2) If the top-level connective of ϕ (that is, the top-most connective that appears in the syntax parse tree) is a Boolean connective ($\wedge$, $\vee$, $\neg$ etc.), the satisfaction question is answered by definition of the ordinary truth table and next recursive step of ϕ.

(3) If the top-level connective is an operator starting with A, then the relationship is satisfied. If all paths from s satisfy the "LTL formula" obtained by removing the symbol A, then the next state points of s

定义 14.3.2　一个 CTL 公式 φ 的子公式是这样的公式 ψ，其语法分析树是 φ 的语法分析树的子树.

14.3.2　分支时态逻辑的语义

给定模型 $M=(S,\rightarrow,L)$, $s\in S$, ϕ 是 CTL 公式. $M,s\models\phi$ 是否成立的定义是就 ϕ 的结构进行递归的，可以通过下面内容帮助理解：

(1) 若 ϕ 是原子的，由 L 确定满足的关系.

(2) 若 ϕ 的顶级连接词（即出现在的语法分析树中最顶层的连接词）是一个布尔连接词（$\wedge$, $\vee$, $\neg$ 等），则满足问题由普通的真值表定义以及 ϕ 的下一递归步骤来回答.

(3) 若顶级连接词是一个从 A 开始的算子，则满足关系成立. 若从 s 出发的所有路径均满足移去符号 A 后所得到的"LTL 公式"，则 s 的所有下一个

all satify ϕ.

(4) Similarly, if the top-level connective starts at E, then the relationship is satisfied. If a certain path from s satisfies the "LTL formula" obtained by removing symbol E, the form of $M,s\models\phi$ is defined as follows:

Definition 14.3.3 Let $M=(S,\rightarrow,L)$ be a model of CTL, s belongs to S, and ϕ is a CTL formula. Inductively define $M,s\models\phi$ based on the structure of ϕ as follows:

- $M,s\models\top$
- $M,s\models\top$
- $M,s\models p$ if and only if $p\in L(s)$
- $M,s\models\neg\phi$ if and only if $M,s\not\models\phi$
- $M,s\models\phi\wedge\psi$ if and only if $M,s\models\phi$ and $M,s\models\psi$
- $M,s\models\phi\vee\psi$ if and only if $M,s\models\phi$ and $M,s\models\psi$
- $M,s\models\phi\rightarrow\psi$ if and only if $M,s\not\models\phi$ and $M,s\models\psi$
- $M,s\models AX\phi$ if and only if for all $s\rightarrow s_1$, there is $M,s_1\models\phi$. Among them, $M,s_0\models AX\phi$ can be understood as: all the next state points of s_0 satisfy ϕ, as shown in Fig. 14-5.

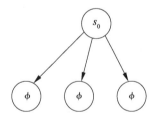

Fig. 14-5 $M,s\models AX\phi$

- $M,s\models EX\phi$ if and only if for some s_1 that makes $s\rightarrow s_1$, there is $M,s_1\models\phi$. Among them, $M,s_0\models EX\phi$ can be understood as: there exists a next state of s_0 satisfying ϕ, as shown in Fig. 14-6.

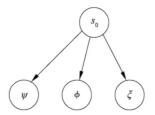

图 14-6　$M, s \models EX\phi$
Fig. 14-6　$M, s \models EX\phi$

- $M, s \models AG\phi$ if and only if for all paths $s_0 \to s_1 \to s_2 \to \ldots$ (where s_0 is s) from s, for all s_i on these paths, there is $M, s_i \models \phi$. Among them, $M, s_0 \models AG\phi$ can be understood as: every state on the path starting from s_0 satisfies ϕ, as shown in Fig. 14-7.

- $M,s \models AG\phi$ 当且仅当对于从 s 出发的所有路径 $s_0 \to s_1 \to s_2 \to \cdots$（其中 s_0 就是 s），对于在这些路上的所有 s_i，都有 $M,s_i \models \phi$. 其中 $M,s_0 \models AG\phi$ 可以理解为：所有从 s_0 出发的路径上每一个状态都满足 ϕ，具体如图 14-7 所示.

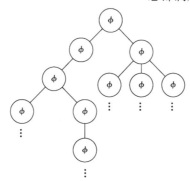

图 14-7　$M, s \models AG\phi$
Fig. 14-7　$M, s \models AG\phi$

- $M, s \models EG\phi$ if and only if there is a path $s_0 \to s_1 \to s_2 \to \ldots$ (where s_0 is s) from s, for all s_i on these paths, there is $M, s_i \models \phi$.

- $M, s \models AF\phi$ if and only if for all paths $s_0 \to s_1 \to s_2 \to \ldots$ (where s_0 is s) from s, for each path, there is s_i, and there is $M, s_i \models \phi$.

- $M, s \models EF\phi$ if and only if there is a path $s_0 \to s_1 \to s_2 \to \ldots$ (where s_0 is s) from s, there exists s_i on this path, and there is $M, s_i \models \phi$. Among them, $M, s_0 \models EF\phi$ can be understood as: the path starting from s_0 must have a state that satisfies ϕ, as shown in Fig. 14-8.

- $M,s \models EG\phi$ 当且仅当存在一条从 s 出发的路径 $s_0 \to s_1 \to s_2 \to \cdots$（其中 s_0 就是 s），对于在这些路上的所有 s_i，都有 $M,s_i \models \phi$.

- $M,s \models AF\phi$ 当且仅当对于从 s 出发的所有路径 $s_0 \to s_1 \to s_2 \to \cdots$（其中 s_0 就是 s），对于每条路径，都存在 s_i，有 $M,s_i \models \phi$.

- $M,s \models EF\phi$ 当且仅当存在一条从 s 出发的路径 $s_0 \to s_1 \to s_2 \to \cdots$（其中 s_0 就是 s），该路径上存在 s_i，有 $M, s_i \models \phi$. 其中 $M,s_0 \models EF\phi$ 可以理解为：从 s_0 出发的路径，一定会存在一个满足 ϕ 的状态，如图 14-8 所示.

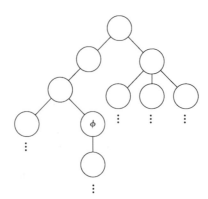

图 14-8　$M, s \models EF\phi$
Fig. 14-8　$M, s \models EF\phi$

- $M, s \models A[\phi U\psi]$ if and only if for the path $s_0 \to s_1 \to s_2 \to \ldots$ (where s_0 is s) from s, there exists $i \geq 0$ such that $M, s_i \models \psi$, and for all $0 \leq j \leq i$, there is $M, s_j \models \phi$.

- $M, s \models E[\phi U\psi]$ if and only if there is a path $s_0 \to s_1 \to s_2 \to \ldots$ (where s_0 is s) from s, there exists $i \geq 0$, such that $M, s_i \models \psi$, and for all $0 \leq j \leq i$, there is $M, s_j \models \phi$.

Now look at the Kripke structure shown in Fig. 14-9.

- $M, s \models A[\phi U\psi]$ 当且仅当对于从 s 出发的路径 $s_0 \to s_1 \to s_2 \to \cdots$（其中 s_0 就是 s）,存在 $i \geq 0$ 使得 $M, s_i \models \psi$, 且对于所有的 $0 \leq j \leq i$ 都有 $M, s_j \models \phi$.

- $M, s \models E[\phi U\psi]$ 当且仅当存在从 s 出发的路径 $s_0 \to s_1 \to s_2 \to \cdots$（其中 s_0 就是 s）,存在 $i \geq 0$, 使得 $M, s_i \models \psi$, 且对于所有的 $0 \leq j \leq i$ 都有 $M, s_j \models \phi$.

现在看看图 14-9 所示的 Kripke 结构.

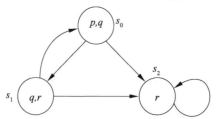

图 14-9　Kripke 结构
Fig. 14-9　Kripke structure

The following conclusions are drawn:

(1) $M, s_0 \models EX(q \wedge r)$: In model M, there exists a state s_i such that $s_i \to s_0$ and $s_i \models (q \wedge r)$.

(2) $M, s_0 \models A(pUr)$: In model M, for all paths starting from $s_0 \to s_1 \to s_2 \to \ldots$ or $s_0 \to s_2 \to s_2 \to s_2 \to \ldots$ and other paths, there exists that all states on the path starting from s_0 satisfy p until a state that satisfies r is encountered.

得出如下结论：

(1) $M, s_0 \models EX(q \wedge r)$: 在模型 M 中,存在一个状态 s_i, 使得 $s_i \to s_0$ 且 $s_i \models (q \wedge r)$.

(2) $M, s_0 \models A(pUr)$: 在模型 M 中,对于从 s_0 出发的所有路径 $s_0 \to s_1 \to s_2 \to \cdots$ 或者 $s_0 \to s_2 \to s_2 \to s_2 \to \cdots$ 等路径,都存在从 s_0 开始的路径上的所有状态一直满足 p, 直到遇到满足 r 的状态.

(3) $M, s_0 \models \neg EF(q \wedge r)$: In model M, for all paths starting from s_0, it is impossible to pass through nodes that satisfy the states p and r at the same time.

(4) $M, s_0 \models EGr$: In model M, for all paths starting from s_0, there must exist that from a certain node, and all subsequent nodes satisfy the state r.

Model satisfiability: Let $M = (S, \rightarrow, L)$ be the model, and ϕ is the CTL formula. If there is $M, s \models \phi$ for any $s \in S$, then model M is said to satisfy the CTL formula ϕ, denoted as $M \models \phi$.

14.4 System verification technologies based on FSM

14.4.1 Verification methods for computation tree logic

The semantic definitions of linear temporal logic and computation tree logic given above allow detecting whether the initial state of a given system satisfies the linear temporal logic and computation tree logic formulas. This is the basic problem of model checking. Although the specification designers prefer to use linear temporal logic, we still start with computation tree logic model checking, which has a much simpler algorithm.

It is easy to see that it is easier to do model checking after specifying an initial state to expand the model into an infinite tree, because then all possible paths are clearly visible. However, if one wants to implement a model checker on a computer, the transition system cannot be expanded into an infinite tree, one needs to check the finite data structures. Therefore, we want to develop new ideas in the semantics of computation tree logic. This deeper understanding will provide the basis for an efficient algorithm for the following problem: Given $M, s \in S$ and ϕ, compute whether $M, s \models \phi$ holds. In the case where ϕ is not satisfied, this algorithm can be augmented to produce an actual path of the system to prove that M cannot satisfy ϕ. This allows us to debug the system by trying to find out what makes the run reject ϕ.

There are various ways to consider the following problems as computational problems:

$M, s_0 \models \phi$, is it valid?

For example, one can take the model M, the formula ϕ, and the state s_0 as input and expect an output formed as "yes" ($M, s_0 \models \phi$ holds) or "no" ($M, s_0 \models \phi$ does not hold). Alternatively, the inputs are just M and ϕ, and the outputs are all states s of the model M that satisfy ϕ.

It turns out that it is easier to provide an algorithm for solving the second problem. This will automatically give the solution to the first problem, since it is only necessary to check whether s_0 is an element in the output set.

Algorithm 14-1: Labeling Algorithm

An algorithm is proposed in which a model and a computation tree logic formula are known, and the algorithm outputs all states of the model that satisfy the formula. This algorithm does not need to deal explicitly with each conjunction of the CTL, since the propositional conjunctions $\perp$, $\neg$ and $\wedge$ form the appropriate set; and AF, EU and EX form the appropriate set of temporal conjunctions. Given any CTL formula ϕ, in order to write it in equivalent form using the appropriate set of connectives, we can simply preprocess ϕ and then call the model checking algorithm. The specific algorithm is as follows:

Input: a CTL model $M = (S, \rightarrow, L)$ and a CTL formula ϕ

Output: the set of states of M satisfying ϕ

First, turn ϕ into the output of TRANSLATE (ϕ), i.e., denote ϕ by the connectives AF, EU, EX, $\perp$, $\neg$ and $\wedge$ using the equivalent formula given earlier in this chapter. Second, starting from the smallest subformula of ϕ, label the states of M by the

这样我们可以通过尝试找出使运行拒绝 ϕ 的原因来调试系统.

有各种方式把以下问题视为计算问题:

$M, s_0 \models \phi$ 是否成立？

例如,可以将模型 M、公式 ϕ 和状态 s_0 作为输入,然后期待一个形成为"yes"($M, s_0 \models \phi$ 成立)或"no"($M, s_0 \models \phi$ 不成立)的输出. 或者,输入只是 M 和 ϕ,而输出是模型 M 的满足 ϕ 的所有状态 s.

原来为解决第二个问题提供算法更容易些. 这将自动给出第一个问题的解,因为只需要检测 s_0 是否为输出集合中的元素即可.

算法 14-1: 标记算法

提出一算法,已知一个模型和一个分支时态逻辑公式,该算法输出满足公式的模型的所有状态. 此算法不需要明确地处理 CTL 的每个连接词,因为命题连接词 $\perp$, $\neg$ 和 $\wedge$ 形成适当集合;而 AF, EU 和 EX 形成时态连接词适当集合. 给定任意 CTL 公式 ϕ,为了用适当连接词的集合,将其写为等价的形式,可以对 ϕ 进行简单预处理,然后调用模型检测算法. 具体算法如下:

输入: 一个 CTL 模型 $M = (S, \rightarrow, L)$ 和一个 CTL 公式 ϕ

输出: 满足 ϕ 的 M 的状态的集合

首先,把 ϕ 变成 TRANSLATE(ϕ) 的输出,即用本章前面给出的等价公式将 ϕ 用连接词 $AF, EU, EX, \perp, \neg$ 和 $\wedge$ 表示. 其次,从 ϕ 的最小子公式开始,用满足 ϕ 的子公式来标记 M 的状

subformulas that are satisfied of ϕ, and gradually extend ϕ from the inside out.

Assume that ψ is a subformula of ϕ and that the states of all direct subformulas satisfying ψ have been labeled. We determine which states are labeled with ψ by situation analysis. If ψ is

- $\bot$: then no state is labeled with $\bot$.
- p: if $p \in L(s)$, then s is labeled with p.
- $\psi_1 \wedge \psi_2$: if s has been labeled with ψ_1 and ψ_2, then s is labeled with $\psi_1 \wedge \psi_2$.
- $\neg \psi_1$: if s is not yet labeled with ψ_1, then s is labeled with $\neg \psi_1$.
- $AF\psi_1$: if any state s is labeled with ψ_1, label it with $AF\psi_1$.

Repeat: label any state with $AF\psi_1$ if all succeeding states are labeled with $AF\psi_1$, until no change occurs. Fig. 14-10 illustrates this step.

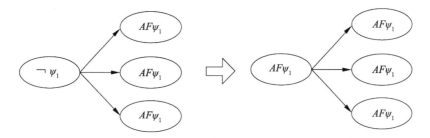

图 14-10 用形式为 $AF\psi_1$ 的子公式标记状态过程的迭代步骤

Fig. 14-10 Iterative steps for labeling a state process with a subformula of the form $AF\psi_1$

- $E[\psi_1 \cup \psi_2]$: if any state s is labeled with ψ_1, label it with $E[\psi_1 \cup \psi_2]$.

Repeat: label any state with $E[\psi_1 \cup \psi_2]$ if it is labeled with ψ_1 and at least one of its successor states is labeled with $E[\psi_1 \cup \psi_2]$, until no change occurs. Fig. 14-11 illustrates this step.

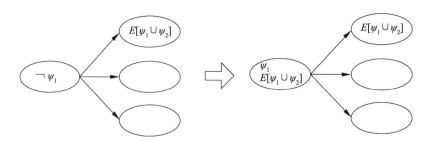

图 14-11 用形式为 $E[\psi_1 \cup \psi_2]$ 的子公式标记状态过程的迭代步骤
Fig. 14-11 Iterative steps for labeling a state process with a subformula of the form $E[\psi_1 \cup \psi_2]$

- $EX\,\psi_1$: label any state with $EX\,\psi_1$ if one of its successors is labeled with ψ_1.

After completing the labeling for all subformulas of ϕ (including ϕ itself), output the state labeled with ϕ.

The complexity of this algorithm is $O(f \cdot V \cdot (V+E))$, where f is the number of connectives in a formula, V is the number of states, and E is the number of transitions. The algorithm is linear with respect to the size of the formula and quadratic with respect to the size of the model.

Algorithm 14-2: Direct Processing of EG

Instead of using the appropriate set of minimal connectives, similar routines can be written for other connectives. In fact, this may be more efficient. However, connectives AG and EG require a slightly different approach than other connectives. The algorithm for processing $EG\,\psi_1$ directly is as follows.

- $EG\,\psi_1$: label all states with $EG\,\psi_1$.

If state s is not labeled with ψ_1, delete label $EG\,\psi_1$.

Repeat: remove label $EG\,\psi_1$ from any state if all successors of that state are not labeled with $EG\,\psi_1$; until no change occurs.

Here, we label all states with the subformula $EG\,\psi_1$ and then weaken this set of labels, instead of going from nothing to start building, as we did for EU. In fact, as far as the final result is concerned, there is no

- $EX\,\psi_1$: 如果其后继之一用 ψ_1 标记, 用 $EX\,\psi_1$ 标记任何状态.

对 ϕ 的所有子公式 (包括 ϕ 本身) 完成标记后, 输出标记为 ϕ 的状态.

这个算法的复杂度为 $O(f \cdot V \cdot (V+E))$, 此处 f 是公式中连接词的个数, V 是状态的个数, 而 E 是迁移的个数. 该算法关于公式的大小为线性的, 而关于模型的大小是二次的.

算法 14-2: 直接处理 EG

不使用最小连接词的适当集合, 也可以为其他连接词写类似的例行程序. 事实上, 这样也许更有效. 然而, 连接词 AG 和 EG 需要一种与其他连接词略微不同的方法. 下面是一个直接处理 $EG\,\psi_1$ 的算法:

- $EG\,\psi_1$: 用 $EG\,\psi_1$ 标记所有的状态.

若未用 ψ_1 标记状态 s, 则删除标记 $EG\,\psi_1$.

重复: 如果该状态的所有后继都没有用 $EG\,\psi_1$ 标记, 从任何状态删除标记 $EG\,\psi_1$; 直到不发生改变为止.

这里, 我们用子公式 $EG\,\psi_1$ 标记所有的状态, 然后削弱这个标记集合, 而不像对 EU 所做的那样, 从一无所有到开始建立. 实际上, 就最终结果而言,

real difference between this process with respect to $EG\ \psi_1$ and doing it after transforming it into $\neg AF\ \neg \psi$.

Algorithm 14-3: A More Efficient Variant

The efficiency of the labeling algorithm can be improved by using a smarter way of handling EG. Instead of using EX, EU and AF as proper sets, replace them with EX, EU and EG. For EX and EU, do the same as before (but be careful to search the model by the reverse breadth-first search method, which guarantees not to pass any node twice). Regarding the case of $EG\ \psi$:

● Limit the graph to states satisfying ψ, i.e., removing all other states and their transitions.

● Find maximum strongly connected branches; they are maximal regions of the state space in which each state is connected to all other states in the region (i.e., there is a finite path).

● Find any state that can reach an SCC (Strongly Connected Component) by applying the reverse breadth-first search algorithm on a restricted graph, see Fig. 14-12.

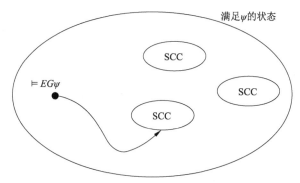

Fig. 14-12　A better way to handle EG

The complexity of this algorithm is $O(f \cdot (V+E))$, i.e., it is linear with respect to the dimensions of both the model and the formula.

Pseudo-code of the CTL model checking algorithm: We give the pseudo-code of the basic labeling algorithm. The main function SAT (denote "satisfy") takes the CTL formula as input. The program SAT expects a syntax parse tree of CTL formulas constructed according to the syntax defined by CTL as input. This expectation reflects an important precondition regarding the correctness of the SAT algorithm. The program simply does not know how to do with the input of non-CTL formulas.

Pseudo-code written for SAT is somewhat like C or Java code snippets. A function with the keyword "return" is used to indicate the result that the function will return. We also use the natural language for the analysis of the case of the root node in the syntax parse tree of ϕ. "local var" declares some new variables that are local with respect to the current instance of the considered program; and "repeat until" executes the subsequent commands repeatedly until the condition becomes true. Also, use some constructive notation about set operations (such as intersection, complement, etc.). In reality, abstract data types may be needed, as well as implementations about these operations, but for now we are only interested in the principle mechanisms of the SAT algorithm. Assume that SAT has access to all relevant parts of the model: S, $\rightarrow$ and L. In particular, we ignore the fact that SAT requires to take a description of M as input and simply assume that SAT operates directly on any given model. Note how SAT translates ϕ into an equivalent formula for the appropriate set chosen.

Fig. 14-13 shows the SAT algorithm and its subfunctions. The subfunctions use the program variables X, Y, V and W representing the set of states. The SAT procedure can handle simple cases directly, while passing more complex cases to special procedures, and the latter call SAT recursively on the subexpressions. These special procedures depend on the implementation of functions in Fig. 14-14–Fig. 14-16.

$\text{pre}_\exists(Y) = \{s \in S \mid \text{existence of } s', (s \rightarrow s' \text{ and } s' \in Y)\}$

入. 程序 SAT 期望一个根据 CTL 定义的语法所构造的 CTL 公式的语法分析树作为输入. 这种期望反映了关于 SAT 算法正确性的一个重要前提. 程序根本不知道怎样处理非 CTL 公式的输入.

为 SAT 写的伪代码有点像 C 或 Java 代码片段. 使用带有关键词"return"的函数,用来指示函数将返回的结果. 我们也使用自然语言对 ϕ 的语法分析树中根结点的情况分析. "local var"声明一些相对于所考虑程序的当前实例,为局部的新变量;而"repeat until"重复地执行后续的命令,直到条件变成真. 此外,使用关于集合运算(如交集、补集等)的一些建设性记号. 在现实中,可能需要抽象数据类型,以及关于这些运算的实现,但目前我们仅对 SAT 算法的原则性机制感兴趣. 假设 SAT 可以访问模型的所有相关部分:S, $\rightarrow$ 和 L. 特别地,我们忽略 SAT 要求将 M 的描述作为输入这个事实,并简单地假设 SAT 在任意给定的模型上直接操作. 注意 SAT 是如何将 ϕ 翻译成所选择的适当集的等价公式的.

图 14-13 显示了算法及其子函数. 子函数使用代表状态集合的程序变量 X, Y, V 和 W. SAT 的程序可直接处理简单情况,而将更复杂的情况传递给特殊的程序,后者对子表达式递归地调用 SAT. 这些特殊过程依赖于图 14-14–图 14-16 中的函数实现.

$\text{pre}_\exists(Y) = \{s \in S \mid 存在 s', (s \rightarrow s' 且 s' \in Y)\}$

$pre_{\forall}(Y) = \{s \in S \mid$ for all s', $(s \to s'$ implies $s' \in Y)\}$

$pre_{\forall}(Y) = \{s \in S \mid$ 对所有 s', $(s \to s'$蕴含 $s' \in Y)\}$

```
function SAT(φ)
/* determines the set of states satisfying φ */
begin
    case
        φ is T : return S
        φ is ⊥ : return ∅
        φ is atomic: return {s ∈ S | φ ∈ L(s)}
        φ is ¬φ1: return S−SAT(φ1)
        φ is φ1 ∧ φ2: return SAT(φ1) ∩ SAT(φ2)
        φ is φ1 ∨ φ2: return SAT(φ1) ∪ SAT(φ2)
        φ is φ1→φ2: return SAT(¬φ1 ∨ φ2)
        φ is AXφ1: return SAT(¬EX¬φ1)
        φ is EXφ1: return SAT_EX(φ1)
        φ is A[φ1 ∪ φ2] : return SAT(¬(E[¬φ2 ∪ (¬φ1 ∨ ¬φ2)] ∨ EG¬φ2))
        φ is E[φ1 ∪ φ2] : return SAT_EU(φ1, φ2)
        φ is EFφ1: return SAT(E(T ∪ φ1))
        φ is EGφ1: return SAT(¬AF¬φ1)
        φ is AFφ1: return SAT_AF(φ1)
        φ is AGφ1: return SAT(¬EF¬φ1)
    end case
end function
```

图 14-13 函数 SAT(将 CTL 公式作为输入并返回满足该公式的状态集合. 如果 *EX*,*EU* 或 *AF* 是输入分析树的根,它分别调用函数 SAT_{EX},SAT_{EU} 和 SAT_{AF})

Fig. 14-13 Function SAT (Take a CTL formula as input and return the set of states that satisfy that formula. If *EX*, *EU* or *AF* is the root of the input parse tree, it calls the functions SAT_{EX}, SAT_{EU} and SAT_{AF}, respectively)

```
function SAT_EX(φ)
/* determines the set of states satisfying EXφ */
local var X, Y
begin
    X := SAT(φ);
    Y := pre_∃(X);
    return Y
end
```

图 14-14 函数 SAT_{EX} (它通过调用 SAT 计算满足 ϕ 的状态,然后沿→向后寻找满足 *EX* ϕ 的状态)

Fig. 14-14 Function SAT_{EX} (The states satisfying ϕ are computed by calling SAT. It then looks backwards along→for states satisfying *EX* ϕ)

```
function SAT_AF(φ)
/* determines the set of states satisfying AF φ */
local var X, Y
begin
    X := S;
    Y := SAT(φ);
    repeat until X = Y
    begin
        X := Y;
        Y := Y ∪ pre_∀(Y)
    end
    return Y
end
```

图 14-15 函数 SAT_{AF}（通过调用 SAT 计算满足 ϕ 的状态.然后,按照标记算法中所描述的方式将满足 $AF\phi$ 的状态累积在一起）

Fig. 14-15 Function SAT_{AF} (The states satisfying ϕ are computed by calling SAT. Then, the states satisfying $AF\phi$ are accumulated according to the method described in the labeling algorithm)

```
function SAT_EU(φ, ψ)
    /* determines the set of states satisfying E[φ∪ψ] */
local var W, X, Y
begin
    W := SAT(φ);
    X := S;
    Y := SAT(ψ);
    repeat until X = Y
    begin
        X := Y;
        Y := Y ∪ (W ∩ pre_∃(Y))
    end
    return Y
end
```

图 14-16 函数 SAT_{EU}（通过调用 SAT 计算满足 ϕ 的状态.然后,按照标记算法中所描述的方法将满足 $E[\phi\cup\psi]$ 的状态累积在一起）

Fig. 14-16 Function SAT_{EU} (The states satisfying ϕ are computed by calling SAT. Then, the states satisfying $E[\phi\cup\psi]$ are accumulated according to the method described in the labeling algorithm) "pre" denotes moving backwards along the transition relation. Two functions compute the original image of the set of states. The function $\text{pre}_\exists$ (mechanism in SAT_{EX} and SAT_{EU}) takes a subset Y of states as input and returns the set of states that can be transited into Y. The function $\text{pre}_\forall$ used in SAT_{AF} takes the set Y as input and returns the set of states that can only be transited into Y. Note that $\text{pre}_\forall$ can be denoted in terms of complement and $\text{pre}_\exists$ as follows:

$$\text{pre}_\forall(Y) = S\text{-pre}_\exists(S\text{-}Y)$$

Here the set of all $s \in S$ that are not in Y is written as $S\text{-}Y$.

Moving on to the "state explosion" problem, although the labeling algorithm (including the clever way of handling EG) is linear in the size of the model, unfortunately the model size itself is usually exponential in the number of variables and the number of system components executing in parallel. This means that, for example, adding a Boolean variable to a program doubles the complexity of its property verification.

The tendency to make the state space very large is called the state explosion problem. A great deal of research has been done to find ways to overcome this problem as well, including the use of the following methods:

- Effective data structures are called ordered binary decision diagrams (OBDD), which denote sets of states rather than individual states. SMV is implemented using OBDD.
- Abstraction: We can interpret a model abstractly, consistently, or with respect to a particular property.
- Partial order reduction: For asynchronous systems, some intertwined component traces may be equivalent with respect to the satisfiability of the formula to be checked. This can often significantly reduce the size of the model checking problem.

"pre"表示沿迁移关系向后移动. 两个函数计算状态集合的原像. 函数 $\text{pre}_\exists$(SAT_{EX} 和 SAT_{EU} 中的机制)以一个状态子集 Y 作为输入,并且返回可迁移进入 Y 的状态集合. 用在 SAT_{AF} 中的函数 $\text{pre}_\forall$ 将集合 Y 作为输入并且返回只能迁移进 Y 的状态集合. 注意, $\text{pre}_\forall$ 可以用补和 $\text{pre}_\exists$ 表达如下:

$$\text{pre}_\forall(Y) = S\text{-pre}_\exists(S\text{-}Y)$$

这里将不在 Y 中的所有 $s \in S$ 的集合写为 $S\text{-}Y$.

接下来讨论"状态爆炸"问题,尽管标记算法(包括处理 EG 的巧妙方式)关于模型的规模为线性的,遗憾的是,模型规模本身通常关于变量个数以及并行执行的系统组件的数目是指数的. 这意味着:若在程序中增加一个布尔变量,则使其性质验证的复杂度加倍.

使状态空间变得非常大的趋势称为状态爆炸问题. 为了找出克服这个问题的方法已经做了大量的研究,包括使用以下方法:

- 有效的数据结构称为有序二元决策图(OBDD),它表示状态集合而不是单个状态. SMV 就是用 OBDD 实现的.
- 抽象:可以抽象地、一致地或者关于一个特殊性质来解释一个模型.
- 偏序归约:对于异步系统,就待测公式的满足性而言,一些相互交织的组件迹可能是等价的. 这经常能够显著缩小模型检测问题的规模.

- Generalization: Model checking for systems with a large number of identical or similar components can often be achieved by "generalizing" the number of components.

- Compounding: Decompose the verification problem into several simpler verification problems.

14.4.2 Verification method of linear temporal logic

Basic strategy of LTL algorithm: Let $M = (S, \rightarrow, L)$ be a model, $s \in S$, and Φ is an LTL formula. We have to determine whether $M, s \models \Phi$, that is, whether Φ is satisfied for all paths starting from s along M. Almost all LTL verification algorithms follow the following three steps.

(1) Construct an automaton for the formula $\neg \Phi$, also called a **tableau**. The automaton of Ψ is called A_Ψ. Therefore, we construct $A_{\neg \Psi}$. The automaton has a concept of acceptance trace. A trace is an assignment sequence of propositional atoms. Starting from a path, its traces can be abstracted. The structure has the following properties: for all paths π: $\pi \models \Psi$ if and only if the trace of π is accepted by A_Ψ. In other words, the automaton A_Ψ accurately encodes the trace that satisfies Ψ.

Therefore, the automaton $A_{\neg \Phi}$ we constructed for $\neg \Phi$ has following properties: encoding all traces satisfying $\neg \Phi$; that is, all traces not satisfying Φ.

(2) Combine the automaton $A_{\neg \Phi}$ with the model M of the system. Combining the results of the calculations produces a transition system, the path of which is both the path of the automaton and the path of the system.

(3) Search in the combined transition system to see if there is a path starting from the state derived from s. If there is one, it can be interpreted as a path in M that does not satisfy Φ starting with s.

If there is no such path, output "Yes, $M, s \models \Phi$". Otherwise, if there is such a path, output "No, $M, s \models \Phi$". In the latter case, counterexamples can be extracted from the path found.

- 归纳:有大量相同或相似组件系统的模型检测经常可以通过对组件数目做"归纳"来加以实现.

- 复合:将验证问题分解成若干个更简单的验证问题.

14.4.2 线性时态逻辑的验证方法

LTL 算法基本策略:设 $M = (S, \rightarrow, L)$ 是一个模型,$s \in S$,Φ 是一个 LTL 公式.要确定是否有 $M, s \models \Phi$,即沿 M 的从 s 出发的所有路径,Φ 是否满足.几乎所有的 LTL 的验证算法都遵循以下三步进行.

(1) 为公式 $\neg \Phi$ 构造一个自动机,也叫作**布景**.关于 Ψ 的自动机称为 A_Ψ.于是,构造 $A_{\neg \Psi}$.该自动机有一个接受迹的概念.迹是命题原子的赋值序列.从一条路径出发,可以抽象出它的迹.该构造有如下性质:对所有路径 π:$\pi \models \Psi$ 当且仅当 π 的迹为 A_Ψ 所接受.换言之,自动机 A_Ψ 精确地编码满足 Ψ 的迹.

因此,为 $\neg \Phi$ 所构造的自动机 $A_{\neg \Phi}$ 有下列性质:编码满足 $\neg \Phi$ 的所有迹;即所有不满足 Φ 的迹.

(2) 将自动机 $A_{\neg \Phi}$ 与系统的模型 M 结合.结合运算的结果产生一个迁移系统,其路径既是自动机的路径又是系统的路径.

(3) 在结合的迁移系统中搜寻,看看是否存在由 s 导出的状态出发的路径.如果存在一条,可以解释为 M 中以 s 开始不满足 Φ 的一条路径.

如果没有这样的路径,则输出 "Yes, $M, s \models \Phi$".否则,如果存在这样的的路径,输出 "No, $M, s \models \Phi$".在后一种情形中,可从找到的路径中提取出反例.

Let us consider an example. As shown in Fig. 14-17, the system is described by the SMV program and its model M. Consider the formula $\neg(a \cup b)$. Because not all paths of M satisfy the formula (for example, paths $q_3, q_2, q_2, \ldots$ do not satisfy it), we expect the model checking to be unsuccessful.

According to step 1, construct an automaton $A_{a \cup b}$, which accurately describes the trace that satisfies $a \cup b$. (Use the fact that $\neg\neg(a \cup b)$ is equivalent to $a \cup b$). Such an automaton is shown in Fig. 14-18. We will consider how to construct it later; for now, we are only trying to understand how and why it works.

The trace t is accepted by the automaton in Fig. 14-18. If there is a path π through the automaton, such that:

考虑一个例子. 如图 14-17 所示, 该系统由 SMV 程序及其模型 M 描述. 考虑公式 $\neg(a \cup b)$. 因为并不是 M 的所有路径都满足公式(例如, 路径 q_3, $q_2, q_2, \cdots$ 不满足它), 所以, 我们预期模型检测是不成功的.

根据步骤1, 构造一个自动机 $A_{a \cup b}$, 它精确地刻画出满足 $a \cup b$ 的迹(使用 $\neg\neg(a \cup b)$ 等价于 $a \cup b$ 的事实). 这样的自动机如图 14-18 所示. 稍后将考虑如何构造它;目前, 我们仅试图理解它是怎样以及为什么可以工作.

迹 t 为如图 14-18 的自动机所接受, 如果存在一条通过自动机的路径 π, 使得:

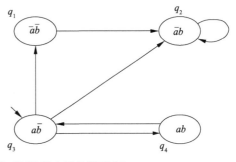

图 14-17 一个 SMV 程序及其模型 M
Fig. 14-17 An SMV program and its model M

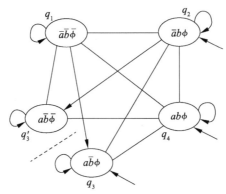

图 14-18 恰好接受满足 $\varphi \stackrel{\text{def}}{=} a \cup b$ 的自动机(无箭头的迁移可以双向迁移. 接受条件是该自动机的路径不能不确定地通过 q_3 循环)

Fig. 14-18 Exactly accept an automaton that meets $\varphi \stackrel{\text{def}}{=} a \cup b$ (Transition without arrows can transit in both directions. The acceptance condition is that the path of the automaton cannot pass through the q_3 cycle indefinitely)

t is the trace of π, its transition matches the corresponding state of π;

This path maintains a certain "acceptance condition". For the automaton of Fig. 14-18, the acceptance condition is: the path should not end with q_3, $q_3, \ldots$ indefinitely.

For example, assume t is $a\bar{b}$, $a\bar{b}$, $a\bar{b}$, ab, ab, $\bar{a}\bar{b}$, $a\bar{b}$, $a\bar{b}$, $\ldots$, and finally repeats the state $a\bar{b}$ forever. Then, we choose the path q_3, q_3, q_3, q_4, q_4, q_1, q'_3, $q'_3, \ldots$. Starting from q_3, because the first state is $a\bar{b}$, it is an initial state. The next state we choose is nothing more than an assignment that follows the π state. For example, in q_1, the next assignment is $a\bar{b}$, and transition allows the choice of q_3 or q'_3. We choose q'_3 and loop there forever. This path satisfies the condition, so the trace t is accepted. Observe that this definition states that "there is a path". In the above example, there are also paths that do not meet this condition:

Any path starting with q_3, q'_3 does not satisfy the condition that the transition relationship must be maintained.

The paths q_3, q_3, q_3, q_4, q_4, q_1, q_3, $q_3, \ldots$ does not satisfy the condition that the cycle must not end of q_3.

These paths will not cause us trouble, because to declare that π is acceptable, it is enough to find a path that satisfies the conditions.

Why does the automaton in Fig. 14-17 work as expected? To understand this, observe that it has enough states to distinguish the value of the proposition, that is, there is one state for all assignments $\{\bar{a}\bar{b}, \bar{a}b, a\bar{b}, ab\}$, in fact, there are two states for assignment $a\bar{b}$. It is intuitive enough for each of $\{\bar{a}\bar{b}, \bar{a}b, ab\}$ to have a state, because these assignments determine whether $a \cup b$ holds. But in $a\bar{b}$, $a \cup b$ may

t 是 π 的迹,其迁移与 π 的对应状态相匹配;

该路径保持一定的"接受条件". 对图 14-18 中的自动机而言,接受条件是:该路径不应该不确定地以 $q_3, q_3, \cdots$ 结束.

例如,假设 t 是 $a\bar{b}$, $a\bar{b}$, $a\bar{b}$, ab, ab, $\bar{a}\bar{b}$, $a\bar{b}$, $a\bar{b}$, $\cdots$,最终永远重复状态 $a\bar{b}$. 然后,我们选择路径 $q_3, q_3, q_3, q_4, q_4, q_1, q'_3, q'_3, \cdots$. 从 q_3 开始,因为第一个状态是 $a\bar{b}$,是一个初始状态. 我们选择的下一个状态不过是遵循 π 状态的赋值. 例如,在 q_1,下一个赋值是 $a\bar{b}$,迁移允许选择 q_3 或 q'_3. 我们选择 q'_3,并在那永远循环. 这条路径满足条件,因此迹 t 被接受. 观察这个定义并规定"存在一条路径". 在上面的例子中,也存在不满足该条件的路径:

以 q_3, q'_3 开始的任何路径不满足必须保持迁移关系的条件.

路径 $q_3, q_3, q_3, q_4, q_4, q_1, q_3, q_3, \cdots$ 不满足必须不以 q_3 的循环结束的条件.

这些路径不会给我们添麻烦,因为要声明 π 是可接受的,只要找到一条满足条件的路径就够了.

为什么图 14-17 中的自动机能如预期工作呢? 为了理解这一点,观察到它有足够的状态来区分命题的值,即对所有赋值 $\{\bar{a}\bar{b}, \bar{a}b, a\bar{b}, ab\}$ 有一个状态,事实上,对赋值 $a\bar{b}$ 有两个状态. 对 $\{\bar{a}\bar{b}, \bar{a}b, ab\}$ 中的每一个都有一个状态足够直观,因为这些赋值确定了 $a \cup b$ 是否成立. 但是在 $a\bar{b}$ 中,$a \cup b$ 可能为

be false or true. Therefore, two cases must be considered. The appearance of $\Phi \stackrel{\text{def}}{=} a \cup b$ in a state shows that we still expect Φ to become true, or we just got it. And $\overline{\Phi}$ points out that we no longer expect Φ, and do not get it either. The transition of automata make the only way to start from q_3 is to get b, that is, move to q_2 or q_4. In addition, the transition is not restricted, allowing any path to follow. Each of q_1, q_2, and q_3 can be transited to any assignment, so q_3 and q'_3 can be taken together, as long as we carefully choose the correct one to enter. The acceptance condition allows any path, except for the path that loops on q_3 indefinitely, to ensure that the promise of $a \cup b$ to release b is eventually satisfied.

Use this automaton $A_{a \cup b}$ to proceed to step 2. In order to combine the automaton $A_{a \cup b}$ with the system model M shown in Fig. 14-19, for convenience, two versions of q_3 are used to redraw M, as shown in Fig. 14-19a. This is an equivalent system. Now all the ways to enter q_3 choose q_3 or q'_3 non-deterministically, whichever choice results in the same successor. But it allows us to add it to $A_{a \cup b}$, and choose the common transition between the two, to get the combined system of Fig. 14-19b.

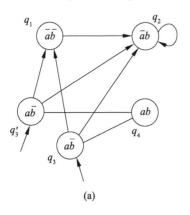
(a)

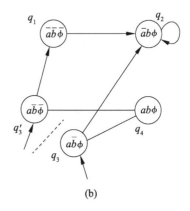
(b)

图 14-19 系统模型 M

Fig. 14-19 System model M

Does step 3 exist a path from q in the combined automaton? We can see that there are two types of paths in the combined system: q_3, $(q_4, q_3)^* q_2$, q_2, ... and q_3, q_4, $(q_3, q_4)^* q_3'$, q_1, q_2, q_2, ..., where $(q_3, q_4)^*$ is recorded as an empty string, or as q_3, q_4, or as q_3, q_4, q_3, q_4, etc. Then, according to step 3, as expected, all paths in the original system M do not satisfy $\neg(a \cup b)$.

Constructing automata: Let's take a more detailed look at how to construct automata. Given an LTL formula Φ, we hope to construct an automaton A_Φ so that A_Φ just accepts those operations that make Φ hold on it. Assume that Φ only contains temporal connectives $\cup$ and X; remember that other temporal connectives can be denoted by these two connectives.

The closure $C(\Phi)$ of the formula Φ is defined as a collection of coins and their complementary sub-formulas, and $\neg\neg \Psi$ and Ψ are regarded as equivalent. For example, $C(a \cup b) = \{a, b, \neg a, \neg b, a \cup b, \neg(a \cup b)\}$. The state of A_Φ (denoted as q, q', etc.) is the largest subset of $C(\Phi)$ that satisfies the following conditions:

For all (non-negative) $\Psi \in C(\Phi)$, $\Psi \in q$ or $\neg \Psi \in q$, but the two cannot be true at the same time.

$\Psi_1 \vee \Psi_2 \in q$ is established, if and only if when $\Psi_1 \vee \Psi_2 \in C(\Phi)$, there is $\Psi_1 \in q$ or $\Psi_2 \in q$.

The conditions for other Boolean combinations are similar.

If $\Psi_1 \cup \Psi_2 \in q$, then $\Psi_2 \in q$ or $\Psi_1 \in q$.

If $\neg(\Psi_1 \cup \Psi_2) \in q$, then $\neg \Psi_2 \in q$.

Intuitively speaking, these conditions imply that the state of A_Ψ can explain which sub-formulas of Φ are true.

The initial states of A_Ψ are those states that contain Φ. Regarding the transition relation δ of A_Φ, there is $(q, q') \in \delta$, if and only if all the following conditions are true:

If $X_\Psi \in q$, then $\Psi \in q'$;

If $\neg X_\Psi \in q$, then $\neg \Psi \in q'$;

If $\Psi_1 \cup \Psi_2 \in q$, and $\Psi_2 \notin q$, then $\Psi_1 \cup \Psi_2 \in q'$;

If $\neg(\Psi_1 \cup \Psi_2) \in q$, and $\Psi_1 \in q$, then $\neg(\Psi_1 \cup \Psi_2) \in q'$.

The rationality of the latter two conditions can be explained by the following recursive law:

$$\Psi_1 \cup \Psi_2 = \Psi_2 \vee (\Psi_1 \wedge X(\Psi_1 \cup \Psi_2))$$
$$\neg(\Psi_1 \cup \Psi_2) = \neg \Psi_2 \wedge (\neg \Psi_1 \vee X \neg(\Psi_1 \cup \Psi_2))$$

In particular, they guarantee that as long as a certain state contains $\Psi_1 \cup \Psi_2$, subsequent states will contain Ψ_1, as long as they do not contain Ψ_2.

So far, the A_Φ we have defined does not satisfy Φ for all paths through A_Φ. We use additional acceptance conditions to guarantee the "ultimate possibility" Ψ promised by the formula $\Psi_1 \cup \Psi_2$, that is, A_Φ cannot stay in the state that satisfies Ψ_1 forever without getting Ψ_2. Recall that for the automaton of $a \cup b$ in Fig. 14-18, we stipulated the acceptance condition that the path through the automaton does not end with $q_3, \cdots$.

The acceptance condition of A_Φ is defined in this way to ensure that all states containing a certain formula $X \cup \Psi$ will eventually follow a state containing Ψ. Let $X_1 \cup \Psi_1, \ldots, X_k \cup \Psi_k$ be all sub-formulas of this form in $C(\Phi)$. We stipulate the following acceptance conditions: If for all i satisfying $1 \leq i \leq k$, the operation has an infinite number of states satisfying $\neg(X_i \cup \Psi_i) \vee \Psi_i$, then one operation is acceptable. In order to understand why this condition achieves the desired effect, imagine a situation where it does not hold. Assume that there is one operation and only a finite number of states satisfy $\neg(X_i \cup \Psi_i) \vee \Psi_i$. Cross all these finite states, take the latter stage of the operation, all states do not satisfy $\neg(X_i \cup \Psi_i) \vee \Psi_i$, that is, all states satisfy $(X_i \cup \Psi_i) \wedge \neg \Psi_i$. This is exactly the kind of operation we want to eliminate.

How to use NuSMV to realize LTL model checking? Knowing the LTL formula Φ, the system M, and a state s of M, by constructing an automaton $A_{\neg\Phi}$, combining M with it, and checking whether there is a path that satisfies the acceptance condition of $A_{\neg\Phi}$ in the resulting system, we can check whether $M, s \models \Phi$ holds.

Program verification of linear temporal logic can consider programs and calculations from two different perspectives: deterministic programs and non-deterministic programs.

Deterministic program: A deterministic program is a quadrtuple (W, R, W_0, π). Among them, W is the state set, R is the global transformation function, W_0 is the only initial state, and $\pi: W \to \Sigma$ is the labeling function. We regard the state label as the property of the state and the actions that the program completes when entering the state. A deterministic program has a unique calculation (execution), which is an infinite sequence of states generated from the initial state $W_0, R(W_0), R^2(W_0), \ldots$.

Our idea of program verification is to associate programs and calculations with a certain temporal logic model and model constructor. A deterministic program has the same structure as an LTL formula model, and the calculation of the program is also the same.

To obtain a verification method, the key is to choose an appropriate method to describe the procedure and the property of the procedure. The program can be characterized by explicit states and transitions, or it can be described by a formula. Since we consider deterministic procedures, the formula has only one model in the equivalent sense. Even if the program is deterministic, the property is not necessarily deterministic because it can be true for a class of programs. The most natural way to describe properties is the LTL formula. Model generators (such as Büchi sequential automata) can also be used. Any program or

如何用 NuSMV 实现 LTL 模型检测？已知 LTL 公式 Φ，系统 M，以及 M 的一个状态 s，通过构造自动机 $A_{\neg\Phi}$，将 M 与其结合，并在结果的系统中检测是否存在一条满足 $A_{\neg\Phi}$ 的接受条件的路径，可以检测 $M, s \models \Phi$ 是否成立.

线性时态逻辑的程序验证可以从两个不同的角度考虑程序和计算：确定性程序和非确定性程序.

确定性程序：一个确定性程序是一个 4 元组 (W, R, W_0, π). 其中，W 是状态集，R 是全局变换函数，W_0 是唯一的初始状态，$\pi: W \to \Sigma$ 是标记函数. 我们把状态标号看成该状态具有的性质以及程序进入该状态完成的动作. 一个确定性程序具有唯一的计算（执行），它是从初始状态产生的一个无穷状态序列 $W_0, R(W_0), R^2(W_0), \ldots$.

我们进行程序验证的思路是将程序和计算与某种时态逻辑的模型和模型构造子联系起来. 一个确定性程序恰与一个 LTL 公式的模型有相同的结构，程序的计算也是相同的.

获得一种验证方法，关键在于选择一种合适的方法描述程序和程序的性质. 程序可以通过显式的状态和转换刻画，也可用一个公式描述. 由于我们考虑的是确定性程序，因此公式在等价意义上仅有一个模型. 即使程序是确定性的，性质也不一定是确定性的，因为它可以对一类程序均为真. 描述性质最自然的方式是 LTL 公式，也可用模型生成子（如 Büchi 序列自动机），任何 LTL 公式描述的程序或性质也可用 Büchi 序列自动机描述. 严格地说，反之不成立.

property described by LTL formulas can also be described by Büchi sequential automata. Strictly speaking, the opposite is not true.

Based on the description of the program and its property, Tab. 14-1 shows the verification method of the deterministic program.

基于程序及其性质的描述,表 14-1 给出了确定性程序的验证方法.

表 14-1 确定性程序的验证方法
Tab. 14-1 Verification methods for deterministic program

程序 Program	性质 Nature	
	LTL 公式 LTL formula	Büchi 序列自动机 Büchi sequential automata
状态和转换 Status and replacement	模型检测 Model checking	自动机验证 Automata verification
LTL 公式 LTL formula	公理化验证 Axiomatic verification	逆模型检验 Inverse model checking

Non-deterministic programs: The most common used area of temporal logic is to verify non-deterministic programs or concurrent programs described in non-deterministic terms.

A non-deterministic program is a quadrtuple (W, Q, W_0, π), where W is the state set, Q is the transformation relationship, W_0 is the only initial state, and $\pi: W \rightarrow \Sigma$ is the labeling function. The calculation of the program is an infinite sequence of states $W_0 W_1 \ldots W_i \ldots$, satisfying that W_0 is the initial state of the program, and for any $i \geq 0$, $(W_i, W_{i+1}) \in Q$, a program can have many calculations.

The method of applying temporal logic to non-deterministic programs: The program is regarded as the model generator of LTL (i.e. Büchi sequential automata), and the calculation is regarded as the model of LTL. Note: The program is different from the Büchi automata in two aspects: one is that the label in the program is in the state rather than on the edge; the other is that the program does not accept the state. These differences are not substantial. Taking the label of the edge leaving a certain state as the label of the state and setting all states to the accepting state, a non-deterministic program is converted into a Büchi

非确定性程序:时态逻辑最常用的领域就是验证非确定性程序或以非确定性描述的并发程序.

一个非确定性程序是一个 4 元组 (W, Q, W_0, π),其中,W 是状态集,Q 是变换关系,W_0 是唯一的初始状态,$\pi: W \rightarrow \Sigma$ 是标记函数.程序的计算是无穷状态序列 $W_0 W_1 \cdots W_i \cdots$,满足 W_0 是程序的初始状态,且对任意 $i \geq 0$,$(W_i, W_{i+1}) \in Q$,一个程序可以有许多计算.

将时态逻辑应用于非确定性程序的方法:将程序看成 LTL 的模型生成子(即 Büchi 序列自动机),计算看成 LTL 的模型.注意:程序与 Büchi 自动机有两方面不同:一是程序中标号在状态而非边上;二是程序没有接受状态.这些区别并非实质性的.将离开某种状态的边的标号作为该状态的标号并将所有状态置为接受状态,就把一个非确定性程序转换为一个 Büchi 自动机,这样得到的 Büchi 自动机接收的字恰为程序计算的集合.事实上,程序

343

automaton, so that the literal received by the Büchi automaton obtained is exactly the set of program calculations. In fact, programs can have acceptance conditions, which is a description form of the fairness of program execution. In this way, the set of program calculations is limited to those calculations that meet the acceptance conditions, that is, a certain receiving state passes infinitely many times. Considering the structural interpretation of the acceptance conditions in Büchi style, the state formula $\exists g(\forall g)$ can be interpreted as "g is true on some (all) paths satisfying the acceptance conditions". Although this is syntactically equivalent to BTL, there is a different set of effective formulas in semantics, and the judgment process is naturally different.

Tab. 14-2 shows the verification methods for non-deterministic programs. We regard the LTL-based verification approach, that is, the program as the LTL model generator, and the calculation as the LTL model, which is called the linear approach.

The basis of applying temporal logic to program verification is to establish the correspondence between temporal logic models and model generators, programs and calculations. Different verification methods depend on various descriptions of programs and properties.

可以有接受条件,这是程序执行的公平性的一种描述形式,这样,程序计算的集合被限定为那些满足接受条件的计算,即某一接收状态通过无限多次. 考虑在 Büchi 风格的接收条件的结构上的解释,状态公式 $\exists g(\forall g)$ 可解释为"g 在某(所有)满足接受条件的路径上为真". 尽管这在语法上等同于 BTL,但语义上有不同的有效公式集,判定过程也自然不同了.

表 14-2 给出了非确定性程序的验证方法. 我们把基于 LTL 的验证途径,即程序看成 LTL 模型生成子,并将计算看成 LTL 模型,称为线性途径.

将时态逻辑应用于程序验证的基础是建立时态逻辑模型和模型生成子与程序和计算的对应联系. 不同的验证方法取决于程序和性质的各种描述方式.

表 14-2 非确定性程序的验证方法
Tab. 14-2 Verification methods for non-deterministic program

程序 Program		性质 Nature	
		基于 LTL Based on LTL	
		LTL 公式 LTL formula	Büchi 序列自动机 Büchi sequential automata
基于 LTL Based on LTL	状态转换接收条件 LTL 公式 State transition receiving condition LTL formula	模型检测 公理化验证 Model checking Axiomatic verification	自动机验证 逆模型检测 Automata verification Inverse model checking

Exercise 14

1. List all sub-formulas of the LTL formula $\neg p \cup (Fr \vee G \neg q \to qW \neg r)$.

2. In terms of "morality", there should be a duality of W. Explain what its meaning should be, and then choose a symbol for it based on the first letter of the meaning.

3. Take traffic lights as an example (gr: green; ye: yellow; re: red), and answer the questions below:

(a) If the order change of light accords to "red, yellow, green, red, yellow, green ...", please write the corresponding LTL.

(b) If the light cannot have more than two colors at the same time, please write the corresponding LTL.

4. Write the syntax parse tree of the following CTL formula:

(a) EGr;

(b) $AG(q \to EGr)$;

(c) $A[p \cup EFr]$.

5. Please explain why the following formula is not the well-formed CTL formula:

(a) FGr;

(b) XXr;

(c) $AEFr$;

(d) $EXXr$.

6. Consider the model M shown in Fig. 14-20 below. For the following CTL formula ϕ, check whether $M, s_0 \models \phi$ and $M, s_2 \models \phi$ are true:

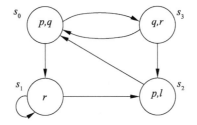

Fig. 14-20 A model with 4 states

(a) AFq;
(b) $AG(EF(p \wedge r))$;
(c) $EX(EXr)$;
(d) $AG(AFq)$.

7. Let $M=(S, \rightarrow, L)$ be an arbitrary model of CTL, and $[[\phi]]$ denotes the set of all $s \in S$ such that $M, s \models \phi$. By examining the sentences in Definition 14.3.3, prove the following set equations.

(a) $[[\bot]] = \phi$;
(b) $[[\neg \phi]] = S - [[\phi]]$;
(c) $[[\varphi_1 \wedge \varphi_2]] = [[\varphi_1]] \cap [[\varphi_2]]$;
(d) $[[AX\varphi]] = S - [[EX\neg\varphi]]$.

8. Write the syntactic parse tree for the following CTL formulas:

(a) EGr;
(b) $AG(q \rightarrow EGr)$;
(c) $A[p \cup EFr]$;
(d) $EF\,EGp \rightarrow AFr$.

9. Prove that the CTL formula ϕ is true for infinite states on a computational path $s_0 \rightarrow s_1 \rightarrow s_2 \rightarrow \ldots$ when and only when for all $n \geq 0$, there exists some $m \geq n$ such that $s_m \models \phi$.

10. Consider the set of LTL/CTL formulas $F = \{Fp \rightarrow Fq, AFp \rightarrow AFq, AG(p \rightarrow AFq)\}$.

(a) Is there a model that makes all formulas in the model hold?

(b) For all $\Phi \in F$, is there a model such that Φ is the only formula in F that satisfies this model?

(c) Find a model in which all formulas in F do not hold.

11. The meanings of the temporal operators F, G and U in LTL and the temporal operators AU, EU, AG, EG, AF and EF in CTL make "the present includes the future". For example, EFp is true about a state, if p is already true about that state. People often need corresponding operators so that "the pres-

ent does not include the future". Using appropriate connectives, define such (six) modified connectives as the derivation operator in CTL.

包含将来". 使用合适的连接词,定义这样的(六个)修饰连接词作为 CTL 中的导出算子.

习题 14 答案
Key to Exercise 14

参考文献
References

[1] 王晓东. 算法设计与分析[M]. 4版. 北京:清华大学出版社,2018.

[2] 迈克尔·西普塞. 计算理论导引[M]. 段磊,唐常杰,等译. 3版. 北京:机械工业出版社,2015.

[3] MICHAEL S. 计算理论导引[M]. 唐常杰,陈鹏,向勇,等译. 2版. 北京:机械工业出版社,2006.

[4] 郑宗汉,郑晓明. 算法设计与分析[M]. 北京:清华大学出版社,2005.

[5] MICHAEL H,MARK R. 面向计算机科学的数理逻辑:系统建模与推理[M]. 何伟,樊磊,译. 2版. 北京:机械工业出版社,2007.

[6] LEWIS H R,PAPADIMITRIOU C H. 计算理论基础[M]. 张立昂,刘田,译. 2版. 北京:清华大学出版社,2000.

[7] DREW J H, EVANS D L, GLEN A G, et al. Computational probability[M]. Springer: International Publishing, 2017.

[8] STOUGHTON A. An introduction to formal language theory that integrates experimentation and proof[Z]. Manhattan: Kansas State University, 2004.

[9] CUTLAND N. Computability: an introduction to recursive function theory[M]. Cambridge: Cambridge University Press, 1980.

[10] GURARI E. An introduction to the theory of computation[M]. New York: Computer Science Press, 1989.

[11] GOLDREICH O. Computational complexity: a conceptual perspective[J]. ACM SIGACT News, 2008, 39(3): 35-39.

[12] ARORA S, BARAK B. Computational complexity: a modern approach[M]. Cambridge: Cambridge University Press, 2009.

[13] KRIPKE S. Elementary recursion theory and its applications to formal systems[M]. New York: New York University Press, 2011.

[14] LEWIS F D. Essentials of theoretical computer science[Z]. Lexington: University of Kentucky, 1996.

[15] GOLDREICH O. Introduction to complexity theory(Lecture Note)[Z]. Department of Computer Science and Applied Mathematics, Weizmann Institute of Science, Israel, 1999.

[16] Fernández M. Models of computation: an introduction to computability theory[M]. London: Springer London, 2009.

[17] HEDMAN S. A first course in logic: an introduction to model theory, proof theory, computability, and complexity[M]. Oxford: Oxford University Press, 2004.

[18] PARBERRY I. Parallel complexity theory[M]. London: Pitman, 1987.

[19] ROTHE J. Complexity theory and cryptology: an introduction to cryptocomplexity[M].

Berlin: Springer, 2005.

[20] DREW J H, EVANS D L, GLEN A G, et al. Computational probability: algorithms and applications in the mathematical sciences[J]. The Journal of the Operational Research Society, 2009,60(7):1041.

[21] WEGENER I. The complexity of symmetric Boolean functions[M]//Computation Theory and Logic. Berlin, Heidelberg: Springer Berlin Heidelberg, 1987: 433-442.

[22] KEARNS M J. The computational complexity of machine learning[M]. Cambridge, Mass.: MIT Press, 1990.

[23] SIPSER M. Introduction to the theory of computation[J]. ACM SIGACT News, 1996, 27(1): 27-29.